Morris Kline has been a college professor, researcher, and writer for over forty years. He received his doctorate in mathematics from New York University in 1936. His teaching career was interrupted for two years, 1936–1938, during which he served as a research assistant at the Institute for Advanced Study in Princeton, New Jersey, and was again interrupted from 1942 to 1945, when he served as a physicist for the United States Army Signal Corps. Though most of his teaching has been at New York University, he has taught at Stanford University, the Technische Hochschule of Aachen, Germany, and at other institutions. For twenty years he was Director of the Division of Electromagnetic Research at the Courant Institute of Mathematical Sciences of New York University, and for eleven years he was Chairman of Undergraduate Mathematics in the main centre of that University. He has been a John Simon Guggenheim Memorial Fellow and a Fulbright lecturer in Germany. He is the author of six books and numerous articles devoted to research, pedagogy, and exposition, and holds several patents for inventions in radio engineering.

MORRIS KLINE

—

MATHEMATICS
IN WESTERN
CULTURE

PENGUIN BOOKS

Penguin Books Ltd, Harmondsworth, Middlesex, England
Penguin Books, 625 Madison Avenue, New York, New York 10022, U.S.A.
Penguin Books Australia Ltd, Ringwood, Victoria, Australia
Penguin Books Canada Ltd, 2801 John Street, Markham, Ontario, Canada L3R 1B4
Penguin Books (N.Z.) Ltd, 182–190 Wairau Road, Auckland 10, New Zealand

—

First published in the U.S.A. 1953
Published in Great Britain by George Allen & Unwin 1954
Published in Pelican Books 1972
Reprinted 1977, 1979, 1982

—

—

Printed and bound in Great Britain by
Cox & Wyman Ltd, Reading
Set in Intertype Plantin

TO ELIZABETH AND JUDITH

Contents

8 *Contents*

Plates

When first I applied my mind to Mathematics I read straight away most of what is usually given by the mathematical writers, and I paid special attention to Arithmetic and Geometry because they were said to be the simplest and so to speak the way to all the rest. But in neither case did I then meet with authors who fully satisfied me. I did indeed learn in their works many propositions about numbers which I found on calculation to be true. As to figures, they in a sense exhibited to my eyes a great number of truths and drew conclusions from certain consequences. But they did not seem to make it sufficiently plain to the mind itself why these things are so, and how they discovered them. Consequently I was not surprised that many people, even of talent and scholarship, should, after glancing at these sciences, have either given them up as being empty and childish or, taking them to be very difficult and intricate, been deterred at the very outset from learning them. . . . But when I afterwards bethought myself how it could be that the earliest pioneers of Philosophy in bygone ages refused to admit to the study of wisdom any one who was not versed in Mathematics . . . I was confirmed in my suspicion that they had knowledge of a species of Mathematics very different from that which passes current in our time.

RENÉ DESCARTES

Foreword

AFTER an unbroken tradition of many centuries, mathematics has ceased to be generally considered as an integral part of culture in our era of mass education. The isolation of research scientists, the pitiful scarcity of inspiring teachers, the host of dull and empty commercial textbooks and the general educational trend away from intellectual discipline have contributed to the anti-mathematical fashion in education. It is very much to the credit of the public that a strong interest in mathematics is none the less alive.

Various attempts have recently been made to satisfy this interest. Together with H. Robbins I attempted in *What Is Mathematics?* to discuss the meaning of mathematics. Our book was, however, addressed to readers with a certain background of mathematical knowledge. More should be done on a less technical level for the large number of people who do not have this background, but still wish to acquire knowledge of the significance of mathematics in human culture.

For some time I have followed with great interest Professor Morris Kline's work on the present book. I believe that it will prove a major contribution and help to bring the mathematical sciences closer to people who have not as yet appreciated the fascination and scope of the subject.

R. COURANT

Preface

Now, when all these studies reach the point of inter-
communion and connection with one another, and come
to be considered in their mutual affinities, then, I think,
but not till then, will the pursuit of them have a value
for our objects; otherwise there is no profit in them.

Plato

THE object of this book is to advance the thesis that mathematics
has been a major cultural force in Western civilization. Almost
everyone knows that mathematics serves the very practical purpose of dictating engineering design. Fewer people seem to be
aware that mathematics carries the main burden of scientific
reasoning and is the core of the major theories of physical
science. It is even less widely known that mathematics has determined the direction and content of much philosophic thought,
has destroyed and rebuilt religious doctrines, has supplied substance to economic and political theories, has fashioned major
painting, musical, architectural, and literary styles, has fathered
our logic, and has furnished the best answers we have to fundamental questions about the nature of man and his universe. As
the embodiment and most powerful advocate of the rational
spirit, mathematics has invaded domains ruled by authority,
custom, and habit, and supplanted them as the arbiter of thought
and action. Finally, as an incomparably fine human achievement
mathematics offers satisfactions and aesthetic values at least equal
to those offered by any other branch of our culture.

Despite these by no means modest contributions to our life and
thought, educated people almost universally reject mathematics
as an intellectual interest. This attitude towards the subject is, in
a sense, justified. School courses and books have presented
'mathematics' as a series of apparently meaningless technical
procedures. Such material is as representative of the subject as an

account of the name, position, and function of every bone in the
human skeleton is representative of the living, thinking, and
emotional being called man. Just as a phrase either loses meaning
or acquires an unintended meaning when removed from its context, so mathematics detached from its rich intellectual setting in
the culture of our civilization and reduced to a series of techniques has been grossly distorted. Since the layman makes very
little use of technical mathematics, he has objected to the naked
and dry material usually presented. Consequently, a subject that
is basic, vital, and elevating is neglected and even scorned by
otherwise highly educated people. Indeed, ignorance of mathematics has attained the status of a social grace.

In this book we shall survey mathematics primarily to show
how its ideas have helped to mould twentieth-century life and
thought. The ideas will be in historical order so that our material
will range from the beginnings in Babylonia and Egypt to the
modern theory of relativity. Some people may question the
pertinence of material belonging to earlier historical periods.
Modern culture, however, is the accumulation and synthesis of
contributions made by many preceding civilizations. The Greeks,
who first appreciated the power of mathematical reasoning,
graciously allowing the gods to use it in designing the universe,
and then urging man to uncover the pattern of this design, not
only gave mathematics a major place in their civilization but
initiated patterns of thought that are basic in our own. As succeeding civilizations passed on their gifts to modern times, they
handed on new and increasingly more significant roles for mathematics. Many of these functions and influences of mathematics
are now deeply embedded in our culture. Even the modern contributions of mathematics are best appreciated in the light of
what existed previously.

Despite the historical approach, this book is not a history of
mathematics. The historical order happens to be most convenient
for the logical presentation of the subject and is the natural way
of examining how the ideas arose, what the motivations for investigating these ideas were, and how these ideas influenced the
course of other activities. An important by-product is that the
reader may get some indication of how mathematics as a whole

has developed, how its periods of activity and quiescence have been related to the general course of the history of Western civilization, and how the nature and contents of mathematics have been shaped by the civilizations that contributed to our modern Western one. It is hoped that new light will be shed on mathematics and on the dominant characteristics of our age by this account of mathematics as a fashioner of modern civilization.

We cannot, unfortunately, do more in one volume than merely illustrate the thesis. Limitations of space have necessitated a selection from a vast literature. The interrelationships of mathematics and art, for example, have been confined to the age of the Renaissance. The reader acquainted with modern science will notice that almost nothing has been said about the role of mathematics in atomic and nuclear theory. Some important modern philosophies of nature, notably Alfred North Whitehead's, have hardly been mentioned. Nevertheless, it is hoped that the illustrations chosen will be weighty enough to prove convincing as well as interesting.

The attempt to highlight a few episodes in the life of mathematics has also necessitated an over-simplification of history. In intellectual, as well as political, enterprises numerous forces and numerous individual contributions determine the outcomes. Galileo did not fashion the quantitative approach to modern science single-handed. Similarly, the calculus is almost as much the creation of Eudoxus, Archimedes, and a dozen major lights of the seventeenth century as it is that of Newton and Leibniz. It is especially true of mathematics that, while the creative work is done by individuals, the results are the fruition of centuries of thought and development.

There is no doubt that in invading the arts, philosophy, religion, and the social sciences the author has rushed in where angels – mathematical ones, of course – would fear to tread. The risk of errors, hopefully minor, must be undertaken if we are to see that mathematics is not a dry, mechanical tool but a body of living thought inseparably connected with, dependent on, and invaluable to other branches of our culture.

Perhaps this account of the achievements of human reason may serve in some small measure to reinforce those ideals of our civilization which are in danger of destruction today. The burning problems of the hour may be political and economic. Yet it is not in those fields that we find evidence of man's ability to master his difficulties and build a desirable world. Confidence in man's power to solve his problems and indications of the method he has employed most successfully thus far can be gained by a study of his greatest and most enduring intellectual accomplishment – mathematics.

It is a pleasure to acknowledge help and favours received from many sources. I wish to thank numerous colleagues in the Washington Square College of Arts and Science of New York University for many helpful discussions, Professor Chester L. Riess of the Brooklyn College of Pharmacy for general criticism and particular suggestions concerning the literature of the Age of Reason, and Mr John Begg of Oxford University Press for advice on the preparation of the figures and plates. Mrs Beulah Marx is to be credited with the excellent illustrations. My wife, Helen, has aided me immeasurably by critical readings and preparation of the manuscript. I am especially grateful to Mr Carroll G. Bowen and Mr John A. S. Cushman for their advocacy of the idea of this book and for guiding the manuscript through the process of publication at Oxford.

I am indebted to the following publishers and individuals for permission to use the material indicated below. The quotation from Alfred North Whitehead in the last chapter is taken from *Science and the Modern World*, The Macmillan Co., N.Y., 1925. Permission to use the graphs of actual sounds, by Dayton C. Miller, was granted by the Case Institute of Technology of Cleveland, Ohio. The quotation from Edna St Vincent Millay is from 'Sonnet XXII' of *The Harp Weaver and Other Poems* published by Harper & Bros., N.Y., copyright 1920 and 1948 by Edna St Vincent Millay. The quotations from Bertrand Russell appeared in *Mysticism and Logic* published by W. W. Norton & Co., Inc., N.Y., and George Allen & Unwin Ltd, London. The quotation from Theodore Merz is taken from Volume II of *A History of European Thought in the Nineteenth Century* published by William Blackwood & Sons Ltd, Edinburgh and London.

New York City　　　　　　　　　MORRIS KLINE
August 1953

I
Introduction:
True and False Conceptions

Stay your rude steps, or e'er your feet invade
The Muses' haunts, ye sons of War and Trade!
Nor you, ye legion fiends of Church and Law,
Pollute these pages with unhallow'd paw!
Debased, corrupted, grovelling, and confin'd,
No definitions touch your senseless mind;
To you no Postulates prefer their claim,
No ardent Axioms your dull souls inflame;
For you no Tangents touch, no Angles meet,
No Circles join in osculation sweet!

John Hookham Frere, George Canning,
and George Ellis

THE assertion that mathematics has been a major force in the moulding of modern culture, as well as a vital element of that culture, appears to many people incredible or, at best, a rank exaggeration. This disbelief is quite understandable and results from a very common but erroneous conception of what mathematics really is.

Influenced by what he was taught in school, the average person regards mathematics as a series of techniques of use only to the scientist, the engineer, and perhaps the financier. The reaction to such teachings is distaste for the subject and a decision to ignore it. When challenged on this decision a well-read person can obtain the support of authorities. Did not St Augustine say: 'The good Christian should beware of mathematicians and all those who make empty prophecies. The danger already exists that the mathematicians have made a covenant with the devil to darken the spirit and to confine man in the bonds of Hell.' And did not the Roman jurists rule, 'concerning evil-doers, mathematicians, and the like', that, 'To learn the art of geometry and to take part in public exercises, an art as damnable as mathematics, are forbidden.' No less a personage than the distinguished modern philosopher, Schopenhauer, described arithmetic as the lowest activity of the spirit, as is shown by the fact that it can be performed by a machine.

Despite such authoritative judgements and despite common opinion, justified as it may be in view of the teachings in the schools, the layman's decision to ignore mathematics is wrong. The subject is not a series of techniques. These are indeed the least important aspect, and they fall as far short of representing mathematics as colour mixing does of painting. The techniques are mathematics stripped of motivation, reasoning, beauty, and significance. If we acquire some understanding of the nature of mathematics, we shall see that the assertion of its importance in modern life and thought is at least plausible.

Let us, therefore, consider briefly at this point the twentieth-century view of the subject. Primarily, mathematics is a method of inquiry known as postulational thinking. The method consists in carefully formulating definitions of the concepts to be discussed and in explicitly stating the assumptions that shall be the basis for reasoning. From these definitions and assumptions conclusions are deduced by the application of the most rigorous logic man is capable of using. This characterization of mathematics was expressed somewhat differently by a famous seventeenth-century writer on mathematics and science: 'Mathematicians are like lovers. . . . Grant a mathematician the least principle, and he will draw from it a consequence which you must also grant him, and from this consequence another.'

To describe mathematics as only a method of inquiry is to describe da Vinci's 'Last Supper' as an organization of paint on canvas. Mathematics is, also, a field for creative endeavour. In divining what can be proved, as well as in constructing methods of proof, mathematicians employ a high order of intuition and imagination. Kepler and Newton, for example, were men of wonderful imaginative powers, which enabled them not only to break away from age-long and rigid tradition but also to set up new and revolutionary concepts. The extent to which the creative faculties of man are exercised in mathematics could be determined only by an examination of the creations themselves. While some of these will appear in subsequent discussion it must suffice here to state that there are now some eighty extensive branches of the subject.

If mathematics is indeed a creative activity, what driving force causes men to pursue it? The most obvious, though not necessarily the most important, motive for mathematical investigations has been to answer questions arising directly out of social needs. Commercial and financial transactions, navigation, calendar reckoning, the construction of bridges, dams, churches, and palaces, the design of fortifications and weapons of warfare, and numerous other human pursuits involve problems that can best be resolved by mathematics. It is especially true of our engineering age that mathematics is a universal tool.

Another basic use of mathematics, indeed one that is especially prominent in modern times, has been to provide a rational organization of natural phenomena. The concepts, methods, and conclusions of mathematics are the substratum of the physical sciences. The success of these fields has been dependent on the extent to which they have entered into partnership with mathematics. Mathematics has brought life to the dry bones of disconnected facts and, acting as connective tissue, has bound series of detached observations into bodies of science.

Intellectual curiosity and a zest for pure thought have started many mathematicians in pursuit of properties of numbers and geometric figures and have produced some of the most original contributions. The whole subject of probability, important as it is today, began with a question arising in a game of cards, namely, the proper division of a gambling stake in a game interrupted before its close. Another most decisive contribution in no way connected with social needs or science was made by the Greeks of the classical period who converted mathematics into an abstract, deductive, and axiomatic system of thought. In fact, some of the greatest contributions to the subject matter of mathematics – projective geometry, the theory of numbers, the theory of infinite quantities, and non-Euclidean geometry, to mention only those that will be within our purview – constitute responses to purely intellectual challenges.

Over and above all other drives to create is the search for

beauty. Bertrand Russell, the master of abstract mathematical thought, speaks without qualification:

Mathematics, rightly viewed, possesses . . . supreme beauty – a beauty cold and austere, like that of sculpture, without appeal to any part of our weaker nature, without the gorgeous trappings of painting or music, yet sublimely pure, and capable of a stern perfection such as only the greatest art can show. The true spirit of delight, the exaltation, the sense of being more than man, which is the touchstone of the highest excellence, is to be found in mathematics as surely as in poetry.

In addition to the beauty of the completed structure, the indispensable use of imagination and intuition in the creation of proofs and conclusions affords high aesthetic satisfaction to the creator. If insight and imagination, symmetry and proportion, lack of superfluity, and exact adaption of means to ends are comprehended in beauty and are characteristic of works of art, then mathematics is an art with a beauty of its own.

Despite the clear indications of history that all of the factors above have motivated the creation of mathematics there have been many erroneous pronouncements. There are the charges – often made to excuse neglect of the subject – that mathematicians like to indulge in pointless speculations or that they are silly and useless dreamers. To these charges a crushing reply can readily be made. Even purely abstract studies, let alone those motivated by scientific and engineering needs, have proved immensely useful. The discovery of the conic sections (parabolas, ellipses, and hyperbolas) which, for two thousand years, amounted to no more than 'the unprofitable amusement of a speculative brain', ultimately made possible modern astronomy, the theory of projectile motion, and the law of universal gravitation.

On the other hand, it is a mistake to assert, as some 'socially minded' writers do rather sweepingly, that mathematicians are stimulated entirely or even largely by practical considerations, by the desire to build bridges, radios, and airplanes. Mathematics has made these conveniences possible, but the great mathematicians rarely have them in mind while pursuing their ideas. Some were totally indifferent to the practical applications, possibly because these were made centuries later. The idealistic mathematical musings of Pythagoras and Plato have led to far more significant contributions than the purposeful act of the warehouse clerks whose introduction of the symbols + and − convinced one writer that 'a turning point in the history of mathematics arose from the common social heritage . . .' It is no doubt true that almost every great man occupies himself with the problems of his age, and that prevailing beliefs condition and limit his thinking. Had Newton been born two hundred years earlier he would most likely have been a masterful theologian. Great thinkers yield to the intellectual fashions of their times as women do to fashions in dress. Even those creative geniuses to whom mathematics was purely an avocation pursued the problems that were agitating the professional mathematicians and scientists. Nevertheless, these 'amateurs' and mathematicians generally have not been concerned primarily with the utility of their work.

Practical, scientific, aesthetic, and philosophical interests have all shaped mathematics. It would be impossible to separate the contributions and influences of any one of these forces and weigh it against the others, much less assert sweeping claims to its relative importance. On the one hand, pure thought, the response to aesthetic and philosophical interests, has decisively fashioned the character of mathematics and made such unexcelled contributions as Greek geometry and modern non-Euclidean geometry. On the other hand, mathematicians reach their pinnacles of pure thought not by lifting themselves by their bootstraps but by the power of social forces. Were these forces not permitted to revitalize mathematicians, they would soon exhaust themselves; thereafter they could merely sustain their subject in an isolation which might be splendid for a short time but which would soon spell intellectual collapse.

Another important characteristic of mathematics is its symbolic language. Just as music uses symbolism for the representation and communication of sounds, so mathematics expresses quantitative relations and spatial forms symbolically. Unlike the usual language of discourse, which is the product of custom, as well as of social and political movements, the language of mathematics is carefully, purposefully, and often ingeniously designed. By virtue of its compactness, it permits the mind to carry and work with ideas which, expressed in ordinary language, would be unwieldy. This compactness makes for efficiency of thought. Jerome K. Jerome's need to resort to algebraic symbolism, though for non-mathematical purposes, reveals clearly enough the usefulness and clarity inherent in this device:

When a twelfth-century youth fell in love he did not take three paces backward, gaze into her eyes, and tell her she was too beautiful to live. He said he would step outside and see about it. And if, when he got out, he met a man and broke his head – the other man's head, I mean – then that proved that his – the first fellow's – girl was a pretty girl. But if the other fellow broke *his* head – not his own, you know, but the other fellow's – the other fellow to the second fellow, that is, because of course the other fellow would only be the other fellow to him, not the first fellow who – well, if he broke his head, then *his* girl – not the other fellow's, but the fellow who *was* the – Look here, if A broke B's head, then A's girl was a pretty girl; but if B broke A's head, then A's girl wasn't a pretty girl, but B's girl was.

While clever symbolism enables the mind to carry complicated ideas with ease, it also makes it harder for the layman to follow or understand a mathematical discussion.

The symbolism used in mathematical language is essential to distinguish meanings often confused in ordinary speech. For example, the word *is* is used in English, in many different senses. In the sentence *He is here,* it indicates a physical location. In the sentence *An angel is white,* it indicates a property of angels that has nothing to do with location or physical existence. In the sentence *The man is running,* the word gives the tense of the verb. In the sentence *Two and two are four,* the form of *is* used denotes numerical equality. In the sentence *Men are the two-legged thinking mammals,* the form of *is* involved asserts the identity of two groups. Of course, for the purposes of ordinary discourse it is superfluous to introduce different words for all these meanings of *is.* No mistakes are made on account of these ambiguities. But the exactions of mathematics, as well as of the sciences and philosophy, compel workers in these fields to be more careful.

Mathematical language is precise, so precise that it is often confusing to people unaccustomed to its forms. If a mathematician should say, 'I did not see one person today,' he would mean that he either saw none or saw many. The layman would mean simply that he saw none. This precision of mathematics appears as pedantry or stiltedness to one who does not yet appreciate that it is essential to exact thinking, for exact thinking and exact language go hand in hand.

Mathematical style aims at brevity and formal perfection. It sometimes succeeds too well and sacrifices the clarity its precision seeks to guarantee. Let us suppose we wish to express in general

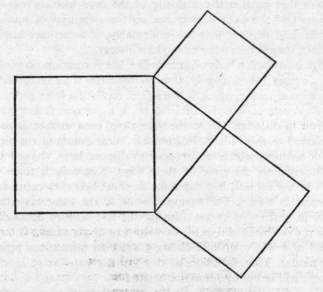

Figure 1. The Pythagorean theorem

terms the fact illustrated in fig. 1. We might be tempted to say: 'We have a right triangle. If we construct two squares each having an arm of the triangle as a side and if we construct a square having the hypotenuse of the triangle for its side, then the area of the third square is equal to the sum of the areas of the first two.' But no mathematician would deign to express himself that way. He prefers: 'The sum of the squares on the arms of a right triangle equals the square on the hypotenuse.' This economy of words makes for deftness of presentation, and mathematical writing is remarkable because it does encompass much in few words. Yet there are times when any reader of mathematical literature finds his patience sorely tried by what he would call a miserliness with ink and paper.

Mathematics is more than a method, an art, and a language. It is a body of knowledge with content that serves the physical and social scientist, the philosopher, the logician, and the artist; content that influences the doctrines of statesmen and theologians; content that satisfies the curiosity of the man who surveys the heavens and the man who muses on the sweetness of musical sounds; and content that has undeniably, if sometimes imperceptibly, shaped the course of modern history.

Mathematics is a body of knowledge. But it contains no truths. The contrary belief, namely, that mathematics is an unassailable collection of truths, that it is like a final revelation from god such as religionists believe the Bible to be, is a popular fallacy most difficult to dislodge. Up to the year 1850, even mathematicians subscribed to this fallacy. Fortunately, some events of the nineteenth century, which we propose to discuss later, showed the mathematicians the error of their ways. Not only is there no truth in the subject, but theorems in some branches contradict theorems in others. For example, some of the theorems established in geometries created during the last century contradict those proved by Euclid in his development of geometry. Though devoid of truth, mathematics has given man miraculous power over nature. The resolution of this greatest paradox in human thought will be one of our major concerns.

Because the twentieth century must distinguish mathematical knowledge from truths it must also distinguish between mathematics and science, for science does seek truths about the physical world. Mathematics has indeed been a beacon light to the sciences and has continually helped them in reaching the position they occupy in our present civilization. It is even correct to assert that modern science triumphs by virtue of mathematics. Yet we shall see that the two fields are distinct.

In its broadest aspect mathematics is a spirit, the spirit of rationality. It is this spirit that challenges, stimulates, invigorates, and drives human minds to exercise themselves to the fullest. It is this spirit that seeks to influence decisively the physical, moral, and social life of man, that seeks to answer the problems posed by our very existence, that strives to understand and control nature, and that exerts itself to explore and establish the deepest and utmost implications of knowledge already obtained. To a large extent our concern in this book will be with the operation of this spirit.

One more characteristic of mathematics is most pertinent to our story. Mathematics is a living plant which has flourished and languished with the rise and fall of civilizations. Created in some prehistoric period it struggled for existence through centuries of pre-history and further centuries of recorded history. It finally secured a firm grip on life in the highly congenial soil of Greece and waxed strong for a brief period. In this period it produced one perfect flower, Euclidean geometry. The buds of other flowers opened slightly and with close inspection the outlines of trigonometry and algebra could be discerned; but these flowers withered with the decline of Greek civilization, and the plant remained dormant for one thousand years.

Such was the state of mathematics when the plant was transported to Europe proper and once more embedded in fertile soil. By A.D. 1600 it had regained the vigour it had possessed at the very height of the Greek period and was prepared to break forth with unprecedented brilliance. If we may describe the mathematics known before 1600 as elementary mathematics, then we may state that elementary mathematics is infinitesimal compared to what has been created since. In fact, a person possessed of the knowledge Newton had at the height of his powers would not be considered a mathematician today for, contrary to popular belief, mathematics must now be said to begin with the calculus and not to end there. In our century, the subject has attained such vast proportions that no mathematician can claim to have mastered the whole of it.

This sketch of the life of mathematics, however brief, may nevertheless indicate that its vitality has been very much dependent on the cultural life of the civilization which nourished it. In fact, mathematics has been so much a part of civilizations and cultures that many historians see mirrored in the mathematics of an age the characteristics of the other chief works of that age. Consider, for example, the classical period of Greek culture, which lasted from about 600 B.C. to 300 B.C. In emphasizing the rigorous reasoning by which they established their conclusions, the Greek mathematicians were concerned not with guaranteeing applicability to practical problems but with teaching men to reason abstractly and with preparing them to contemplate the ideal and the beautiful. It should be no surprise to learn, then, that this age has been unsurpassed in the beauty of its literature, in the supremely rational quality of its philosophy, and in the ideality of its sculpture and architecture.

It is also true that the absence of mathematical creations is indicative of the culture of a civilization. Witness the case of the Romans. In the history of mathematics the Romans appear once and then only to retard its progress. Archimedes, the greatest Greek mathematician and scientist, was killed in 211 B.C. by Roman soldiers who burst in upon him while he was studying a geometrical diagram drawn in sand. To Alfred North Whitehead,

> The death of Archimedes by the hands of a Roman soldier is symbolical of a world change of the first magnitude; the theoretical Greeks, with their love of abstract science, were superseded in leadership of the European world by the practical Romans. Lord Beaconsfield, in one of his novels, has defined a practical man as a man who practises the errors of his forefathers. The Romans were a great race, but they were cursed with the sterility which waits upon practicality. They did not improve upon the knowledge of their forefathers, and all their advances were confined to the minor technical details of engineering. They were not dreamers enough to arrive at new points of view, which could give a more fundamental control over the forces of Nature. No Roman lost his life because he was absorbed in the contemplation of a mathematical diagram.

In fact, Cicero bragged that his countrymen, thank the gods, were not dreamers, as were the Greeks, but applied their study of mathematics to the useful.

Practical-minded Rome, which devoted its energies to administration and conquest, symbolized best perhaps by the stolid if not graceful arches under which victorious troops celebrated their homecomings, produced little that was truly creative and original. In short, Roman culture was derivative; most of the contributions made during the period of Roman supremacy came from the Greeks of Asia Minor, who were under the political domination of Rome.

These examples show us that the general character of an age is intimately related to its mathematical activity. This relationship is especially valid in our times. Without belittling the merits of our historians, economists, philosophers, writers, poets, painters, and statesmen, it is possible to say that other civilizations have produced their equals in ability and accomplishments. On the other hand, though Euclid and Archimedes were undoubtedly supreme thinkers and though our mathematicians have been able to reach farther only because, as Newton put it, they stood on the shoulders of such giants, nevertheless, it is in our age that mathematics has attained its range and extraordinary applicability. Consequently, present-day Western civilization is distinguished from any other known to history by the extent to which mathematics has influenced contemporary life and thought. Perhaps we shall come to see in the course of this book how much the present age owes to mathematics.

II

The Rule of the Thumb
in Mathematics

Do not imagine that mathematics is hard and crabbed
and repulsive to common sense. It is merely the eth-
erealization of common sense.

Lord Kelvin

THE cradle of mankind, as well as of Western culture, was the
Near East. While the more restless abandoned this birthplace to
roam the plains of Europe, their kinsmen remained behind to
found civilizations and cultures. Many centuries later the wise
men of the East had to assume the task of educating their still
untutored relatives. Of the knowledge which these sages imparted to Western man the elements of mathematics were an integral
part. Hence, to trace the impress of mathematics on modern
culture, we must turn to the major Near Eastern civilizations.

We should mention in passing that simple mathematical steps
were made in primitive civilizations. Such steps were no doubt
prompted by purely practical needs. The barter of necessities,
which takes place in even the most primitive types of human
society, requires some counting.

Since the process of counting is facilitated by the use of the
figures and toes, it is not surprising that primitive man, like a
child, used his fingers and toes as a tally to check off the things he
counted. Traces of this ancient way of counting are embedded in
our own language, the word *digit* meaning not only the numbers
1, 2, 3 ... but a finger or a toe as well. The use of the fingers
undoubtedly accounts for the adoption of our system of counting
in tens, hundreds (tens of tens), thousands (tens of hundreds),
and so forth.

Even primitive civilizations developed special symbols for
numbers. In this way, these civilizations showed cognizance of the

fact that three sheep, three apples, and three arrows have much in common, namely the quantity *three*. This appreciation of number as an abstract idea, abstract in the sense that it does not have to relate to particular physical objects, was one of the major advances in the history of thought. Each of us in his own schooling goes through a similar intellectual process of divorcing numbers from physical objects.

Primitive civilizations also invented the four elementary operations of arithmetic, that is, addition, subtraction, multiplication, and division. That these operations did not come readily to man can be learned even from a study of contemporary backward peoples. When sheep owners of many primitive tribes sell several animals, they will not take a lump sum for the lot but must be paid for each one separately. The alternative of multiplying the number of sheep by the price per sheep confuses them and leaves them suspecting that they have been cheated.

There is little question that geometry, like the number system, was fostered in primitive civilizations to satisfy man's needs. Fundamental geometric concepts came from observation of figures formed by physical objects. It is likely that the concept of angle, for example, first came from observation of the angles formed at the elbows and knees. In many languages, including modern German, the word for the side of an angle is the word for leg. We ourselves speak of the arms of a right triangle.

The major Near Eastern civilizations from which our culture and our mathematics sprang were the Egyptian and the Babylonian. In the earliest records of these civilizations we find well-developed number systems, some algebra, and very simple geometry. For the numbers from 1 to 9 the Egyptians used simple strokes thus: |, ||, |||, etc. For 10 they introduced the special symbol ∩, and there were special symbols for 100, 1000, and other large numbers. For intermediate numbers they combined these symbols in a very natural manner. Thus 21 was written ∩∩|.

The Babylonian method of writing quantity deserves more attention. For 1 they wrote 𒀹 ; 2 was represented by 𒀹𒀹 ; 4, by 𒀹𒀹𒀹 ; and so forth up to nine. The symbol ⟨ was used for 10. Thus 33 was ⟨⟨⟨𒀹𒀹𒀹 . The number 𒀹⟨⟨𒀹

is especially significant. Here the first 𝕐 meant not 1 but 60, and the whole group represented 60 + 10 + 10 + 1 or 81. Thus the same symbol represented different values depending on its position in the number. The principle involved here is that of place value and is precisely the one we use today. In the number 569, the 9 represent 9 units but the 6 means 6 times 10 and the 5 means 5 times 100 or 5 times 10^2. In other words, the position of a digit in the number determines the value it represents, and this value is a multiple of 10, of the square of 10, or of the cube of 10 and so on, depending on the position of the digit. The number *ten* is called the base of our system.

Because the Babylonians introduced place value in connection with the base sixty, the Greeks and Europeans used this system in all mathematical and astronomical calculations until the sixteenth century and it still survives in the division of angles and hours into 60 minutes and 60 seconds. Base ten was developed by the Hindus and introduced into Europe during the late Middle Ages.

The principle of place value is so important that it merits a bit of discussion. Taken in conjunction with base ten, ten symbols suffice to represent any quantity no matter how large. The representation is systematic and compact compared to other methods such as the Egyptian. Even more important is the fact that the principle permitted the development of our modern efficient methods of computation.

We should notice too that it is not necessary to use ten as a base. Any whole number would do as well in principle. Suppose, for example, that a person were to use five. He would then need just five symbols, say 1, 2, 3, 4, and 0. To indicate the quantity five he would write 10, the 1 this time meaning 1 times five, just as the 1 in the familiar 10 means 1 times ten. To write six in the base five he would write 11. Seven would be 12. Eleven would be 21. Twenty-five would be 100 or 1 times 5^2 + 0 times 5 + 0 units. To use the base five systematically he would of course have to learn the relevant addition and multiplication tables. Thus 3 + 4 would be 12; 13 + 14, the two numbers being in base five and representing eight and nine, respectively, would be 32; and so on.

The question, what is the best base, has been seriously considered and there are good reasons to favour twelve. Custom rules in favour of ten, however, as far as ordinary uses of numbers are concerned.

To use the principle of place value to best advantage a zero is required, for there must be some way to distinguish 503, say, from 53. The Babylonians used a special symbol to separate the 5 and 3 in the former case but failed to recognize that this symbol could also be treated as a number; that is, they failed to see that zero indicates quantity and can be added, subtracted, and used generally like other numbers. The number *zero* must be carefully distinguished from the concept of nothing. A student's grade in a mathematics course is nothing if he never took that course. If, on the other hand, he did take the course and his work was judged worthless, his grade would be zero.

For early civilizations computation with fractions was not a simple matter. The Babylonians lacked adequate notation. Thus ⪪⪪⪪ meant 36/60, as well as 30; the correct value had to be understood from the context. The Egyptians found it necessary to reduce a fraction to a sum of fractions in each of which the numerator was unity. Thus they would express ⅝ as ½ + ⅛ before computing with it. Though modern methods of handling fractions are much more efficient they still give trouble to many adults.

The ancient civilizations of Babylonia and Egypt carried their arithmetic beyond the use of integers and fractions. We know they were able to solve some problems involving unknown quantities, although by methods cruder and less general than we learn in our secondary schools. Babylonia is, in fact, considered to be the source of some of Euclid's knowledge of algebra.

Whereas the Babylonians developed a superior arithmetic and algebra, the Egyptians are generally considered to have surpassed them in geometry. There is much speculation about why this was so. One reason offered by historians is that the Egyptians never developed convenient methods of working with numbers, particularly fractions, and consequently were prevented from going further in the field of algebra. Instead they emphasized geometry. Another view is that geometry is a 'gift of the Nile'. Herodotus relates that in the fourteenth century B.C. King Sesostris had so divided the land among all Egyptians that each received a rectangle of the same size and was taxed accordingly. If a man lost any of his land by the annual overflow of the Nile, he had to report the loss to the Pharaoh who would then send an overseer to measure the loss and make a proportionate abatement of the tax. Thus from the soil of Egypt the science of geometry – *geo* meaning earth, *metron* meaning measure – arose and flourished. Herodotus may have correctly selected the reason for the emphasis on geometry in Egypt but seems to have overlooked its existence for millenniums preceding the fourteenth century B.C.

Egyptian and Babylonian geometry was of the rule-of-thumb or practical variety. Straight lines meant no more than stretched pieces of cord; the Greek word 'hypotenuse', in fact, means 'stretched against', presumably against the two arms of the right angle. A plane was merely the surface of a piece of flat land. Their formulas for volume of granaries and areas of land were arrived at by trial and error. As a consequence, many of these formulas were definitely faulty. For example, an Egyptian formula for the area of a circle was 3·16 times the square of the radius. This is not correct though close enough for the uses the Egyptians made of it.

The Egyptians and Babylonians made numerous practical applications of their mathematics. Their papyri and clay tablets show promissory notes, letters of credit, mortgages, deferred payments, and the proper apportionment of business profits. Although arithmetic and algebra were used in such commercial transactions, geometrical formulas produced the areas of fields and the amounts of grain stored in cylindrical and pyramidal granaries. In addition, the Babylonians and Egyptians were indefatigable builders. Even in this age of skyscrapers their temples and pyramids appear to us to be admirable engineering achievements. The Babylonians were also highly skilled irrigation engineers. Through cleverly dug canals, the Tigris and Euphrates Rivers, the life's blood of these people, fertilized the land and made possible in that dry, hot climate the support of thriving and populous cities such as Ur and Babylon.

But it is a mistake – no matter how often it is repeated – to believe that mathematics in Egypt and Babylonia was confined just to the solution of practical problems. This belief is as false for those times as it is for our own. Instead we find, upon closer investigation, that the exact expression of man's thoughts and emotions, whether artistic, religious, scientific, or philosophical, involved then, as today, some aspect of mathematics. In Babylonia and Egypt the association of mathematics with painting, architecture, religion, and the investigation of nature was no less intimate and vital than its use in commerce, agriculture, and construction.

Those writers who believe that mathematics possesses only utilitarian value often read into history a practical motivation for mathematical activity that logically could not have existed. Their argument runs like this: mathematics was applied to calendar reckoning and navigation; hence the creation of mathematics was motivated by these practical problems much as the need to count led to the number system. This *post hoc ergo propter hoc* type of argument has no history and very little probability in its favour. No mariner lost at sea suddenly decided that the stars were the answer to his navigation problem; nor did some Egyptian farmer, concerned about the number of days until the annual flood of the Nile, decide that he would thereafter watch the course of the sun.

Preceding the use of astronomy and of mathematics for navigation and calendar reckoning there must have been centuries during which men filled with instinctive wonder and awe of nature, men with irrepressible philosophical drives, patiently observed the movement of the sun, moon, and stars. These seers, obsessed by the mystery of nature, overcame the handicaps of lack of instruments and woefully inadequate mathematics to distill from their observations the patterns the heavenly bodies describe. These are the men who very early in the Egyptian civilization learned that the solar year, the year of the seasons, consists of about 365 days.

Their patience and persistence accomplished even more. They observed that the star Sirius appeared in the sky at sunrise on that day of the year when the annual flood of the Nile reached Cairo. This observation must have been made for many years before it was decided to chart Sirius' path in the heavens in order to predict the flood. More than that, since the calendar year of 365 days was a quarter of a day short of the true solar year, after several years the calendar no longer told when Sirius would appear in the sky at dawn. Only after 1,460 years, that is, 4 × 365, would the calendar and the position of Sirius in the sky agree once more. This period of 1,460 years, called the Sothic cycle, was also known to the Egyptian astronomers. Surely the existence of such regularities in the heavens had to be recognized before anyone could think of applying them.

Once the astronomical and mathematical studies revealed these regularities, the Babylonian and Egyptian learned to watch the face of the sky. He hunted, fished, sowed, reaped, danced, and performed religious ceremonies at the times the heavens dictated. Soon particular constellations received the names of the activities their appearance sanctioned. Sagittarius, the hunter, and Pisces, the fish, are still in the sky.

The heavens decided the time of events. But such imperious masters would brook no tardiness in compliance with their orders. The Egyptian, who made his living by tilling the soil which the Nile covered with rich silt during its annual overflow of the country, had to be well prepared for the flood. His home, equipment, and cattle had to be temporarily removed from the area, and arrangements made for sowing immediately afterwards. Hence the coming of the flood had to be predicted. Not only in Egypt but in all lands it was necessary to know beforehand the time for planting and the coming of holidays and days of sacrifice.

Prediction was not possible, however, by merely keeping count of the passing days and nights. For the calendar year of 365 days soon lost all relation to the seasons just because it was short by a quarter of a day. Prediction of a holiday or the Nile flood even a few days in advance required an accurate knowledge of the motions of the heavenly bodies and of mathematics that was possessed only by the priests. These votaries, knowing the importance of the calendar for the regulation of daily life and for provident preparation, capitalized on this knowledge to secure power over the uninformed masses. In fact, it is believed that the

Egyptian priests knew the solar year, that is, the year of the seasons, to be 365¼ days in length but deliberately withheld this knowledge from the people. Knowing also when the flood was due, the priests could pretend to bring it about with their rites while making the poor farmer pay for the performance. Knowledge of mathematics and science was power then as it is today.

Though wonder about the heavens led to mathematics through its respectable relation astronomy, religious mysticism, itself an expression of wonder about life, death, wind, rain, and the panorama of nature, soon fastened on mathematics through its now disreputable relation astrology. Of course, the importance of astrology in ancient religions must not be judged by its discredited position today. In almost all these religions, the heavenly bodies, the sun especially, were gods who ruled over events on the Earth. The will and plans of these gods might be fathomed by studying their activities, their regular comings and goings, the sudden visitations of meteors, and the occasional eclipses of the sun and moon. It was as natural for the ancient priests to work out formulas for the divination of the future based on the motions of the planets and star constellations as it is for the modern scientist to study and master nature with his techniques.

Even if the heavenly bodies had not been gods, a scientifically immature people would have had good reasons to associate the positions of the sun, moon, and stars with human affairs. The dependence of crops upon the sun and upon weather in general, the mating of animals at definite seasons of the year, the periodicity in women, which even Aristotle and Galen believed to be controlled by the action of the moon, and numerous other similar associations, all lent strong credence to such a doctrine. To the Egyptians, in particular, the coming of the Nile flood on just the day that Sirius appeared in the sky at sunrise meant one thing: Sirius caused the flood.

Religious mysticism expressed itself directly *de more geometrico* in the construction and orientation of beautiful temples and pyramids. Every major Babylonian city built a *ziggurat*, a temple in the form of a tower. This was an imposing edifice erected on top of a succession of terraces, approached by broad flights of steps, and clearly visible for miles around. The

Egyptian pyramids and temples are, of course, well known. The pyramids in particular were constructed with special care because they were royal tombs, and the Egyptians believed that construction according to exact mathematical prescriptions was essential for the future life of the dead. The orientation of these religious structures in relation to the heavenly bodies is well illustrated by the famous temple of Amon-Ra, the sun god, at Karnak. The building faced the setting sun at the summer solstice, and on that day the sun shone directly into the temple and illuminated the rear wall.

Nor did religious mysticism overlook the intriguing properties of numbers as a vehicle for expressing its ideas. The numbers three and seven attracted special attention. Since the universe was evidently constructed in a definite period of time, why not utilize a desirable number like seven? That it should be a matter of days seemed a good compromise between the power of God and the complexity of nature.

The science of the cabala illustrates how far religionists were willing to go to explain the mystery of the universe in terms of number. Tradition credits the Babylonian priests with the invention of this mystic and demoniac science of numbers, which the Hebrews later expanded. This pseudo-science was based on the following idea. Each letter of the alphabet was associated with a number. In fact, the Greeks and Hebrews used the letters of the alphabet as their number symbols. With each word was associated the number that was the sum of the numbers attached to the letters spelling the word. Two words having the same associated number were believed to be related, and this connection was used to make predictions. Thus a man's death might be prophesied because the numbers attached to the name of some enterprise he planned to undertake and to the word death were the same.

Man's artistic interests vied with his religious feelings to discover and utilize mathematical knowledge. While the architects studied and applied geometry to the design and construction of beautiful public buildings, temples, and royal palaces, the painters were attracted by geometrical figures as a means of expressing their conceptions of beauty. Artists of the city of Susa, in Persia, used geometrical forms six thousand years ago in a con-

ventionalized artistic style as sophisticated as that of modern
abstract art. Goats, whose fore and hind quarters were triangles
and whose horns were sweeping semi-circles, and storks, whose
bodies and heads were drawn as large and small triangles, decor-
ated their pottery. Geometry was not, as Herodotus claimed, the
gift of the Nile alone. The artists too presented this gift to civilization.

The Egyptian and Babylonian civilizations drew inspiration
for mathematical activity from many human needs and interests,
but they fell short of greatness both in their understanding of
mathematics and in their actual contributions to the subject.
They accumulated simple formulas and numerous elementary
rules and techniques, all of which answered questions arising in
particular situations. There was, however, no general development of a subject nor do the texts enunciate any general principles. The Ahmes papyrus, from which we derive most of our
knowledge of Egyptian mathematics, merely works out specific
problems; no explanations or reasons for the operation are furnished. It has been suggested that the Babylonian and Egyptian
priests may have possessed general mathematical principles and
may have kept that knowledge secret. This is largely speculation,
supported partly by the title of the Ahmes papyrus: *Directions
for Obtaining Knowledge of All Dark Things*, and partly by the
general character of the Egyptian theocracy with its oral transmission of knowledge and its attempt to develop in the people a
reverence for the ruling class.

The failure to build a major scientific body of knowledge or to
encompass details in some broad synthesis is noticeable also in
Egyptian and Babylonian astronomy. During thousands of years
of observation no theory was ever developed to correlate and
illuminate the observations.

Too much has been made of the mathematics used in the construction of pyramids and temples as evidence of the profundity
of ancient mathematics. It is pointed out by some writers that the
sides of a pyramid are almost exactly the same length and that
the right angles are very close to 90°. Not mathematics, however,
but care and patience were required to obtain such results. Accurate computers are not necessarily great mathematicians and

neither were the pyramid builders. What is amazing about their work was the organization and engineering of such large-scale efforts.

From the modern point of view Egyptian and Babylonian mathematics was defective in another very important respect: the conclusions were established empirically. It will profit us shortly if we examine the method by which the Egyptian and Babylonian acquired his formulas.

Suppose a farmer wished to enclose 100 square feet of area as cheaply as possible and desired to have the area rectangular in shape. To keep the cost of fencing low he would want the perimeter to be as small as possible. Now he can lay out a rectangle with 100 square feet of area by using dimensions such as 50 by 2 feet, 20 by 5 feet, 8 by 12½ feet, and many other combinations. The perimeters of these various rectangles, however, are not the same despite the fact that the areas are all 100 square feet. For example, the dimensions 2 by 50 require a perimeter of 104 feet; the dimensions 5 by 20 require a perimeter of only 50 feet; and so forth. From our few calculations we can readily see that the differences in perimeter for different dimensions can be considerable.

Now the farmer is in a plight. If he knows some arithmetic he can try various dimensions which yield an area of 100 square feet and take those which yield the smallest perimeter. But since the possibilities are infinite he can never try all of them; hence he cannot determine the best choice. An alert farmer might notice that the more nearly equal the two dimensions are the smaller the perimeter required. He might suspect, then, that the square with dimensions 10 feet by 10 feet requires the smallest perimeter. But he could not be sure. His trial-and-error procedure, however, has led to a likely conclusion, namely, that of all rectangles with a given area, the square has the least perimeter.

The farmer would no doubt use this conjecture and, because arithmetic and continued experience with rectangular areas support this conclusion, it would be handed down to posterity as a reliable mathematical fact. Of course, the conclusion is by no means established and no modern mathematics student would be permitted to 'prove' it in this manner. About the best that can be said for this ancient approach to mathematical knowledge is that it substitutes patience for brilliance.

One other aspect of the mathematics of ancient times deserves our attention. The priests monopolized all learning, mathematics included, in order to use it for their own ends. Knowledge gave them power, and by restricting knowledge they reduced the likelihood that anyone would be able to challenge that power. Moreover, ignorance begets fear and people who are afraid turn to leaders who will guide and reassure them. In this way, the priests reinforced their position and were able to maintain their rule over the people. The theocracies of Babylonia and Egypt compare very unfavourably with civilizations in which there was no dominant priestly class. We shall see that the few hundred years during which the Greeks flourished and the last few hundred years of our modern era produced infinitely more knowledge and progress than the millenniums of the two ancient civilizations.

III

The Birth of the Mathematical Spirit

Whatever we Greeks receive, we improve and perfect.

Plato

THERE is a story told of Thales that once during an evening walk he became absorbed in observation of the stars and fell into a ditch. A woman accompanying him exclaimed, 'How canst thou know what is doing in the heavens, when thou seest not what is at thy feet?' Thales, however, did do many things simultaneously and successfully. During one lifetime he not only founded Greek mathematics, observed the stars, and took nature walks with congenial companions, but also fathered Greek philosophy, contributed a major cosmological theory, travelled extensively, made notable contributions to astronomy, and realized enormous success in business.

Thales, along with most of the early Greek mathematicians, learned the elements of algebra and geometry from the Egyptians and the Babylonians. In fact, many of these scholars came from Asia Minor, which inherited the Babylonian culture. Others, born on the Greek mainland, went to Egypt and studied there. Despite the unquestioned influence of Egypt and Babylonia on Greek minds, the mathematics produced by the Greeks differed radically from that which preceded it. Indeed, from the point of view of the twentieth century, mathematics and, it may well be added, modern civilization began with the Greeks of the classical period, which lasted from about 600 to 300 B.C.

The mathematics that existed before Greek times has already been characterized as a collection of empirical conclusions. Its formulas were the accretion of ages of experience much as many medical practices and remedies are today. Though experience is no doubt a good teacher, in many situations it would be a most inefficient way of obtaining knowledge. Who would erect a mile-long bridge to determine whether a particular steel cable could support it? The method of trial and error may be direct but it may also be disastrous.

Is experience the only way of obtaining knowledge? Not for beings endowed with a reasoning faculty. Reasoning can follow many routes, among which is the commonly travelled one of analogy. The Egyptians, for example, believed in immortality and so they buried their dead with clothes, utensils, jewellery, and other things that might be of use in the next world. Their reasoning was that since life on Earth required these articles, the after-life would also.

Reasoning by analogy is useful, but it also has its limitations. There may not be an analogous situation at all; airplanes, radios, and submarines could hardly have been invented by reasoning by analogy. Or, there may be an analogous situation that differs slightly but enough to matter a great deal. Though human beings resemble apes, some conclusions about humans cannot be drawn from a study of the apes.

A more commonly used method of reasoning is known as induction. A farmer may observe that heavy rains during several successive springs were followed by excellent crops. He concludes that heavy rains are beneficial to crops. Again, because a person may have had unfortunate experiences in dealing with lawyers, he concludes that all lawyers are undesirable people. Essentially, the inductive process consists in concluding that something is *always* true on the basis of a limited number of instances.

Induction is the fundamental method of reasoning in experimental science. Suppose a scientist heats a given quantity of water from 40° to 70° and sees that the volume occupied by the water increases. If he is a good scientist, he will draw no conclusion as yet but will repeat the experiment many times. Let us suppose that he observes the same expansion each time. He will then declare that water expands as it is heated from 40° to 70°. This conclusion is obtained by inductive reasoning.

Though the conclusions obtained by inductive reasoning seem warranted by the facts, they are not established beyond all doubt. Logically these conclusions are not any better established than the generalization drawn from the observation of four hundred million Chinese that all human beings are yellow-skinned. In other words, we cannot be certain of any conclusion obtained by inductive reasoning. There are other limitations to this type of reasoning. We cannot conclude inductively what the effect on society of an untried law may be. Nor can we conclude inductively, as one uncritical observer did, that all Indians walk single file by seeing one do so!

The several methods of obtaining conclusions, each undoubtedly useful in a variety of situations, possess a common limitation: even if the facts of experience, or the facts on which reasoning by analogy or induction are based, are entirely correct, the conclusion obtained is not certain, and where certainty is vital these methods are practically useless.

Fortunately, there is a method of reasoning that does guarantee the certainty of the conclusions it produces. The method is known as deduction. Let us consider some examples. If we accept the facts that all apples are perishable and that the object before us is an apple, we *must* conclude that this object is perishable. As another example, if all good people are charitable and if I am good, then I must be charitable. And if I am not charitable I am not good. Again, we may argue deductively from the premises that all poets are intelligent and that no intelligent people deride mathematics, to the inevitable conclusion that no poet derides mathematics.

It does not matter, in so far as the reasoning is concerned, whether we agree with the premises. What is pertinent is that if we accept the premises we must accept the conclusion. Unfortunately, many people confuse the acceptability or truth of a conclusion with the validity of the reasoning that leads to this conclusion. From the premises that all intelligent beings are humans and that readers of this book are human beings, we might conclude that all readers of this book are intelligent. The conclusion is undoubtedly true but the purported deductive reasoning is invalid because the conclusion does not necessarily follow from the premises. A moment's reflection shows that even though all intelligent beings are humans there may be human beings who are not intelligent, and nothing in the premises tells us to which group of human beings the readers of this book belong.

Deductive reasoning, then, consists of those ways of deriving new statements from accepted facts that compel the acceptance of the derived statements. We shall not pursue at this point the question of why it is that we experience this mental conviction. What is important now is that man has this method of arriving at new conclusions and that these conclusions are unquestionable if the facts we start with are also unquestionable.

Deduction, as a method of obtaining conclusions, has many advantages over trial and error or reasoning by induction and analogy. The outstanding advantage is the one we have already mentioned, namely, that the conclusions are unquestionable if the premises are. Truth, if it can be obtained at all, must come from certainties and not from doubtful or approximate inferences. Second, in contrast to experimentation, deduction can be carried on without the use or loss of expensive equipment. Before the bridge is built and before the long-range gun is fired, deductive reasoning can be applied to decide the outcome. Sometimes deduction has the advantage of being the only available method. The calculation of astronomical distances cannot be carried out by applying a yardstick. Moreover, whereas experience confines us to tiny portions of time and space, deductive reasoning may range over countless universes and aeons.

With all of its advantages, deductive reasoning does not supersede experience, induction, or reasoning by analogy. It is true that 100 per cent certainty can be attached to the conclusions of deduction when the premises can be vouched for 100 per cent. But such unquestionable premises are not necessarily available. No one, unfortunately, has been able to vouchsafe the premises from which a cure for cancer could be deduced. For practical purposes, moreover, the certainty deduction grants is sometimes superfluous. A high degree of probability may suffice. For centuries the Egyptians used mathematical formulas drawn from experience. Had they waited for deductive proof the pyramids at Giza would not be squatting in the desert today.

Each of these various ways of obtaining knowledge, then, has its advantages and disadvantages. Despite this fact, the Greeks insisted that *all mathematical conclusions be established only by deductive reasoning*. By their insistence on this method, the Greeks were discarding all rules, formulas, and procedures that had been obtained by experience, induction, or any other non-deductive method and that had been accepted in the body of mathematics for thousands of years preceding their civilization. It would seem, then, that the Greeks were destroying rather than building; but let us withhold judgement for the present.

Why did the Greeks insist on the exclusive use of deductive proof in mathematics? Why abandon such expedient and fruitful ways of obtaining knowledge as induction, experience, and analogy? The answer can be found in the nature of their mentality and society.

The Greeks were gifted philosophers. Their love of reason and their delight in mental activity distinguished them from other peoples. The educated Athenians were as much devoted to philosophy as our smart-set is to night-clubbing; and pre-Christian fifth-century Athens was as deeply concerned with the problems of life and death, immortality, the nature of the soul, and the distinction between good and evil as twentieth-century America is with material progress. Philosophers do not reason, as do scientists, on the basis of personally conducted experimentation or observation. Rather their reasoning centres about abstract concepts and broad generalizations. It is difficult, after all, to experiment with souls in order to arrive at truths about them. The natural tool of philosophers is deductive reasoning, and hence the Greeks gave preference to this method when they turned to mathematics.

Philosophers are, moreover, concerned with truths, the few, immaterial wisps of eternity that can be sifted from the bewildering maze of experiences, observations, and sensations. Certainty is the indispensable element of truth. To the Greeks, therefore, the mathematical knowledge accumulated by the Egyptians and Babylonians was a house of sand. It crumbled to the touch. The Greeks sought a palace built of ageless, indestructible marble.

The Greek preference for deduction was, surprisingly, a facet of the Hellenic love for beauty. Just as the music lover hears music as structure, interval, and counterpoint, so the Greek saw beauty as order, consistency, completeness, and definiteness. Beauty was an intellectual as well as an emotional experience. Indeed, the Greek sought the rational element in every emotional experience. In a famous eulogy Pericles praises the Athenians who died in battle at Samos not merely because they were courageous and patriotic, but because reason sanctioned their deeds. To people who identified beauty and reason, deductive arguments naturally appealed because they are planned, consistent, and complete, while conviction in the conclusions offers the beauty of truth. It is no wonder, then, that the Greeks regarded mathematics as an art, as architecture is an art though its principles may be used to build warehouses.

Another explanation of the Greek preference for deduction is found in the organization of their society. The philosophers, mathematicians, and artists were members of the highest social class. This upper stratum either completely disdained commercial pursuits and manual work or regarded them as unfortunate necessities. Work injured the body and took time from intellectual and social activities and the duties of citizenship.

Famous Greeks spoke out unequivocally about their disdain of work and business. The Pythagoreans, an influential school of philosophers and religionists we shall soon meet, boasted that they had raised arithmetic, the tool of commerce, above the needs of merchants. They sought knowledge, not wealth. Arithmetic, said Plato, should be pursued for knowledge and not for trade. Moreover, he declared the trade of a shopkeeper to be a degradation for a freeman and wished the pursuit of it to be punished as a crime. Aristotle declared that in a perfect state no citizen should practice any mechanical art. Even Archimedes, who contributed extraordinary practical inventions, cherished his discoveries in pure science and considered every kind of skill connected with daily needs ignoble and vulgar. Among the Boeotians there was a very decided contempt for work. Those who defiled themselves with commerce were excluded from state office for ten years.

The Greek attitude towards work might have had little influence on their culture were it not for the fact that they did possess a large slave class to whom they could 'pass the buck'. Slaves ran the businesses and the households, did unskilled and technical work, managed the industries, and practised even the most important professions such as medicine. The slave basis of classical Greek society fostered a divorce of theory from practice and the development of the speculative and abstract side of science and mathematics with a consequent neglect of experimentation and practical applications.

In view of the eschewal of commerce and trade by the Greek upper class – certainly a contrast to the preoccupation of our highest social class with finance and industry – it is not hard to understand the preference for deduction. If a person does not 'live' in the world about him, experience teaches him very little. Similarly, in order to reason inductively or by analogy he must be willing to go about and observe the real world. Experimentation would certainly be alien to thinkers who frowned upon the use of the hands. Since the Greeks were not idlers they fell quite naturally into the mode of inquiry that suited their tastes and social attitudes.

Jonathan Swift observed and ridiculed this isolation of Greek culture, as well as its influence on the abstract nature of what he believed to be the pseudo-science of his own day. When Gulliver is led on a tour of inspection of Laputa, he observes:

Their houses are very ill built, the walls bevil, without one right angle in any apartment, and this defect ariseth from the contempt they bear to practical geometry, which they despise as vulgar and mechanic, those instructions they give being too refined for the intellectuals of their workmen, which occasions perpetual mistakes. And although they are dexterous enough upon a piece of paper in the management of the rule, the pencil, and the divider, yet in the common actions and behaviour of life, I have never seen a more clumsy, awkward, and unhandy people, nor so slow and perplexed in their conceptions upon all other subjects, except those of mathematics and music.

Nevertheless, Greek insistence on deductive reasoning as the sole method of proof in mathematics was a contribution of the first magnitude. It removed mathematics from the carpenter's tool box, the farmer's shed, and the surveyor's kit, and installed it as a system of thought in man's mind. Man's reason, not his senses, was to decide thenceforth what was correct. By this very decision reason effected an entrance into Western civilization, and thus the Greeks revealed more clearly than in any other manner the supreme importance they attached to the rational powers of man.

The exclusive use of deduction has, moreover, been the source of the surprising power of mathematics and has differentiated that subject from all other fields of knowledge. In particular, therein lies one sharp distinction between mathematics and science, for science also uses conclusions obtained by experiment-ation and induction. Consequently, the conclusions of science occasionally need revision and sometimes must be thrown overboard entirely, whereas the conclusions of mathematics have stood for thousands of years even though the reasoning in some cases has had to be supplemented.

Had the Greeks done no more to the character of mathematics than to convert it from an empirical science into a deductive system of thought their influence on history would still have been enormous. But their contributions only began there.

A second vital contribution of the Greeks consisted in their having *made mathematics abstract*. Earlier civilizations learned to think about numbers and operations with numbers somewhat abstractly, but only in the unconscious manner in which we as children learned to think about and manipulate them. Geo-metrical thinking, before Greek times, was even less advanced. To the Egyptians, for example, a straight line was quite literally no more than either a stretched rope or a line traced in sand. A rectangle was a fence bounding a field.

With the Greeks not only was the concept of number con-sciously recognized but also they developed *arithmetica*, the higher arithmetic or theory of numbers; at the same time mere computation, which they called *logistica* and which involved hardly any appreciation of abstractions, was deprecated as a skill in much the same way as we look down upon typing today. Similarly in geometry, the words *point, line, triangle,* and the like became mental concepts merely suggested by physical objects but differing from them as the concept of wealth differs from land, buildings, and jewellery and as the concept of time differs from a measure of the passage of the sun across the sky.

The Greeks eliminated the physical substance from mathematical concepts and left mere husks. They removed the Cheshire cat and left the grin. Why did they do it? Surely it is far more difficult to think about abstractions than about concrete things. One advantage is immediately apparent – the gain in generality. A theorem proved about the abstract triangle applies to the figure formed by three match sticks, the triangular boundary of a piece of land, and the triangle formed by the Earth, sun, and moon at any instant.

The Greeks preferred the abstract concept because it was, to them, permanent, ideal, and perfect whereas physical objects are short-lived, imperfect, and corruptible. The physical world was unimportant except in so far as it suggested an ideal one; man was more important than men. The strong preference for abstractions will be evident from a brief glance at the leading doctrine of Greece's greatest philosopher.

Plato was born in Athens about 428 B.C. of a distinguished and active Greek family, at a time when that city was at the height of her power. While still a youth he met Socrates and later supported him in the defence of the aristocracy's leadership of Athens. When the democratic party took power, Socrates was sentenced to drink poison and Plato became *persona non grata* in Athens. Convinced that there was no place in politics for a man of conscience – of course, politics was different in those days – he decided to leave the city. After travelling extensively in Egypt and visiting the Pythagoreans in lower Italy, he returned to Athens about 387 B.C. where he founded his academy for philosophy and scientific research. Plato devoted the latter forty of his eighty years of life to teaching, writing, and the making of mathematicians. His pupils, friends, and followers were the greatest men of his age and of many succeeding generations, and among them could be found every noteworthy mathematician of the fourth century B.C.

There is, Plato maintained, the world of matter, the Earth and the objects on it, which we perceive through our senses. There is also the world of spirit, of divine manifestations, and of ideas such as Beauty, Justice, Intelligence, Goodness, Perfection, and the State. These abstractions were to Plato as the Godhead is to the mystic, the Nirvana to the Buddhist, and the spirit of God to the Christian. Whereas our senses grasp the passing and the concrete, only the mind can attain the contemplation of these eternal ideas. It is the duty of every intelligent man to use his mind towards this end, for these ideas alone, and not the daily affairs of man, are worthy of attention. These idealizations, which are the core of Plato's philosophy, are on exactly the same mental level as the abstract concepts of mathematics. To learn how to think about the one is to learn how to think about the other. Plato seized upon this relationship.

In order to pass from a knowledge of the world of matter to the world of ideas, he said, man must prepare himself. Light from the highest realities, which reside in the divine sphere, blinds the person who is not trained to face it. He is, to use Plato's own famous figure, like one who lives continually in the deep shadows of a cave and is suddenly brought out into the sunlight. To make the transition from darkness to light, mathematics is the ideal means. On the one hand, it belongs to the world of the senses, for mathematical knowledge pertains to objects on this Earth. It is, after all, the representation of properties of matter. On the other hand, considered solely as idealization, solely as an intellectual pursuit, mathematics is indeed distinct from the physical objects it describes. Moreover, in the making of proofs, physical meanings must be shut out. Hence mathematical thinking prepares the mind to consider higher forms of thought. It purifies the mind by drawing it away from the contemplation of the sensible and perishable to the eternal. The path to salvation, then, to the understanding of Truth, Beauty, and Goodness, led through mathematics. This study was an initiation into the Mind of God. In Plato's words, '. . . geometry will draw the soul towards truth, and create the spirit of philosophy. . . .' For geometry is concerned not with material things but with points, lines, triangles, squares, and so on, as objects of pure thought.

Arithmetic, too, said Plato, 'has a very great and elevating effect, compelling the soul to reason about abstract numbers, and

rebelling against the introduction of visible or tangible objects
into the argument.' He advised 'the principal men of our State to
go and learn arithmetic, not as amateurs, but they must carry on
the study until they see the nature of numbers with the mind
only.'

To sum up Plato's position: a modicum of geometry and calculation suffice for practical needs; however, the higher and
more advanced portions tend to lift the mind above mundane
considerations and enable it to apprehend the final aims of philosophy, the idea of the Good. For this reason Plato recommended that the future philosopher-kings be trained for ten years,
from the age of twenty to the age of thirty, in the study of the
exact sciences: arithmetic, plane geometry, solid geometry, astronomy, and harmonics. In his stress on mathematics as a preparation for philosophy, Plato spoke not merely for his followers
and for his generation but for the whole classical Greek age.

The Greek preference for idealizations and abstractions expressed itself in philosophy and mathematics. It showed itself
just as clearly in art. Greek sculpture of the classical period dwelt
not on particular men and women but on ideal types (Plates i and
ii). This idealization extended to standardization of the ratios of
the parts of the body to each other. No finger or toenail was
overlooked in Polyclitus' prescriptions of these ratios. The
modern practice in beauty contests of awarding the prize to the
girl whose measurements most closely approximate an established
standard is a continuation of the Greek interest in an ideal figure.

The faces and postures of the classical Greek draped and un-
draped figures, at least until the decadent 'Laocoön', show no
emotion or concern. Judged by their facial expressions the Greek
gods and the Greek people neither thought, nor laughed, nor
worried. Their demeanour is calm even in pieces of sculpture depicting dramatic action. The faces are as serene as we could
expect those of man in the abstract to be. Particular emotions are,
after all, a matter of the moment, whereas these sculptors were
depicting the eternal in the nature of man. This epic style of
sculpture contrasts sharply with what is found in the numerous
busts and statues of military and political leaders done in the
Roman period (Plate iii).

The Greeks standardized their architecture as they did their sculpture. Their simple and austere buildings were always rectangular in shape; even the ratios of the dimensions were fixed. The Parthenon at Athens (Plate IV) is an example of the style and proportions found in almost all Greek temples. The insistence on ideal dimensions is, incidentally, closely related to the Greek insistence on form, form in the abstract, a concept not alien to our day, in which art and abstraction are practically synonymous.

The insistence on deductive and abstract mathematics created the subject as we know it. Both of these characteristcs were imparted by philosophers. Despite the fact that mathematics was born of Greek philosophy, many great mathematicians and some of the not so great have been extremely scornful of all philosophic speculation. Of course this attitude is no more than an expression of narrowness. These mathematicians are in their chosen field like mighty rivers that wear down mountains to reach the sea but whose paths are then confined to narrow gorges. Their power has enabled them to penetrate deeply below the surface they started to explore but has also enclosed and entrapped them in high walls over which they can no longer see. These disdainful mathematicians overlook the fact that the deepest and mightiest rivers are continually fed by tenuous, vaguely defined clouds. So, too, do the clouds of philosophic thought distil their essence into mathematical streams.

The Greeks put their stamp on mathematics in still another way that has had a marked effect on its development, namely, by their emphasis on geometry. Plane and solid geometry were thoroughly explored. A convenient method of representing quantities, however, was never developed nor were efficient methods of reckoning with numbers. Indeed, in computational work they even failed to utilize techniques the Babylonians had created. Algebra in our present sense of a highly efficient symbolism and numerous established procedures for the solution of problems was not even envisioned. So marked was this disparity of emphasis that we are impelled to seek the reasons for it. There are several.

We mentioned earlier that in the classical period industry, commerce, and finance were conducted by slaves. Hence the educated people, who might have produced new ideas and new methods for handling numbers, did not concern themselves with such problems. Why worry about the use of numbers in measurement if one doesn't measure, or in trading if one dislikes trade? Nor do philosophers need the numerical dimensions of even one rectangle to speculate about the properties of all rectangles.

Like most philosophers the Greeks were star-gazers. They studied the heavens to penetrate the mysteries of the universe. But the use of astronomy in navigation and calendar reckoning hardly concerned the Greeks of the classical period. For their purposes, shapes and forms were more relevant than measurements and calculations, and so geometry was favoured. Of these forms, the circle and sphere, suggested of course by superficial observation of the sun, moon, and planets, received the major share of attention. Hence their astronomical interests, too, led the classical Greeks to favour geometry.

The twentieth century seeks reality by breaking matter down – witness our atomic theories. The Greeks preferred to build matter up. For Aristotle and other Greek philosophers the form of an object is the reality to be found in it. Matter as such is primitive and shapeless; it is significant only when it has shape. It is no wonder, then, that geometry, the study of forms, was the special concern of the Greeks.

Finally, it was the solution of a vital mathematical problem that drove the Greek mathematicians into the camp of the geometers. We have already spoken of the fact that the Babylonian civilization, as well as earlier ones, used integers and fractions. The Babylonians were familiar also with a third type of number which arose through the application of a theorem on right triangles.

First, let us examine the theorem. If a right triangle has arms of lengths 3 and 4, the hypotenuse, or side opposite the right angle (AB in fig. 2) has length 5. Now the square of 5, namely 25, is the sum of the squares of 3 and 4; i.e., $5^2 = 3^2 + 4^2$. This relationship among the sides of a right triangle, that is, that the square of the length of the hypotenuse is equal to the sum of the squares of the lengths of the other two sides, is commonly known as the Pythagorean theorem. To the Babylonians and Egyptians the fact, if not a proof, of this relationship was known.

Suppose now that the arms of a right triangle both have length 1 (fig. 3). What would be the length of the hypotenuse? Let us call the hypotenuse x. Then according to the Pythagorean theorem its length must be such that

$$x^2 = 1^2 + 1^2 = 2.$$

Hence x, the length of the hypotenuse, must be a number whose square is 2. We indicate the number whose square is 2 by $\sqrt{2}$ and

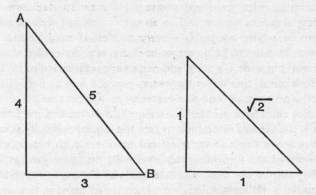

Figures 2 and 3. Two right triangles

call it the square root of 2. But what number equals $\sqrt{2}$? That is, what number multiplied by itself gives 2?

The answer, as the Pythagorean school of mathematicians discovered to its great dismay, is that there is no whole number or fraction whose square is 2. $\sqrt{2}$ is a new kind of number, and they called it *irrational* because it could not be expressed exactly as a ratio of whole numbers, as 4/3 or 3/2. By contrast, whole numbers and fractions are called *rational* numbers. These terms are in use today.

The irrational number is a much neglected topic in the history of thought and a troublesome member of our number system. We have just seen that such numbers must be used in order to represent lengths and they are, moreover, explicitly and implicitly involved in almost all of mathematics. Yet how can we add, subtract, multiply, or divide such numbers ? For example, how can we add 2 and $\sqrt{2}$? How do we divide $\sqrt{7}$ by $\sqrt{2}$?

The Babylonians had a makeshift, though practical, solution of these difficulties. They approximated the value of $\sqrt{2}$. For example, since the square of 14/10 or 1·4 is 1·96, and since 1·96 is nearly equal to 2, 1·4 must be nearly equal to $\sqrt{2}$. An even better approximation to $\sqrt{2}$ is 1·41 because the square of 1·41 is 1·988.

The Babylonian approximation to $\sqrt{2}$ does not permit exact reasoning with irrational numbers, for no matter how many decimal places we are willing to use we cannot write a rational number whose square is *exactly* 2. Yet, if mathematics is to merit its claim to being an exact study, it must evolve a method of working with $\sqrt{2}$ itself and not an approximation of it. To the Greek mind, this difficulty was as genuine and as prepossessing as the problem of food to a castaway on a coral reef.

Not content to use the less scrupulous method of the Babylonians, the Greeks undertook to face the logical difficulty squarely. In order to think about irrational numbers with exactness they conceived the idea of working with all numbers geometrically. They started out this way. A length was chosen to represent the number 1. Other numbers were then represented in terms of this length. To represent $\sqrt{2}$, for example, they used a length equal to the hypotenuse of a right triangle whose sides were one unit in length. The sum of 1 and $\sqrt{2}$ was a length formed by adjoining a unit segment to the length representing $\sqrt{2}$. In this geometrical form the sum of a whole number and an irrational one is no more difficult to conceive than the sum of one and one.

Similarly the product of two numbers, 3 and 5 for example, was expressed geometrically as the area of the rectangle with dimensions 3 and 5. In the case of 3 and 5 the use of area as a way of thinking about the product may be no great advantage. But one can also think of the product of 3 and $\sqrt{2}$ as an area. To think about this second rectangle is no more difficult than to think of the first one; yet it provides an exact way of working with the product of an integer and an irrational number or, for that matter, two irrational numbers.

The Greeks not only operated with numbers in the geometric

manner but went so far as to solve equations involving unknowns by series of geometrical constructions. The answers to these constructions were line segments whose lengths were the unknown values. The thoroughness of their conversion to geometry may be judged from the fact that the product of four numbers was unthinkable in classical Greece because there was no geometric figure to represent it in the manner that area and volume represented the product of two and three numbers respectively. Incidentally, we still speak of a number such as 25 as the *square* of 5 and 27 as the *cube* of 3 in conformity with Greek thought.

The preference of the Greeks for geometry was so marked that during his travel in Laputa, Gulliver was again forced to comment:

The knowledge I had in mathematics gave me great assistance in acquiring their phraseology, which depended much upon that science and music; and in the latter I was not unskilled. Their ideas are perpetually conversant in lines and figures. If they would, for example, praise the beauty of a woman, or any other animal, they describe it by rhombs, circles, parallelograms, ellipses, and other geometrical terms, or by words of art drawn from music, needless here to repeat. I observed in the King's kitchen all sorts of mathematical and musical instruments, after the figures of which they cut up the joints that were served to his Majesty's table.

Because the Greeks converted arithmetical ideas into geometrical ones and because they devoted themselves to the study of geometry, that subject dominated mathematics until the nineteenth century, when the difficulties in treating irrational numbers on an exact, purely arithmetical basis were finally resolved. In view of the clumsiness and complexity of arithmetical operations geometrically performed, this conversion was, from a practical standpoint, a highly unfortunate one. The Greeks not only failed to develop the number system and algebra which industry, commerce, finance, and science must have, but they also hindered the progress of later generations by influencing them to adopt the more awkward geometrical approach. Europeans became so habituated to Greek forms and fashions that Western civilization had to wait for the Arabs to bring a number system from far-off India.

Unfortunate as this Greek perversion of the number system and of algebra may appear to us with our understanding of progress, it still should not invoke on Greek heads the condemnation that has sometimes been heaped there. The one backward step the Greeks took was in itself thoroughly reasonable; moreover, the damage done is heavily outweighed by the incomparable good of their other accomplishments.

When most people describe the Greek contributions to modern civilization, they talk in terms of art, philosophy, and literature. No doubt the Greeks deserve the highest praise for what they bequeathed to us in these fields. Greek philosophy is as alive and significant today as it was then. Greek architecture and sculpture, especially the latter, are more beautiful to the average educated person of the twentieth century than the creations of his own age. Greek plays still appear on Broadway. Nevertheless, the contribution of the Greeks that did most to determine the character of present-day civilization was their mathematics. By altering the nature of the subject in the manner we have related, they were able to proffer their supreme gift. This we proceed to examine.

IV

The *Elements* of Euclid

Euclid alone
Has looked on Beauty bare. Fortunate they
Who, though once only and then but far away
Have heard her massive sandal set on stone.

Edna St Vincent Millay

IN a relatively brief period great intellects such as Thales, Pythagoras, Eudoxus, Euclid, and Apollonius produced an amazing
amount of first-class mathematics. The fame of these men spread
to all corners of the Mediterranean world and attracted numerous pupils. Masters and pupils gathered in schools which,
though they had few buildings and no campus, were truly centres
of learning. The teachings of these schools dominated the entire
intellectual life of the Greeks and, therefore, we shall refer to
them in several different connections.

The Pythagorean school was the most influential in determining both the nature and contents of Greek mathematics. Its
leader, the legend-veiled Pythagoras, was born on the island of
Samos about 569 B.C. Through extensive travel in Egypt and
India he absorbed much of mathematics and mysticism. He then
founded in Croton, a Greek colony of Southern Italy, a community which embraced both mystical and rational doctrines. On
the mystical side the group drew inspiration from Greek religion
and considered it necessary to purify the soul from the taint of
the physical and redeem it from the prison of the body. To
achieve these ends the Pythagoreans maintained celibacy and
performed rituals and ceremonial purgations. In addition, they
believed it necessary to observe certain taboos. They would not
wear wool clothing, eat meats or beans except on the occasion of
a religious sacrifice, touch a white cock, sit on a quart measure,
walk on the high roads, use iron to stir a fire, or leave marks of

ashes on a pot. Once released from a body, the soul was re-incarnated in another. Xenophanes says that one day Pythagoras passed a dog being beaten and he cried out, 'Stop, beat no more, it is the soul of a friend; I recognized it, hearing its complaints.'

The community devoted itself primarily to the study of philosophy, science, and mathematics. As if it could foresee the terrible uses to which some of this knowledge might be put, it pledged new members to secrecy and required them to join up for life. Though membership was restricted to men, women were admitted to lectures, for Pythagoras believed that females were of some value. The esoteric character of the group and its mystic and secret observances aroused the suspicion and dislike of the people of Croton, who finally drove the Pythagoreans out and burned their buildings. Pythagoras fled to Metapontum in Southern Italy and, according to one story, was murdered there. His followers, however, scattered to other Greek centres and continued his teachings.

Concerning other mystical and speculative doctrines of the Pythagoreans we shall say more in a later chapter. At the moment we should mention the fact that the Pythagoreans are credited with giving the subject of mathematics special and independent status. They were the first group to treat mathematical concepts as abstractions, and though Thales and his fellow Ionians had established some theorems deductively, the Pythagoreans employed this process exclusively and systematically. They distinguished mathematical theory from practices such as geodesy and calculation, and proved the fundamental theorems of plane and solid geometry and of *arithmetica,* the theory of numbers. To their dismay they also discovered and proved the irrationality of the square root of two.

More widely known than the Pythagoreans was the Academy of Plato, which had Aristotle as its most distinguished student. (The latter founded his own school, the Lyceum, when he left the Academy at the time of Plato's death.) We saw earlier that Plato's pupils were the most famous philosophers, mathematicians, and astronomers of their age. Under Plato's influence they emphasized pure mathematics to the extent of ignoring all practical applications and they added immensely to the body of knowledge. The school retained its pre-eminence in philosophy long after leadership in mathematics and science passed on to Alexandria. When closed by the Emperor Justinian in the sixth century A.D., it had endured nine hundred years.

The work of the many schools and of isolated individuals who lived all over the Mediterranean area from Asia Minor to Sicily and Southern Italy was unified by Euclid in one masterful book called the *Elements*. This most famous account, formulated about 300 B.C., therefore constitutes the mathematical history of an age as well as the logical presentation of geometry. From a few sagaciously chosen axioms, Euclid deduced all the important results of the Greek masters of the classical period, roughly some five hundred theorems. The axioms, the arrangement, the form of presentation, and the completion of partially developed topics were his.

Much of the material in Euclid's *Elements* is familiar to us through our high-school studies. Nevertheless, before we pass on to consider the significance of this mathematics for our culture we should like to review a few features of this most influential, and to some, most revolting, textbook in history. It is the structure of Euclid that concerns us for the moment.

Geometry, we know, deals with points, lines, planes, angles, circles, triangles, and the like. For Euclid and for the Greeks whose work Euclid was presenting, those terms represented not physical objects themselves, but concepts abstracted from physical objects. Actually only a few properties of the physical object are reflected in the mathematical abstraction to which it gives rise. The stretched string gave rise to the mathematical straight line but the colour of the string and the material of which it was made are not properties of the straight line. To be precise about what his abstract terms included Euclid began with some definitions. A straight line he defined as that which lies evenly between its ends. (The abstraction from the stretched string and the mason's level is clearly evident here.) A point, he said, is that which has no parts. And so on to triangles, circles, polygons, and the like.

In his definitions Euclid went to unnecessary and inadvisable lengths. A logical, self-sufficient system must start somewhere. It cannot hope to define every concept it uses, for definition involves describing one concept in terms of others and the latter in terms of still others. Obviously, if the process is not to be circular, a person must start with some undefined terms and define others in terms of these. For example, Euclid's definition of a point as that which has no parts obviously calls for a definition of parts. Other writers in attempting to improve on Euclid have defined a point as pure position. And what, then, is position? No doubt in some social spheres position is everything in life, but the concept of position does not clarify the meaning of point.

Again, we are saying that not all concepts can be defined in a self-contained system. It is true that all the concepts arise from and represent definite physical objects but these physical meanings are of no help in the process of formal definition because they are not part of mathematics. Surprisingly, the inability to define some of the concepts with which geometry deals causes no hardship, as we shall see in a moment.

Having defined, at least to his satisfaction, the concepts with which he was to deal, Euclid proceeded to the all-important task of establishing facts or theorems about them. To undertake the deductive process he needed premises for, as Aristotle points out,

It is not everything that can be proved, otherwise the chain of proof would be endless. You must begin somewhere, and you start with things admitted but undemonstrable. These are first principles common to all sciences which are called axioms or common opinions.

In the selection of axioms Euclid displayed great insight and judgement. The mathematicians of the leading schools had started with axioms acceptable to them. As the contributions increased in number there was a growing danger that many axioms were being employed which not all mathematicians would regard as unquestionably true of the physical world. Also, there was an unnecessary profusion of axioms, a wasteful state of affairs from a logical standpoint, since it is always better to assume as little as possible and to prove those statements which can be deduced from the axioms already accepted. Hence Euclid's task was to find an adequate and universally acceptable set of axioms for geometry. Moreover, since the geometrical investigations of the Greeks were part and parcel of their search for truth, these axioms had to be unquestionable, absolute truths.

The axioms which Euclid proposed state properties of points, lines, and other geometrical figures that are possessed by their physical counterparts. The properties in question appear to be so obviously true of these physical objects that all men have been willing to agree to them as a basis for further reasoning. The

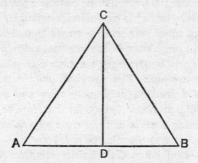

Figure 4. An isosceles triangle

extraordinary merit of Euclid's selection is that though they are immediately acceptable they are no mere superficialities, for they lead to profound consequences. Moreover, he was able to choose a very limited number, ten in all, and yet secure the construction of the whole system of geometry.

Merely to reassure ourselves of the wisdom of Euclid's choice, let us recall one or two of his axioms. He affirmed that, 'It shall be possible to draw a straight line joining any two points'; that, 'It shall be possible to draw a circle with given centre and through a given point'; and that, 'The whole is greater than any of its parts.' Surely these are unassailable and acceptable to all men.

Having selected the concepts with which geometry is to deal and having chosen basic truths about these concepts, Euclid proceeded to establish theorems or conclusions. The method of proof was, of course, strictly deductive. In order to appreciate fully why later generations esteemed the solidity of Euclid's conclusions, let us review one of his proofs.

An early theorem in Euclid asserts that *the base angles of an isosceles triangle are equal*. This theorem has special interest because, despite its elementary character, it marked the limit of the study of geometry in medieval universities. It has been called the '*pons asinorum*' or the 'bridge of asses' because fools could not comprehend this proof and hence, like asses at a bridge, would proceed no further.

Before we review the proof, let us examine the meaning of the theorem. If ABC (fig. 4) is an isosceles triangle, two sides, say AC and BC, are equal. We wish to prove that the base angles A and B, that is the angles opposite the equal sides, are equal.

The proof begins by drawing the line CD which bisects the angle C of the triangle. The justification for this step is the following. Euclid showed previously that *any* angle could be bisected. Since C is an angle, it too can therefore be bisected. The deductive reasoning here is of the form: all apples are red; here is an apple; hence this apple must be red.

Introduction of the line CD divides triangle ABC into two triangles ACD and DCB. Of these triangles we know, first, that AC equals CB, because the original triangle ABC is stated to be isosceles. Second, angle ACD equals angle DCB because CD is the angle bisector. Third, since CD is common to the two smaller triangles, these triangles have this equal side. We may therefore assert that triangle ACD is congruent to triangle DCB because a previous theorem asserts that *any* two triangles which have two sides and the included angle of one equal to two sides and the included angle of the other are congruent. Since the two triangles in question have such equal parts, these two triangles are congruent. We may assert, finally, that angle A equals angle B because, by the very definition of congruent triangles, corresponding parts are equal, and angles A and B are such corresponding parts.

The theorem in question is therefore proved by several deductive arguments, each of which employs unquestionable premises and yields an unquestionable conclusion. Of course, not all the proofs in Euclid are so simple. Nevertheless, each proof, no matter how complex it may seem to be at first glance, consists of no more than a series of simple deductive arguments.

We need not re-examine one by one the theorems that Euclid established. It will be enough to mention that from the axioms some simple theorems are immediately proved, and that these furnish stepping-stones to more elaborate theorems, the whole structure being marvellously and closely knit. Indeed, many students have fumed that so large a number of seemingly involved theorems can be derived from so few self-evident axioms.

Let us see next that Euclid's topics concern fundamental properties of the sizes and shapes of objects. His first major concern was under what conditions two objects are equal in size and shape, that is, under what conditions these objects are congruent. Suppose, for example, a surveyor has two pieces of land, triangular in shape. How can he establish that the two are equal? Must he measure every side, every angle, and even the areas of both to decide the equality? Not with the theorems of Euclid to aid him. Two triangles are equal in all respects if, for example, the sides of one are known to be equal respectively to the sides of the other. This fact hardly seems to be more than a triviality but the reader can see it is not quite that if he asks himself under what conditions he can guarantee the complete equality of two quadrilaterals, that is, two four-sided figures. Such questions and related ones apply, of course, to all sorts of geometric figures.

Euclid asked next: if figures are not equal, what significant relationship may they bear to each other and what geometric properties can they have in common? The relationship he chose is shape. Figures of unequal size but of the same shape, that is, similar figures, have many geometric properties in common. As applied to triangles, for example, similarity means that the angles of one equal the corresponding angles of the other. From this defining property it follows that the ratio of any two corresponding sides is constant. Thus if ABC and $A'B'C'$ are similar triangles (fig. 5), then $AB/A'B'$ equals $BC/B'C'$. Moreover, if the ratio of two corresponding sides is r, say, then the ratio of the two areas is r^2.

If figures have neither shape nor size in common, what can be said about them? They may, of course, have the same area, or, in geometric terms, they may be equivalent. Or they may be inscribable in the same circle. The number of possible relationships and the questions that can be raised about each are endless. Euclid selected fundamental ones.

All of the concepts Euclid studied he applied not merely to figures formed by straight lines but also to circles and spheres. Interest in these figures was considerable, for, to the Greeks, the circle and the sphere were perfect figures.

From the standpoint of aesthetic appeal another class of

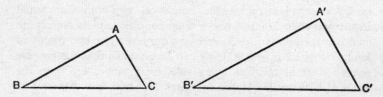

Figure 5. Two similar triangles

figures was equally enticing to them. Among triangles, the equilateral triangle was noteworthy because its sides are equal in length and its angles are of the same size. Among four-sided figures, the square was attractive for the same reason. Plane figures with five, six, and more sides can also be constructed so that the sides of the figure are all equal as are the angles. Such figures are called regular polygons and were studied in detail. Complete surfaces can be formed with regular polygons, any one surface to be built up with only one kind of polygon. For example, the surface of a cube is a complete surface built up by joining six squares along their edges. Such surfaces, of which the cube is but one type, are called regular polyhedra.

One of the first questions raised in connection with regular polyhedra was, how many different types are there? By masterful reasoning, which we shall not repeat here, Euclid showed that there must be exactly five types of regular polyhedra. These are pictured in fig. 6. Plato admired these figures so much that he could not conceive of God not making use of them. He therefore elaborated on one Greek school of thought, which affirmed that

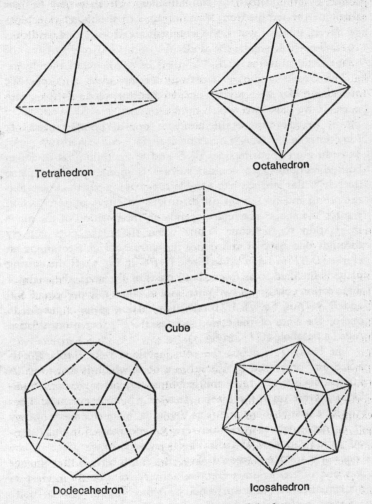

Tetrahedron

Octahedron

Cube

Dodecahedron

Icosahedron

Figure 6. The five regular polyhedra

all objects are composed of four elements, earth, air, fire, and water, by adding that the fundamental particles of fire had the shape of the tetrahedron; those of air had the shape of the octahedron, those of water, the icosahedron, and those of earth, the cube. The fifth shape, the dodecahedron, God reserved for the shape of the universe itself.

The Greeks studied exhaustively another class of curves. We are all familiar with a cone-shaped figure such as an ice cream cone. If we have two such very long cones, placed as shown in fig. 7, we get what mathematicians call a conical surface, or sometimes just a cone. This conical surface consists of two parts, extending on opposite sides of O and to an unlimited extent in both directions. If a conical surface is intersected by a plane (merely a flat surface like a table-top, having no thickness and extending indefinitely in all directions) a curve of intersection results, the shape of which depends on the position of the plane in relation to the cone. Thus when the plane cuts entirely through one part of the cone, the curve of intersection is an ellipse (DEF in fig. 7), or a circle (ABC in fig. 7). If the cutting plane is inclined so as to cut both parts of the cone, the curve of intersection consists of two parts and is called a hyperbola (RST and $R'S'T'$ in fig. 7). If, finally, the cutting plane is parallel to one of the lines of the cone, such as POP', the intersection is called a parabola (GIK in fig. 7).

The basic facts about the conic sections were similarly collected and organized by Euclid in a book which is lost to us. A little after Euclid's time another famous mathematician, Apollonius, wrote on the subject a treatise which is extant and for which he is almost as famous as Euclid is for his *Elements*. Many other mathematical works were created and written in this classical period, but few of these have survived. If we judge by the books and fragments we do have, it is fairly certain that the age was one of tremendous creative activity, of intense interest in mathematics, and of unsurpassed brilliance.

Greek mathematics is as significant for the questions it raised and did not answer as for those it did. Among such questions are three famous ones known to every layman. They are called 'squaring the circle', 'doubling the cube', and 'trisecting the

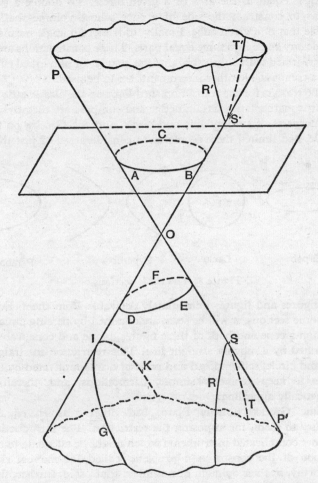

Figure 7. A conical surface and the sections made by intersecting planes

angle'. To square a circle means to construct a square, the area of which is equal to the area of a given circle. To double a cube means to construct the side of a cube whose volume shall be double that of a given cube. Finally, to trisect an angle means to divide *any* angle into three equal parts. These constructions are to be performed with only a straightedge, that is, an unmarked ruler, and a compass. No other instruments are to be used.

The reasons for the restriction shed light on the classic attitude towards mathematics. Straightedge and compass are the physical counterparts of straight line and circle, and the Greeks, on the whole, had limited their geometry to consideration of just these

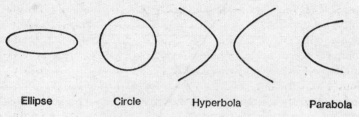

| Ellipse | Circle | Hyperbola | Parabola |

Figure 8. The conic sections

two figures and figures immediately derivable from them. Even the conic sections, it will be seen, are obtained by passing a plane through a cone and both of these figures, plane and cone, can be generated by a moving straight line. This restriction to straight line and circle, self-imposed and arbitrary, was motivated by the desire to keep geometry simple, harmonious, and, therefore, aesthetically appealing.

Some Greeks, notably Plato, had other reasons, equally weighty to them, for imposing the restriction. The introduction of more complicated instruments which might be adequate to the solution of the construction problems called for manual skill unworthy, in their opinion, of a thinker. Plato said, further, that by using complicated instruments 'the good of geometry is set aside and destroyed, for we again reduce it to the world of sense, instead of elevating and imbuing it with the eternal and incorporeal images of thought, even as it is employed by God, for which reason He always is God.'

The three construction problems were very popular in Greece. The first historical reference to them states that the philosopher Anaxagoras passed time in prison trying to square the circle. Despite repeated efforts of the best Greek mathematicians, the problems were not solved. Nor were they to be solved for the next two thousand years. About seventy years ago it was finally proved that the constructions *cannot* be performed under the conditions stated. Despite this fact, people still try and often claim success. We can assert without examining their work that they are in error or have misunderstood the problems.

The long years of labour on these famous problems indicate the care, the rigour, the patience, and the persistence of mathematicians. The questions are not of practical importance, for the constructions can be performed readily by resorting to instruments only slightly more complicated than straightedge and compass. Nevertheless, people with an irrepressible desire to meet intellectual challenges attempted the theoretical constructions.

Actually, the search for iron has often led to gold. The conic sections, which paved the way for modern astronomy, were discovered during attempts to perform the famous constructions, as were hosts of other beautiful and useful mathematical results. In fact, if we were to list those major mathematical ideas arrived at by tackling impractical, 'worthless' problems, we might be led to define mathematics as the development of the trivial. (Many an 'educator', though ignorant of the subject and its history, has not hesitated to render this judgement.) The history of the work on the famous construction problems shows how unfair are the attacks made on the 'impractical' Greeks, for these visionaries did far more for the advancement of our scientific age than did the so-called practical people.

We have already praised the Greeks for having made mathematics abstract. It would be well for our appreciation of the scope of mathematics to see just what this abstractness implies, at least in Euclidean geometry.

Let us consider a rather simple situation. Suppose we select any two fixed points A and B, and a line L not through A or B but in the same plane (fig. 9). Suppose, in addition, we wish to find that point P on line L for which the distance $AP + PB$ is least; that is, if Q is any other point on L then $AP + PB$ must be less than $AQ + QB$. This problem is a purely geometrical one. It is not hard to prove that if P is chosen so that AP and BP make *equal angles with line L*, then the distance $AP + PB$ is least.

Let us take the proof of this theorem for granted, and let us see how the theorem could be applied to practical situations. Sup-

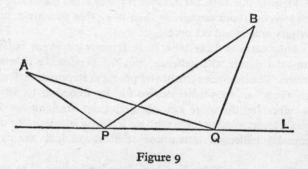

Figure 9

pose A and B are the positions of two towns and L is a river. A pier which will serve both towns is to be built along the river in such a way that the total distance from the pier to town A and from the pier to town B is as short as possible. At what point along the river should the pier be placed? Our general theorem furnishes the answer: At that point P where AP and PB make equal angles with the river.

Let us consider another 'practical' situation. A billiard ball situated at a point A on the table is to be hit so that it will rebound from the side L of the billiard table and hit the ball at B. A billiard ball always behaves so that the angle its path makes in approaching the side of the table equals the angle its path makes in rebounding. That is, in figure 10, angle 1 equals angle 2. Every billiard player knows this fact at least subconsciously and uses it. That is, he directs the ball to the point P so that AP and PB make equal angles with the side. But he undoubtedly does not know that the path he selects, and hopes the ball will take, is the shortest path that a ball can take in going from A to B by rebounding from the side of the table.

Our illustrations show how one mathematical theorem supplies information in two widely different and unrelated situations. There are, in fact, numerous other applications of the same theorem. The fact that a theorem developed to answer a question in one field so often turns out to be vital in a completely different one fills the history of mathematics with surprises. Of course, this broad applicability of mathematics is bought at a price, the price of abstractness, for to achieve theorems about all triangles by working with the ideal triangle, the mathematician must struggle with elusive and sometimes unmanageable thoughts instead of fingering a triangle made of wood.

There is another point about the relation of the abstract theorems of mathematics to their applications that it is very important to keep in mind, namely, that the abstract theorem states the ideal case, whereas the physical situation to which it is applied may fall far short of the ideal. Suppose, for example, we lay out a triangle on the surface of the Earth. Can we apply the theorems of

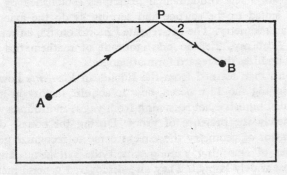

Figure 10

plane geometry to this triangle? In the first place, the Earth is spherical and not flat. Moreover, the Earth's surface is hardly that of a perfect sphere, but is rather irregular. On at least two accounts, then, this triangle on the Earth's surface falls short of the ideal triangle of plane geometry. Hence, there is likely to be some error involved in the use of the mathematical theorem. To the extent that the physical triangle approaches the ideal, the conclusions of mathematics will apply. Failure to recognize this fact can lead to serious error in application.

The creation of Euclidean geometry was more than the contribution of some useful and beautiful theorems. It engendered a rational spirit. No other human creation has demonstrated how much knowledge can be derived by reasoning alone as have the hundreds of proofs in Euclid. The deduction of these numerous and profound results taught the Greeks and later civilizations the power of reason and gave them confidence in what could be achieved by this faculty. Encouraged by this evidence, Western man was inspired to apply reason elsewhere. Theologians, logicians, philosophers, statesmen, and all seekers of truth have imitated the form and procedure of Euclidean geometry.

Even among the Greeks themselves mathematics was set up as the standard for all the sciences. Aristotle, in particular, insisted that each science must consist in the deductive demonstration of truths from a few fundamental principles established by some method suited to the science and serving as do the axioms of Euclidean geometry. The oft-repeated motto on the entrance to Plato's Academy, 'Let no one ignorant of mathematics enter here', epitomizes this regard for mathematics.

Western man learned from the Euclidean *Elements* how perfect reasoning should proceed, how to acquire facility in it, and how to distinguish exact reasoning from vague mouthings which carry merely the pretence of proof. During the course of the development of geometry the Greeks came to recognize general principles of reasoning, among which the syllogistic laws are now most widely known. They also discovered general methods of attack on problems. For example, Plato is credited with devising the analytical attack, which starts with the desired conclusion and deduces consequences until a known fact is reached. The correct proof is then made by reversing the exploratory steps. The reader may recall using this method himself in Euclidean geometry to discover the proof of a theorem. Of course, the method transcends geometry. Greek geometers also discovered and gloried in the power of the indirect method of proof, which pursues the implications of several alternatives in the expectation that all but the correct one will lead to contradictions and hence must be discarded. The logical foundations on which this method rests, known to logicians as the laws of contradiction and excluded middle, were formulated by Aristotle.

The necessity for accurate definition, for clearly stated assumptions, and for rigorous proof also became evident in the course of the work on geometry. Men such as Socrates and Plato not only stressed these needs but in turn contributed to mathematics polish, smoothness, and clear structure. In effect, the grand exercise in logic that geometry afforded to the Greeks led to the construction and systemization by Aristotle of those laws of thought now accepted and applied by all of us. Thus Greek geometry served as the progenitor of the science of logic.

Hundreds of generations since Greek times have also learned how to reason by studying Euclid, a procedure deprecated by many people, who argue that we can learn logic without studying mathematics. The argument is as valid as the statement that we can all conceive of great paintings and so the world would be as well off with the conceptions as with the paintings themselves. Unfortunately, the concept of a painting never stirred a heart.

The importance of Euclid transcends by far its values as a logical exercise and as a model of reasoning. With the development of the beautiful structure and elegant reasoning of geometry, mathematics was transformed from a tool for the advancement of other activities to an art. It was appreciated as such by the Greeks. Arithmetic, geometry, and astronomy were to them music for the soul and the art of the mind.

Indeed, rational and aesthetic as well as moral interests can hardly be separated in Greek thought. Repeatedly we read that the Earth must be spherical because the sphere has the most beautiful shape of all bodies and is therefore divine and good. For the same reason Plato believed that the sun, moon, and stars were each rigidly attached to a sphere that rotated on its own axis about the Earth. Moreover, the path of each body must be a circle, for the circle shared aesthetic appeal with the sphere. The circle and the sphere were the perfect paths which represented the changeless, eternal order of the heavens as contrasted with straight-line motion which prevailed on the imperfect Earth. For

aesthetic and moral reasons, too, it was decided that the heavenly
bodies moved with uniform speed, traversing equal distances in
equal intervals of time. This stately, regular, unhurried motion
befitted heavenly bodies. In fact, argued the Pythagoreans, in-
constant speed for the planets is inadmissible; 'even in the human
sphere, such irregularity is incompatible with the orderly procedure of a gentleman'. The truths of poetry and the truths of
science were one or, to paraphrase Aristotle, nature's purpose and
her deep-seated laws all tend in her multitudinous work to one
form or another of the beautiful.

Geometry, philosophy, logic, and art were all expressions of
one type of mind, one outlook on the universe, and it is fascinating to trace, as some historians have, the existence of common
characteristics in all these phases of the classic Greek culture. For
example, the clear, transparent, and simple structure of Euclidean geometry is a mathematical manifestation of the same
love of clarity and ordered design which the plain, simple forms
of the Greek temple display. Infinitely complex by comparison is
the Gothic cathedral with its multitudinous subordinate interior
and exterior structures. Greek sculpture of the classical period is
also surprisingly simple. No elaborate dress, military decorations,
frills, or furbelows clutter up the statue or detract from the principal theme.

In a like manner, the literary classics of the period were written
in a simple, clear, matter-of-fact style, sparing of imagery and
adjective. We have but to contrast the nightingale whose song has
'charmed magic casements, opening on the foam of perilous seas,
in fairy lands forlorn', with Sophocles' bird who 'sings her clear
note deep in the green glades ivy-grown, sheltered alike from
sunshine and from wind' to perceive the qualities of Greek style.
Lucidity, simplicity, and restraint were the ingredients of beauty.
Greek art is the art of the intellectual, the art of clear thinkers,
and it is, consequently, plain art. Nevertheless, the geometry,
architecture, sculpture, and literature achieve a beauty and
elegance that transcend their simplicity.

Euclidean geometry is often described as being closed and
finite. These adjectives apply in several senses. The subject
matter is limited, as we have seen, to figures that can be con-

structed with straightedge and compass and to theorems that can be derived from a fixed set of axioms. No new axioms are introduced as the reasoning unfolds the subject. Euclidean geometry is finite in the sense, too, that it avoids the infinite. The straight line, for example, is not considered in its entirety by Euclid. Rather, he said, a line segment can be extended as far as necessary in either direction, as though he begrudged the necessity for extension. So too, in treating the whole numbers, the Greeks thought of this set as potentially infinite, that is, infinite only in the sense that more numbers can always be added to any given finite set; they would not deal with the entire collection of whole numbers as an entity in itself.

These characteristics of closure and finiteness are also dominant in Greek architecture. The whole structure of a Greek temple is small, near at hand, completely visible to the observer. It suggests finality, completeness, and definiteness. The eye and mind soon grasp and encompass its proportions and grandeur. The Greek temple may be compared with the Gothic in these respects too. The latter is almost never visualized as a whole. It seems to lose itself in all directions and escape complete comprehension. It suggests great distances and, through its spires, spiritual aspiration. The imagination is stirred and the individual awed by endless vistas of receding arches and by high altars visible in the gloomy interiors as if from a distance, while immense size conjures up impressions of the invisible. The sense of finiteness is vanquished as the high structure swallows the individual and loses him in its dim interiors.

In Greek science the concept of the infinite is scarcely understood and frankly avoided. The simplest type of motion for the Greeks is not, as it is for us, along a straight line, because the straight line is not perceptible in its entirety; straight-line motion is never completed. The Greeks preferred circular motion. The concept of a limitless process frightened them and they shrank before 'the silence of the infinite spaces'.

In philosophy, too, the infinite was avoided. Paradoxes of the infinite, some of which we shall encounter later, proved insurmountable barriers to Greek philosophic thought. Aristotle says the infinite is imperfect, unfinished, and therefore unthinkable.

It is formless and confused. Good and evil were founded, in fact, on the notions of the limited and determinate, for the one, and on the indeterminate and infinite for the other. The limited and definite qualities of objects also gave them character and perfection. Only as objects were distinct and defined did they have a nature and meaning. 'Nothing that is vast enters into the life of mortals without a curse,' says Sophocles.

Another characteristic of Greek mathematics runs throughout the culture. Euclidean geometry is static. The properties of changing figures are not investigated. Rather, figures are given in their entirety and are studied as is. The restful atmosphere of the Greek temple reflects this theme. Mind and spirit are at peace there. So, also, in Greek sculpture the figures are static, aloof, psychologically calm. They are as emotionally aroused as an equilateral triangle. Myron's 'Discus Thrower' (Plate 11), who is about to make a tremendous physical effort, is as calm and unruffled as the proverbial Englishman drinking tea.

The static character of the Greek drama, too, has often been pointed out. There is little or no action. We are presented at the very beginning of the play with a complete account of prior happenings that pose a problem or predicament for the characters involved, and the play concerns itself with mental struggles and minor deeds that eventuate in a denouement almost foreseen.

Linked with the static quality of the Greek drama is another characteristic also found in Euclidean geometry. The Greek tragedies emphasize the workings of fate or necessity. The characters in a play do not seem to have the will or the power to make decisions but are driven by hidden forces. Thus Oedipus is forced relentlessly to incest and patricide. The working of fate has been likened to the compulsion inherent in the use of deductive reasoning, wherein the mathematician is not free to choose the conclusions he may draw from his premises but is forced to accept the necessary consequences.

There is one other major characteristic of Greek art, geometry, and philosophy which, though generally present in such creations, nevertheless seems pre-eminently so with the Greeks. Their works reflect the fact that they strove to view the universe *sub specie aeternitatis*. They sought knowledge of what is universal

and eternal rather than individualistic and fleeting. The mathematical sphere is eternal and its mathematical properties will
hold forever. Hence knowledge of the sphere is most desirable.
The water bubble and the brilliantly coloured balloon, fascinating though they may be, are not worthy of attention, for they
will soon burst. So, too, Greek art of the classical period strove to
evoke and depict the broad, basic qualities not of men but of
man. What mattered in any one person were the qualities he
displayed of mankind in general. Dress, individual relations, and
daily activities, all these were accidental and trivial details. In
their philosophical speculations, the Greeks also sought to define
and understand the perfect form of concepts and qualities, for
the perfect is by its very nature eternal. The perfect state was
worthy of contemplation; the democratization of Greek society
was hardly recognized as a serious problem.

The mathematics we have thus far surveyed and the culture it
reflects belong to the classical Greek period. They do not by any
means exhaust the contributions of the 'morning-land of civilization' to mathematics and to our own lives and thought, for
the significant epoch extending from about 300 B.C. to A.D. 600
still awaits us. Before turning the page let us recall that the age
we are leaving created mathematics in the sense in which we
understand the term today. The insistence on deduction as the
exclusive method of proof, the preference for the abstract as
opposed to the particular, and the selection of a most fruitful and
highly acceptable set of axioms determined the character of
modern mathematics, while the divination and proof of numerous fundamental theorems sent it well on its way. Accompanying
the mathematics and indeed shining forth from it was the
brilliant light of human reason which was first kindled by the
Greeks. Their mathematical documents proclaimed the supremacy
of mind in human affairs and therewith a new concept of civilization.

V

Placing a Yardstick to the Stars

'Tis late; the astronomer in his lonely height
Exploring all the dark, descries from far
Orbs that like distant isles of splendour are,

He summons one dishevelled, wandering star;
'Return ten centuries hence on such a night.'

That star will come. It dare not by one hour
Cheat science, or falsify her calculation;
Men will have passed, but watchful in the tower
Man shall remain in sleepless contemplation;
And should all men have perished there in turn,
Truth in their stead would watch that star's return.

Sully Prudhomme

FOR at least four thousand years the civilization of Egypt followed a rigid pattern. In religion, mathematics, philosophy, commerce, and agriculture each man imitated his forefathers. No external influences disrupted the calm life and fixed ways. Then, about 325 B.C., Alexander the Great conquered this vast land, as well as Greece and the Near East, and proceeded to Hellenize his conquests. He founded the city of Alexandria and moved the capital of the ancient world from Athens to this new city; then the conquering culture was in its turn conquered. From a fusion of cultures, centred at Alexandria, a new civilization appeared and made its very significant and distinctive contribution to mathematics and to Western civilization.

Alexandria became the centre of the entire ancient world, for it was ideally located at the junction of Asia, Africa, and Europe. On the streets of the city native Egyptians met and traded with Greeks, Jews, Persians, Ethiopians, Syrians, Romans, and Araos. Aristocrat, citizen, and slave jostled each other. No city in the world, not even modern New York, has ever embraced a greater variety of peoples.

To this important centre came traders and businessmen from all corners of the world. In the harbour were ships that brought wines from Italy, tin from Wales, and amber from Sweden. Outward-bound ships sailed to the Ganges and Canton. The Alexandrian traders not only spread Greek culture over the world but brought back to Alexandria knowledge that had been acquired in other countries. As a result, the city became truly cosmopolitan while the wealth that was accumulated permitted expansion in many directions. Splendid buildings, statues, obelisks, mausoleums, tombs, temples, and synagogues abounded. To the pleasure-loving, Alexandria offered bazaars, baths, parks, theatres, libraries, a hippodrome, a race course, and homes for the wealthy.

Credit for making Alexandria the intellectual centre of the new world does not go to the founder of the city, who died while still engaged in conquests, but to the very capable Ptolemy the First, the general who took over control of Egypt on the death of Alexander. Aware of the cultural importance of the great Greek schools such as those founded by Pythagoras, Plato, and Aristotle, Ptolemy decided that Alexandria should have such a school and that it should become the centre of Greek culture in this new world. He built, therefore, a home for the Muses to whom scholars were dedicated.

Adjacent to the Museum Ptolemy built a library not only for the preservation of important manuscripts but also for the use of the general public. This famous library was said at one time to contain 750,000 volumes. Together with the Museum, the library resembled a modern university, though no university of today can boast of possessing as many great intellects as were assembled there.

Scholars of all countries were invited to Alexandria by Ptolemy and were supported by grants from him. Consequently, there gathered at this Museum poets, philosophers, philologists, astronomers, geographers, physicians, historians, artists, and the most famous mathematicians of the Alexandrian age. The principal group of the scholars gathered at the Museum was Greek, but distinguished members of many other nations also settled there. Among the non-Greeks the most celebrated was the learned Egyptian astronomer, Claudius Ptolemy.

Two factors seem to have vitally influenced the character of the culture which grew out of the mixture of peoples and scholars and the broadened physical horizons. The commercial interests of the Alexandrians, more extensive than those of the Athenians, brought geographical and navigational problems to the fore and directed attention to materials, methods of production, and the improvement of skills. Second, because the commerce was carried on by free people who were not segregated socially from the scholars, the latter became aware of and involved in the problems facing the people at large. As a result, scholars were induced to unite the flourishing theoretical studies with concrete scientific and engineering investigations. Technical fields were pursued and extended; training schools were established; and mechanics and other sciences were advanced. Also, arts despised or ignored in the classical period were taken up with zest.

The ingenuity of the mechanical devices invented by the Alexandrians in response to the new interests are astonishing even by modern standards. They designed improved water clocks and sun dials and used them to good advantage in the courts to limit lawyers' speeches. Pumps, pulleys, wedges, tackles, geared devices, and a mileage-measuring device no different from the one to be found in the modern automobile were widely employed. Among the mechanical inventions were new instruments for astronomical measurements. To the mathematician and inventor Heron (first century B.C.), the age owed an automatic machine for sprinkling holy water when a five-drachma coin was inserted. Musical organs could be operated in a similar way. The temples mystified the public with doors that opened when coins were deposited.

The study of gases and liquids produced a water-driven organ, a gun powered by compressed air, and a hose for spraying liquid fire. The public gardens were enhanced by water fountains with moving statues driven by water pressure. The generation of steam power was another development of the Alexandrians. It was used to drive automobiles along the city streets in the annual religious parades. When produced by fires maintained under the temple altars, the steam put life into the gods. Awe-struck audiences observed gods who raised their hands to bless the worshippers, gods who shed tears, and statues that poured out libations. Mechanical doves rose up into the air and descended by means of the unobservable action of the steam.

The Alexandrians also applied knowledge of the behaviour of sound and light to practical devices. Most spectacular of these was Archimedes' huge mirror which concentrated the sun's rays on Roman ships besieging his native city of Syracuse. The ships were supposed to have burned under the intense heat.

In contrast to the closely guarded and orally transmitted learning of an earlier Egyptian period, books freely disseminated the new knowledge. Fortunately for the Alexandrians, Egyptian papyrus was cheaper than parchment and so Alexandria became the centre of the book-copying trade of the ancient world. For the first time in the history of science there appeared an excellent work on mechanical and metallurgical knowledge. The principles underlying water- and steam-driven devices were explained in treatises on pneumatics and hydrostatics while other treatises explained the construction of vaults, catapults, and tunnels. Ingenious for those times were Heron's mathematical prescriptions for digging tunnels under a mountain, which made it possible to work from both ends and meet in the middle.

Of course mathematics had a most important place in the Alexandrian world, but it was not the mathematics that the classical Greek scholars knew. No matter what some mathematicians may say about the purity of their thoughts and their indifference to, or elevation above, their environment, the fact of the matter is that the Hellenistic civilization of Alexandria produced a kind of mathematics almost opposite in character to that produced by the classical Greek age. The new mathematics was practical, the earlier entirely unrelated to application. The new mathematics measured the number of grains of sand in the universe and the distance to the farthest stars; the older one refused to measure. The new mathematics enabled men to travel over land and sea; the older one prepared him to sit motionless and to view with his mind's eye the immaterial abstractions of philosophic thought. The great Alexandrian mathematicians, Eratosthenes, Archimedes, Hipparchus, Ptolemy, Heron, Menelaus, Diophantus, and Pappus, though they displayed almost without exception the Greek genius for theoretical abstractions, nevertheless were quite willing to apply their talents to the practical problems necessarily important in their civilization.

Typical of the new Greek was Eratosthenes (275–194 B.C.), director of the library at Alexandria and universal genius. Distinguished in the classical pursuits of mathematics, poetry, philosophy, and history, he also displayed profound learning in geodesy and geography. Not only did Eratosthenes collect and integrate all available historical and geographical knowledge, but he made maps of the entire universe known to the Greeks. He also found a simple way of measuring the radius of the Earth and of surveying large tracts of land. Astronomical measurements and the construction of astronomical instruments added to his fame.

Eratosthenes also improved the calendar. Most early civilizations had difficulty in keeping track of celestial events because they did not know the exact length of the solar year. For example, one early Greek calendar, descending most probably from the Babylonians, was based on a year of twelve months each containing thirty days. The inadequacy of this calendar became clear when dates originally planned to designate particular astronomical events, such as an equinox, occurred too late or too soon. Naturally the gods objected to such mismanagement of their affairs. Aristophanes records their complaint as transmitted through the Clouds:

> The Moon by us to you her greeting sends,
> But bids us say that she's an ill-used moon,
> And takes it much amiss that you should still
> Shuffle her days, and turn them topsy-turvy;
> And that the gods (who know their feast-days well),
> By your false count are sent home supperless,
> And scold and storm at her for your neglect.

Eratosthenes' calendar called for a year of 365 days and an extra day every fourth year. This calendar was later adopted by the Romans and is essentially the one we use today. Eratosthenes also insisted upon dating all events by the calendar as opposed to the earlier Greek practice of dating by the number of Olympiads since the fall of Troy, or the practice common in other civilizations of dating by the number of years in a king's reign. Eratosthenes worked at Alexandria until blindness overtook him in his old age, whereupon he ended his life by starving himself to death.

The man whose work best epitomizes the character of the Alexandrian age is Archimedes, one of the greatest intellects of antiquity. Though born in Syracuse, a Greek settlement in Sicily, Archimedes received his education in Alexandria. He then returned to Syracuse where he spent the rest of his life. Possessed of a lofty intellect, great breadth of interest both practical and theoretical, extraordinary mechanical skill, and a fertile imagination which Voltaire declared to be finer than Homer's, he was greatly respected and admired by his contemporaries.

The most obvious indications of Archimedes' practical interests are his highly original inventions. In his youth he constructed a planetarium which reproduced the motions of the heavenly bodies. He invented a pump for raising water from a river; used compound pulleys to launch a galley for King Hiero of Syracuse; and invented military engines and catapults to protect Syracuse when it was under attack by the Romans. It was at this time that he applied the focusing property of a curved mirror to burn the Roman ships. He also developed the use of the lever to move great weights.

Perhaps the most famous of his scientific discoveries is the hydrostatic principle now named after him. A story has come down in history that tells how Archimedes was led to make this discovery. The king of Syracuse ordered a crown made of gold. When the crown was delivered he suspected that it was filled with baser metals, and so he sent it to Archimedes and asked him to devise some method of testing the contents without, of course, destroying the workmanship. Archimedes pondered the problem and one day while bathing observed that his body was partly buoyed up by the water. He suddenly grasped the principle that enabled him to handle the problem. He had discovered that a body immersed in water is buoyed up by a force equal to the weight of the water displaced. Since the weight of the displaced water as well as the weight of a body in air can be measured, the ratio of the weights is known. This ratio is constant for a given metal no matter what its shape, and differs from metal to metal. Hence Archimedes had but to determine this ratio for a piece of metal known to be gold and compare it with the corresponding ratio for the crown. Unfortunately, history does not record his decision. The principle that Archimedes discovered is one of the first universal laws of science; he incorporated it among others in his book, *On Floating Bodies*.

Even his theoretical work in mathematics was influenced by the spirit of the Alexandrian age, for he devoted a great deal of his time to problems of measurement. He proved that the area of a circle is half the circumference times the radius, which gives the usual πr^2 formula, and then determined the value of π. The result of his computations – that π lies between $3\frac{1}{7}$ and $3\frac{10}{71}$ – was indeed remarkable for his times. He also proved many other formulas for area and volume.

Moved again by the spirit of the age, Archimedes undertook a task repugnant to the Greek of the classical period. He devised a system for expressing large numbers and, in a final account of this work which he entitled *The Sand Reckoner*, showed how it was possible to express the number of all the grains of sand in the universe.

However much Archimedes may have been moved by the practical interests of the times, he nevertheless possessed the classical Greeks' love for basic theory. Of all his accomplishments, he was proudest of a theoretical one. We know this from his request that there be inscribed on his tombstone the figure of a sphere, a cylinder circumscribed about it, and the ratio two-thirds. This refers to his great discovery that the ratio of the volume of a sphere inscribed in a cylinder to the volume of the cylinder is as two is to three, and that the ratio of the surface of the sphere to the surface of the cylinder is similarly two-thirds.

The death as well as the life of Archimedes epitomizes the events of his age. We have already related that he was challenged by one of the Roman soldiers who had just captured Syracuse. Archimedes was so lost in thought that he did not hear the challenge, whereupon the soldier killed him despite the order of the Roman commander, Marcellus, that Archimedes not be harmed. Archimedes was then seventy-five years old and still in full possession of all his powers. By way of compensation the Romans built an elaborate tomb upon which they inscribed the famous theorem referred to above.

In the field of mathematics proper the Alexandrians created and applied methods of indirect measurement. Their simplest contributions to this subject were formulas for areas and volumes of particular geometrical shapes. These formulas, surprisingly, are not in Euclid, for though Euclid lived at the beginning of the Alexandrian age, his subject matter was really the summation and culmination of the mathematics of the classical period. That the areas of two similar triangles are to each other as the squares of corresponding sides was of great interest to Euclid, but that the area of any one single triangle could be found directly by taking the product of its base and half its altitude was made known to us by the Alexandrians.

The contribution of formulas for areas and volumes is often underestimated. How does a person find out the area of a floor? Does he take little squares one foot on a side, lay them out over the entire floor, and thus decide that the area of a floor is 100 square feet, for this indeed is the meaning of area? Of course he does not. He measures the length and width, quantities usually quite simple to measure, and then multiplies the two numbers to obtain the area. This is indirect measurement, for area has been obtained by measuring lengths. The extension of this idea to volumes is obvious. Thus, even the very common formulas of geometry, which we owe to the Alexandrians and which permit us to measure areas and volumes indirectly by expressing these quantities in terms of readily measured lengths, represent an immense practical achievement.

But this type of indirect measurement was really child's play to the Alexandrians. They were eventually able to measure by indirect means the radius of the Earth, the diameters of the sun and moon, and the distances to the moon, the sun, the planets, and the stars. That we can measure such physically inaccessible lengths and do so, moreover, with an accuracy as great as we wish, seems, at first blush, incredible. Not only did the Alexandrians transform seeming impossibility to actuality, but they did so with a simplicity and a finality hardly to be anticipated at this date in the march of mathematical ideas.

It was in the second century B.C. that Hipparchus, the greatest astronomer of the ancient world, created the branch of mathematics that was so ingeniously applied to the charting of the Earth and heavens. The basis of Hipparchus' astute method is a simple theorem of geometry. Before noting the theorem let us recall that two triangles are similar, by definition, if the angles of one equal, respectively, the angles of the other. To show that two triangles are similar it is sufficient to show that two angles of

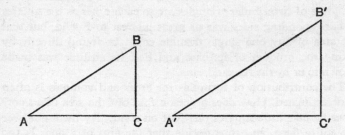

Figure II. Two similar right triangles

one are equal, respectively, to two angles of the other. The reason is simply that the third angles of the two triangles must then be equal, for the sum of the angles in any triangle is 180°. In particular, if we deal with right triangles, since the right angles of the two triangles are equal, it is sufficient to know that an acute angle of one equals an acute angle of the other to conclude that the triangles are similar.

The theorem that Hipparchus applied states that if two triangles are similar, the ratio of the lengths of any two sides of one triangle equals the corresponding ratio of the other. Thus if triangles ABC and $A'B'C'$ (fig. 11) are similar, for example, then BC/AB equals $B'C'/A'B'$. If triangles ABC and $A'B'C'$ are right triangles and if angle A equals angle A', then, in view of the conclusion in the preceding paragraph, we know that the triangles are similar. We may therefore state, with Hipparchus, that *the ratio of the side opposite angle A to the hypotenuse of the triangle must be the same in any right triangle containing angle A.* This ratio of BC to AB is so important that it is given the special name of sine and since the ratio depends on the size of angle A it is written *sine A.* Thus, by definition,

$$sine \ A = \frac{BC}{AB} = \frac{side \ opposite \ angle \ A}{hypotenuse}.$$

The discussion showing that *sine A* is the same in all right triangles containing angle A could be applied to other ratios formed from the sides of a right triangle containing A. For example, the ratios

$$cosine \ A \ = \frac{AC}{AB} = \frac{side \ adjacent \ to \ angle \ A}{hypotenuse}$$

and

$$tangent \ A = \frac{BC}{AC} = \frac{side \ opposite \ angle \ A}{side \ adjacent \ to \ angle \ A}$$

are the same in all right triangles containing angle A.

We may now see how these ratios were used by Hipparchus to measure the Earth and heavens. The first step is to find the height of a mountain. Just to simplify the latter problem somewhat we shall suppose that the mountain has a sheer side, BC in fig. 12, with the point C as the foot of the mountain. We first measure the readily accessible distance AC along the ground, and obtain, let us say, the length ten miles. We also measure angle A, which, for example, might be $17°$. Then we may say, in view of the meaning of *tangent A*, that

$$tangent \ 17° = \frac{BC}{AC}.$$

Since AC is 10,

$$tangent \ 17° = \frac{BC}{10},$$

and, by multiplying both sides of this equality by 10, we learn
that

$$BC = 10.tangent\ 17°.$$

If we knew *tangent* 17° we could immediately obtain *BC*. Now
tangent 17° has the same value in any right triangle containing
this angle. Hence we can choose any convenient triangle for the
purpose of determining this quantity.

A carpenter would obtain this quantity simply. He would construct a small right triangle having an acute angle of 17°,
measure the opposite and adjacent sides, and then compute the
ratio of the two sides. A mathematician would be more soph-

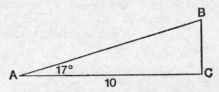

Figure 12. Calculation of the height of a mountain

isticated – and more accurate. Hipparchus, being a mathematician as well as an astronomer, devised a method of
calculating these ratios for any right triangle and listed the results
in famous tables which he passed on to his successors and which
are now incorporated in textbooks.

We need not follow the details of Hipparchus' calculations.
What matters is that it is possible to calculate these ratios as
accurately as we may wish. From these calculations we know that
tangent 17° is ·3057 to four decimal places. Hence *BC*, which is
10.*tangent* 17° is 3·057 miles. Thus the height of the mountain
can be found without having to lay a yardstick along it.

Now let us see how this result can be used to measure the size
of the Earth. We should mention first that the educated Greeks
believed the shape of the Earth to be a perfect sphere. Though
this conclusion was reached through aesthetic and philosophical
arguments rather than by circumnavigation, it was none the less
firmly held. Hence the essential quantity to be measured is the
radius of the sphere.

To measure this length we can proceed as follows. We ascend a mountain, say three miles high, and look towards the horizon.

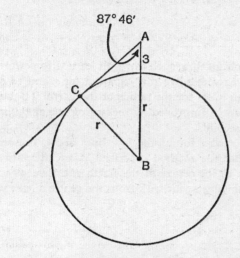

Figure 13. Calculation of the radius of the earth

We then measure with whatever instrument we can command, the angle between our line of sight and the vertical, angle CAB in fig. 13. This angle would turn out to be approximately $87°\ 46'$. With this measurement in hand we complete the diagram as shown, wherein r is the radius of the Earth. In this figure the radius BC is perpendicular to the line of sight AC, for AC is tangent to the Earth's surface and, according to a theorem of Euclidean geometry, a radius of a circle drawn to the point of tangency of a tangent is perpendicular to that tangent. Now, following Hipparchus, let us note the ratio of the side opposite our measured angle to the hypotenuse of the right triangle. In the symbols of our diagram this ratio is $\dfrac{r}{r+3}$. This ratio is also *sine A*, or *sine* $87°\ 46'$. Hence

$$sine\ 87°\ 46' = \frac{r}{r+3}.$$

Since Hipparchus had already calculated the sine ratios, he knew that *sine* 87° 46′ is ·99924 to five decimal places. Hence we see that

$$·99924 = \frac{r}{r+3}.$$

By very simple algebra, such as we all learned to perform in high school, it is easy to solve this equation for r and to obtain the value of 3,944 miles for the radius of the Earth. The accuracy of this result can be improved by measuring the angle involved to the order of seconds.

The reader who found the few lines above tiresome should remember that the method described is an alternative to tunneling down to the centre of the Earth and then measuring the radius by applying a yardstick from the centre all the way to the surface.

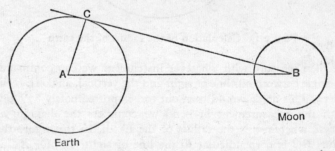

Figure 14. Calculation of the distance from the earth to the moon

And now let us see how Hipparchus found the distance from the Earth to the moon or, more exactly, from the centre of the Earth to the centre of the moon. Our description may be a slight simplification, but it contains the essence of Hipparchus' method. Let us suppose that the calculation is made at a time when the line from the centre of the Earth to the centre of the moon, line AB in fig. 14, cuts the surface of the Earth at a point on the equator. We can then imagine a line drawn from B to the Earth's surface so that it just touches the surface, say at C. Now, ac-

cording to the theorem of geometry cited above, the line AC of
the diagram, which represents the radius of the Earth drawn to
the point of tangency C on the surface, forms a right angle with
the tangent at C. The angle CAB of our figure is the latitude of C.
Incidentally, Hipparchus himself had instituted the system of
locating points on the surface of the Earth by means of latitude
and longitude, the very system used universally today. Hence
Hipparchus knew the latitude of C. He also knew CA, the radius
of the Earth, Hence he could argue that

$$cosine\ A = \frac{AC}{AB}.$$

A reasonable value for A in the diagram above is $89° 3'$; moreover, Hipparchus' tables told him that $cosine\ 89° 3' = ·01658$.
The distance AC, the radius of the Earth, was just found to be
about 3950 miles. Hence

$$·01658 = \frac{3950}{AB}.$$

By multiplying both sides of this equation by AB and then dividing both sides by $·01658$ we obtain

$$AB = \frac{3950}{·01658} = 238,000.$$

That is, the distance from the centre of the Earth to the centre of
the moon is about 238,000 miles.

A glance backward will reveal that by starting with a readily
measurable distance on the surface of the Earth we were able to
calculate successively the height of a mountain, the radius of the
Earth, and the distance to the moon. With this knowledge and
with Hipparchus' method, we could proceed to calculate the
distances to the sun, any planet, and the stars. In fact, Hipparchus did carry out a great number of astronomical calculations. The simplicity and, at the same time, the broad
applicability of his trigonometry should be apparent.

The mathematics that Hipparchus created in order to survey
the Earth and heavens has been used since his times to handle

numerous practical problems. Surveyors, navigators, and map-makers employ it constantly. Indeed, through the power of Hipparchus' methods and other mathematical methods not detailed here, the Alexandrian Greeks set map-making up as a science. Their maps offered the best knowledge of the Earth up to the time of the great explorations in the fifteenth and sixteenth centuries. Very fortunately for later generations the astronomer Ptolemy summarised all of the geographical knowledge amassed by the ancient world in his *Geographia,* a work in eight volumes. This work gave the latitude and longitude of some 8000 places on the Earth and was the world's first atlas and gazetteer.

The subject of trigonometry is an excellent example of a branch of mathematics the investigation of which was motivated by both practical and intellectual interests – surveying, map-making, and navigation on the one hand, and curiosity about the size of the universe on the other. With it the Alexandrian mathematicians triangulated the universe and rendered precise their knowledge about the Earth and the heavens. They then proceeded to capitalize on this work in a manner we shall relate in the next chapter.

VI

Nature Acquires Reason

SOCRATES. Very good; let us begin then, Protarchus, by asking a question.

PROTARCHUS. What question?

SOCRATES. Whether all this which they call the
universe is left to the guidance of unreason and chance
medley, or, on the contrary, as our fathers have declared, ordered and governed by a marvellous intelligence and wisdom.

PROTARCHUS. Wide asunder are the two
assertions, illustrious Socrates, for that which you were
just now saying to me appears to be blasphemy, but
the other assertion, that mind orders all things, is
worthy of the aspect of the world. . . .

Plato: PHILEBUS

PEOPLE who would create must first be willing to dream. Because the philosophically minded Greeks often allowed their
speculations to lapse into dreams, they were rewarded with one of
the greatest prophetic insights that man has ever attained. The
vision proved so extraordinary that it actuated the intellectual
life of all thinking Greeks. For Western civilization the sequel
was most significant.

In part, the vision said that nature is rationally ordered; all
natural phenomena follow a precise and unvarying plan. The
vision disclosed, also, that mind is the supreme power and, therefore, that the pattern of nature can be rendered intelligible by the
application of mind to the affairs of the universe.

The Greeks projected their dream into reality and became the
first people with the audacity and genius to give reasoned explanations of natural phenomena. The Greek urge to understand
had the excitement of a quest and an exploration. And while they

explored they made maps, such as Euclidean geometry, so that others might find their way quickly to the frontiers and help conquer new regions.

Preceding civilizations, notably the Babylonian and Egyptian, had made untold observations and had obtained many useful, empirical formulas. But though they must have discovered some evidence of order in nature, they conceived no embracing theories; and they scarcely dreamed of design. The complex and varied actions and reactions in nature concealed from them any indication of plan, order, and law. Nature appeared and remained capricious, mysterious, and often terrifying. The Greeks thought otherwise. Spurred on by their desire for knowledge and their love of reason, these protagonists of the power of the mind were confident that an examination of nature's ways would reveal the order inherent in the physical world.

The search for a rational interpretation of nature was actively carried on in the earliest days of Greek civilization. Typical was the cosmology of Thales, who argued that everything is ultimately water and that mist and earth are forms of water. He believed that the universe was a mass of water with a bubble in it, the bubble being our world with the Earth floating on the bottom and the rains coming from the water on top. The heavenly bodies were water in an incandescent state; they floated on the water surrounding the bubble. Whereas to the Egyptians and Babylonians the stars were gods, to Thales they were 'steam from a pot'. In offering this theory about the construction of the universe, Thales took an extremely modern point of view. He did not maintain that his explanation necessarily described what literally existed. Rather, he advanced it because it organized observations into a rational pattern.

Such analyses of natural phenomena seem shallow and childish compared with the sophistication and relative profundity of modern scientific theories. Nevertheless, Thales and his Ionian colleagues progressed far beyond the thinking of preceding civilizations. At the very least, these men dared to tackle the universe and they refused any help from gods, spirits, ghosts, devils, angels, or other agents unacceptable to a rational mind. Their material and objective explanations displaced mythical and supernatural accounts, and their reasoned approaches discredited the fanciful and uncritical explanations of the poets. Brilliant intuitions fathomed the nature of the universe and reason defended these insights.

With the advent of the Pythagoreans, the programme for rationalizing nature enlisted the aid of mathematics. The Pythagoreans were struck by the fact that phenomena which are physically most diverse exhibit identical mathematical properties. The moon and a rubber ball share the same shape and all other properties common to spheres. Similarly a garbage can and a cask of fine wine may have the same volume. Is it not apparent, then, that mathematical relations underlie diversity and must be the essence of phenomena?

To be specific, the Pythagoreans found this essence in number and in numerical relations. Number was the first principle in the explanation of nature and was the matter and form of the universe. Said Philolaus, a famous Pythagorean of the fifth century B.C., 'Were it not for number and its nature, nothing that exists would be clear to anybody either in itself or in its relation to other things. . . . You can observe the power of number exercising itself not only in the affairs of demons and gods but in all the acts and the thoughts of men in all handicrafts and music.'

The reduction of music, for example, to simple relationships among numbers became possible for the Pythagoreans when they discovered two facts: first, that the sound caused by a plucked string depends on the length of the string; and second, that harmonious sounds are given off by strings whose lengths are to each other as the ratios of whole numbers. For example, a harmonious sound is produced by plucking two equally taut strings, one twice as long as the other. The musical interval between the two notes is now called an octave. Another harmonious combination is formed by plucking two strings whose lengths are in the ratio of 3 to 2; in this case the shorter one gives forth a note called the fifth above that given off by the longer one. In fact, the relative lengths in every harmonious combination of plucked strings can be expressed as ratios of whole numbers.

The Pythagoreans also reduced the motions of the planets to number relations. They believed that bodies moving in space produce sounds and that a body which moves rapidly gives forth a higher note than one which moves slowly. Perhaps these ideas were suggested by the swishing sound of an object whirled on the end of a string. According to Pythagorean astronomy, the greater the distance of a planet from the Earth the more rapidly it moved. Hence the sounds produced by the planets varied with their distances from the Earth and these sounds all harmonized. But this 'music of the spheres', like all harmony, reduced to no more than number relationships and hence so did the motions of the planets.

In addition to these more 'substantial' elements of their philosophy, the Pythagoreans attached very interesting affinities or interpretations to the individual number. The number *one* they identified with reason, for reason could produce only a consistent whole; *two* was identified with opinion; *four* with justice because it is the first number which is the product of equals (to the Pythagoreans *one* was not a number in the full sense because unity was opposed to quantity); *five* signified marriage because it was the union of the first odd and the first even number; *seven* was identified with health, and *eight* with love and friendship. Because the Pythagoreans thought of four as four dots arranged in a square and identified four with justice, the association of the square and justice continues to this day. The square-shooter is still the man who acts justly.

All the even numbers were regarded as feminine, the odd numbers as masculine. From these associations it followed that the even numbers represented evil and odd, good. The trouble with the even numbers was that they permitted bisection into more and more even numbers as 2 into 1 and 1, 4 into 2 and 2, 8 into 4 and 4, and so on. The process of continued bisection suggested the infinite, a horrible thought to the Greeks who preferred the definite and limited. The odd numbers, on the other hand, prevented the even numbers from pursuing the bisection process indefinitely; they prevented the even numbers from going to pieces. Moreover, they themselves resisted bisection for that led, in the case of an odd number, to vulgar improper fractions.

A number was perfect if it equalled the sum of its divisors, as

$6 = 1 + 2 + 3$. Two numbers were 'friendly' if each was the sum of the divisors of the other. Thus 220 and 284 were amicable as a check on the divisors will show. Such numbers were written on pellets and the latter were eaten as aphrodisiacs. The ideal number was 10 because for one thing it was the sum of consecutive integers 1, 2, 3, and 4. And because 10 was ideal, the moving bodies in the heavens must be 10 in number. The Pythagoreans could readily account for 9 of these bodies, for they believed that the Earth, sun, moon, sphere of stars, and the other 5 planets known at that time moved around a fixed central fire. They asserted the existence of a tenth moving body, which they called the counter-Earth. This body was always on the side of the central fire opposite to the Earth and so it was not visible. The ideality of 10 also required that every object in the universe be describable in terms of 10 pairs of categories such as odd and even, bounded and unbounded, right and left, one and many, male and female, and good and evil.

These speculative vagaries of the Pythagoreans are, to a large extent, idle, unscientific, and useless. Their obsession with the importance of number caused them to build a natural philosophy which certainly had little correspondence with nature. Unfortunately, some of this philosophy was passed on to medieval Europe where it was made sacrosanct by religious mystics. Nevertheless the major thesis of the Pythagoreans, namely, that nature should be interpreted in terms of number and number relations, that number is the essence of reality, dominates modern science. The Pythagorean thesis was revived and refined in the work of Copernicus, Kepler, Galileo, Newton, and their successors, and is represented today by the doctrine that nature must be studied quantitatively. These relatively modern scientists adopted several other Pythagorean beliefs, namely, that the universe is ordered by perfect mathematical laws, that divine reason is the organizer of nature, and that human reason, in probing nature, seeks to discern the divine pattern. We shall see that this philosophy is to be credited with the success of modern science, and that numerical relations ultimately usurped the favoured position that the Greeks had granted to geometry.

The foremost Pythagorean, next to Pythagoras, was Plato,

who shared the belief that the reality and intelligibility of the physical world could be comprehended only through mathematics, for 'God eternally geometrizes'. Plato went further than most Pythagoreans. He wished not merely to understand nature through mathematics but to transcend nature in order to comprehend the ideal, mathematically organized world that he believed to be the true reality. The sensible, the impermanent, and the imperfect were to be replaced by the abstract, eternal, and perfect. He hoped that a few penetrating glances at the physical world would supply basic truths which reason, unaided by further observation, could then develop. From that point on, nature would be replaced entirely by mathematics. Indeed, he criticized the Pythagoreans because they investigated the numbers of the harmonies which are heard but never reached the natural harmonies of numbers themselves. The mere study of sounds as such he declared to be useless, while reflection on harmonious numbers, if sought after with a view to the beautiful and the good, was of the highest value.

Plato's attitude towards astronomy illustrates his attitude towards all natural science. According to Plato, true astronomy is not concerned with the movements of the visible, heavenly bodies. The arrangement of the stars in the heavens and their apparent movements are indeed wonderful and beautiful to behold, but mere observation and explanation of the motions fall far short of true astronomy. Before we can attain to the latter we 'must leave the heavens alone', for true astronomy deals with the laws of motion of true stars in a mathematical heaven of which the visible heaven is but an imperfect expression. He encouraged devotion to a theoretical astronomy whose problems please the mind and not the eye. Navigation, the calendar, and the measurement of time were evidently alien to Plato's astronomy.

There is no doubt that Plato's unwillingness to observe and experiment hindered the development of Greek science and placed too much reliance on the power of the mind to grasp fundamental truths and deduce logical consequences. Nevertheless, the good that resulted from his conception of natural science was of inestimable value. It produced the first master plan of the same heavens that he preferred to leave alone.

The Greeks of this period had observed what anyone can see who cares to chart the motions of the planets. As seen from the Earth their progress in the sky is disorderly with no apparent regulative rhyme or reason. They advance and retrogress. Indeed these vagabonds of the sky (the word *planet* means 'wanderer' in Greek) seemingly refuse to follow any orderly course.

Now it is one thing to observe and carefully chart the motions of the planets as the Egyptians and Babylonians did for centuries. These peoples were merely observers. It is quite another, and indeed a major step forward, to ask for some unifying theory of the motions of the heavenly bodies that will reveal a plan underlying the seeming irregularity. This is the problem Plato set before the Academy, that is, to devise a mathematical scheme which, though calling for systematic motions of the planets, would account for the disorderly motions we see. He described his problem in a now famous phrase as 'saving the appearances'.

The answer to Plato's problem given by Eudoxus, a pupil of Plato, a master in his own right and one of the foremost Greek mathematicians, is the first major astronomical theory known to history and a decided advance in the programme of rationalizing nature.

Eudoxus' scheme employed a series of concentric spheres whose centre is the immovable Earth. To account for the complex motion of any one body Eudoxus supposed first that it was attached to a sphere which rotated at constant speed about some axis through the Earth. Thus the planet P (fig. 15) is attached to the sphere whose cross section is AMB, and this sphere rotates about the axis AB. Eudoxus next imagined this axis AB to project beyond A and B until it terminated on a second sphere, at C and D in figure 15, where it was regarded as rigidly attached. This second sphere was supposed to rotate on an axis of its own, GF in figure 15, and to carry the first axis and rotating sphere along. Since two spheres were not sufficient to describe the motion of any heavenly body, Eudoxus presumed that the axis of the second sphere projected beyond it to terminate on a third sphere, which in turn rotated about an axis of its own. In the cases of the planets, Eudoxus used four such spheres for each. The speeds of rotation and the radii of these spheres were fitted by Eudoxus to the observed motions of the planets.

It is, of course, difficult to visualize the path any body would take under the combined motions of two or three spheres. The

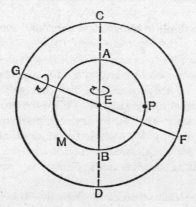

Figure 15. Sketch of the Eudoxian scheme

very complex motions that do result were, however, precisely what Eudoxus needed to describe the paths of the five planets, the sun, moon, and the stars as seen from the Earth. A total of twenty-seven spheres sufficed for the whole system.

Eudoxus' scheme for describing and predicting the motions of the seemingly errant heavenly bodies was, of course, ingenious and it impressed the Greeks immensely. It instituted a mathematical order in nature and, at the same time, it evidenced the power of the human mind to devise such an order. It is also worth noticing that Eudoxus regarded his scheme as a purely mathematical one. He attached no physical meaning to the spheres. They were fictions, and the whole plan just a theory which accounted for the observed motions.

The Eudoxian construction was not the last word in Greek astronomy. We shall see shortly that it was superseded by a superior theory. But before we leave the classical period of Greek culture we should mention other weighty evidence that the age had amassed to establish the rationality of nature. The classical Greek did not have to peer into space to conclude that nature is mathematically designed. He had but to reflect on the significance of Euclidean geometry.

Euclid started his geometry with ten axioms. Some of these, such as that equals added to equals give equals, are immediately acceptable. Others, such as that two points may be joined by one and only one straight line, were suggested by observation of the physical world. Once these axioms were selected, however, the theorems were deduced by the action of the mind alone. Each and every one of the hundreds of theorems contained in the *Elements* could have been deduced by a Euclid sitting with blindfolded eyes in an ivory tower. Nevertheless, when any one of these theorems was applied to a physical situation, it was found that the theorem described the situation perfectly. The theorem gave knowledge as precise and as reliable as if it had been inferred directly from the situation. What should the Greeks have concluded from the fact that a theorem, deduced by pure reasoning involving hundreds of successive deductions from the axioms, applied perfectly? Did it not prove that nature was designed to accord with a rational plan, that nature conformed to a whole body of reasoned knowledge? Was not this overwhelming evidence of design?

The Greeks studied numerous other physical phenomena and found evidence of design and of the mathematical structure of nature. An example from the field of optics will illustrate such successes.

Euclid discovered that the angle at which a light ray strikes a mirror equals the angle at which it is reflected; that is, angle 1 = angle 2 in fig. 16. This fact, often described by the statement that the angle of incidence equals the angle of reflection, reveals law and mathematical design in nature's behaviour.

There is a second mathematical law involved in this phenomenon of optics. We noticed in another connection that if A and B are any two points on one side of a line, then of all the paths leading from point A to the line and then to point B, the shortest path is by way of the point P such that the two line segments AP and PB make equal angles with the line. And this shortest path is exactly the one a light ray takes. Apparently nature is well acquainted with geometry and employs it to full advantage.

If the Greeks of the classical period had excellent evidence of the mathematical design of nature, then the Alexandrian Greeks could, with all justification, assert that they had indisputable proof. The supreme achievement of these people was the creation of the most accurate and most influential astronomical theory of ancient times. The central figure in this work was Hipparchus, the man who demonstrated the use of indirect measurement to calculate the sizes and distances of the heavenly bodies. In the course of his astronomical investigations, he improved instruments for observation, discovered the precession of the equinoxes, determined the angle of the ecliptic, measured irregularities in the motion of the moon, revised earlier estimates of the length of the year (Hipparchus estimated the length of the solar year to be 365 days, 5 hours, and 55 minutes, or about 6½ minutes too much), and catalogued about a thousand stars. These relatively minor contributions were climaxed by the construction of a complete astronomical system.

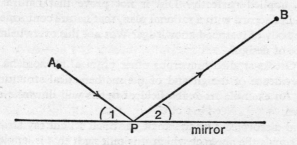

Figure 16. The angle of incidence equals the angle of reflection

Hipparchus recognized that the scheme of Eudoxus, which supposed that the heavenly bodies were attached to rotating spheres with centres at the Earth's centre, did not account for many facts observed by other Greeks and by Hipparchus himself. The Eudoxian theory contained significant errors especially in the motions of Mars and Venus. Instead of Eudoxus' scheme, Hipparchus supposed that a planet *P* (fig. 17) moved in a circle at a constant speed and that the centre of this circle, *Q*, moved at constant speed on another circle with the centre at the Earth. By properly selecting the radii of the two circles and the speeds of *Q* and *P*, he was able to get an accurate description of the motion of many planets. The motion of a planet, according to this

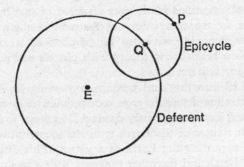

Figure 17. Sketch of Hipparchus' scheme

scheme, is like the motion of the moon according to modern astronomy. The moon revolves around the Earth at the same time that the Earth revolves around the sun. The motion of the moon around the sun is then like the motion of a planet around the Earth in Hipparchus' system.

In the cases of some heavenly bodies Hipparchus found it necessary to use three or four circles, one moving on another. That is, a planet *P* moved in a circle about the mathematical point *Q*, while *Q* moved in a circle about the point *R*, and *R* moved about the Earth; again, each object or point travelled at its own constant speed. In still other cases Hipparchus had to suppose that the centre of the innermost circle or deferent was not at, but near, the centre of the Earth. Motion in accordance with this latter geometrical construction was called eccentric whereas when the centre of the deferent was at the Earth, the motion was called epicyclic. By the use of both types and by the proper choice of radii and speeds of the circles involved, Hipparchus was able to describe quite well the motions of the moon, sun, and planets. With this theory an eclipse of the moon could be predicted to within an hour or two, though solar eclipses were predicted less accurately.

It is worth mentioning that from the modern point of view Hipparchus was taking a step backward, for about a century before his time another famous Alexandrian, Aristarchus, had advanced the theory that all the planets move around the sun. But observations made over a period of one hundred and fifty years by the observatory at Alexandria along with older Babylonian records convinced Hipparchus of what we know today, that a heliocentric theory with planets moving in circles about the sun will not do.

Instead of pursuing and perhaps improving on Aristarchus' idea Hipparchus dismissed it as too speculative. Others rejected Aristarchus' idea because they deemed it impious to identify the corruptible matter of the Earth with the incorruptible heavenly bodies by regarding the Earth as a planet. This distinction between the Earth and the other heavenly bodies was quite solidly established in Greek thought and was defended, though not dogmatically, even by Aristotle. The distinction became a scientific doctrine in Christian theology and the subsequent elimination of this falsity was one of the triumphs of modern mathematics and science.

The development of Greek astronomical theory reached its culmination in the work of Claudius Ptolemy, who was a member of the royal family of mathematicians, if not the political rulers of Egypt. Indeed, Hipparchus' work is known to us because it survived in Ptolemy's *Almagest*, a work almost as important as Euclid's for its influence on succeeding generations. In its mathematical content, the *Almagest* brought Greek trigonometry into the definitive form it was to retain for more than a thousand years. And in the field of astronomy it offered a complete exposition of the geocentric theory of epicycles and eccentrics that has come to be known as the Ptolemaic theory. So accurate was it quantitatively and so long was it accepted that people were lured into regarding it as an absolute truth.

This theory is the final Greek answer to Plato's problem of rationalizing the appearances of the heavens and is the first truly great scientific synthesis. With Ptolemy's completion of Hipparchus' work the evidence for design in the universe was complete to the tenth decimal place. The universe was rational and the principles that determined its motions were mathematical. This astronomical theory, transformed during the Renaissance by Copernicus and Kepler and refounded and refined by Newton, provided the chief evidence for the most important doctrine of modern science – the uniformity and invariability of nature.

More immediate use of the theory was made by a totally different group of thinkers. Since the Ptolemaic system made the Earth the centre of the universe, it was quite natural for Christian theology, proceeding along rational lines, to advance the thesis that man was the most important creation of God and that man's welfare was the most weighty of God's concerns. This theological conclusion was made all important and the mathematical evidence on which it rested was subordinated. Nevertheless, as the Church clearly recognized, the Christian doctrine that man was the most important object in the universe, indeed the one for whom the universe was specially designed, rested heavily on Ptolemaic theory.

The Greeks did not complete the rationalization of nature. We are still engaged in that task today. But they left monuments in astronomy, mechanics,* optics, and the study of space and figures in space. In each of these achievements mathematics was either the essence or the essential tool.

Unfortunately, the intellectual life of the Greeks was cut short by political events beyond the control of mathematicians and philosophers. During the period when Alexandria flourished the Roman Juggernaut rolled over the Italian peninsula and then began to attack other lands bordering the Mediterranean. Through intervention in the family strife between Cleopatra, the last of the Ptolemy dynasty, and her brother, Caesar managed to secure a hold on Egypt. He then attempted to destroy the Egyptian fleet riding at anchor in the harbour by setting fire to it. As a result, there began the most tragic holocaust in the long history of man's battle against barbarism. The fire swept in from the sea and destroyed the great library at Alexandria. Two and a half

*See Chapter XIII.

centuries of book collecting and a half a million manuscripts that represented the bright pageant of ancient culture were wiped out. The Romans withdrew only to return at the death of Cleopatra in the year 31 B.C., and from that time on its interference with, and domination of, the Museum at Alexandria proved more and more destructive to the culture there.

The fire at Alexandria was symbolic of the Roman contempt for abstract knowledge. The history of the Romans co-extends with that of the Greeks, but we should never know the Romans existed from reading even a comprehensive history of mathematics. The Romans were practical people and they boasted of their practicality. They undertook and completed vast engineering projects such as viaducts, magnificent roads which survive even today, bridges, public buildings, and land surveys, but they refused to consider any ideas over and above what were required for the particular concrete applications they were making at the moment. A problem taken from one of their texts illustrates their general attitude. The problem asks for a method of finding the width of a river when the enemy is already on the opposite bank. Cicero admitted that because 'the Greeks held the geometer in the highest honour, accordingly, nothing made more brilliant progress among them than mathematics'. He boasted, however, that 'we have established as the limit of this art its usefulness in measuring and counting'.

Just as the mathematics of the Greeks was seen to be related to the ideality of their art, so the practical interests of the Roman people are manifested in their concrete and rather mundane art. Roman art was purposive, for example, didactic or commemorative, and beauty descended to decoration and ornamentation. The sculptures and portraits were always of individuals and were intended either to honour or deify. Augustus, for example, was sculptured as a soldier with all his armour and his medals (Plate 111), and the little child next to him symbolizes the fertility of Rome. Gone was the contemplation of the ideal and the devotion to perfectly proportioned gods and human figures. In architecture, Rome is represented best by its public buildings, such as baths, all of which served a useful purpose.

The shortsightedness of the Romans produced a one-sided,

imitative, and second-rate culture. For several centuries they were able to supply the deficiencies of inspiration and original thought by relying upon the Greeks. When Augustus undertook a survey of the empire he called in specialists from Alexandria, and when Julius Caesar undertook to reform the calendar, he also called in an Alexandrian. When the wells of learning had almost dried up, the Romans began to realize the error of adding statues to a fountain while neglecting the water supply. But they were too late.

The insignificance of Roman achievements in the fields of mathematics, science, philosophy, and many of the arts is the best answer to those 'practical' people who condemn abstract thought that is not motivated by utilitarianism. Certainly one lesson to draw from the history of the Romans is that people who scorn the highly theoretical work of mathematicians and scientists and decry its uselessness are ignorant of the manner in which practical and important developments have arisen. Indeed, any large corporation knows that today it must spend millions of dollars and years of time on research which offers no promise of immediate usefulness in order to produce new ideas and techniques.

Roman rule of the Greek civilization was destructive for another reason. It enslaved millions of people and kept millions of others subject. Roman officialdom suppressed all social and economic betterment and held education to a minimum. At the same time, enormous wealth was drawn from the subject countries by taxation and taken to Rome. The lot of the multitudes became unbearable. Among these miserable people the Christian appeal with its emphasis on ethics, brotherhood, and rewards in an after-life took hold readily and drew thousands and eventually millions away from the Greek culture. Bloody street riots between 'heathen' and Christian became common. Unfortunately all Greek learning was identified with paganism and, therefore, severely attacked. The scholars at the Museum in Alexandria were persecuted and driven from the city.

The fate of Hypatia, the last mathematician of the Alexandrian school, dramatizes the end of the era. Because she refused to abandon her Greek religion she was set upon by an enraged mob of Christians and torn limb from limb in the streets of Alexandria. Hypatia's fate was also that of Greek thought.

The final blow to the Museum at Alexandria, which destroyed as it were the cover of the great Book of the Ancients, whose pages had already been scattered to the winds, was the burning of the Museum by the Moslems who conquered the city in the year 640. The entire Museum and the remaining manuscripts rescued from other enemies of Hellenistic enlightenment were destroyed on the grounds that if the scrolls contained anything contrary to the writings of Mohammed they were wrong, and if they did not, they were superfluous.

Though the Museum was destroyed and the scholars dispersed, Greek culture did survive and eventually re-emerged to help mould Western civilization. Europe did ultimately learn from the Greeks the potential in human reason, as well as some of its finest products. Europe also inherited the mathematical evidence of design in nature and the confidence to apply reason to all the affairs of man. Western civilization was born when the spirit of reason took hold of man, and this civilization has advanced or retrogressed in accordance with the varying strength of that spirit.

VII

Interlude

Solicit not thy thoughts with matters hid,
Leave them to God, Him serve and fear.

John Milton

THE Earth, wrote a widely travelled merchant of sixth-century
Alexandria, is flat. The inhabited portion has the shape of a
rectangle whose length is double its breadth. Surrounding this
inhabited portion is water, which in turn is surrounded by more
earth. In the north is a high conical-shaped mountain around
which the sun and moon revolve. At night the sun is behind the
mountain; by day, of course, it is in front. The sky is glued to the
edges of the outer earth. Above the sky is heaven which is divided
into two floors, the upper one for the blessed and God and the
lower one for the angels ministering to man.

So wrote a man who breathed the same air as did Euclid,
Archimedes, Hipparchus, and Ptolemy. This merchant, Cosmas
by name, who later became a monk, did not learn these facts
about the universe through his travels. Rather, he said, the topography of the universe is established by demonstration from
Divine Scriptures, which it is not lawful for a Christian to doubt.
In his work *Topographia Christiana*, which was very popular
among both the educated and ignorant until the twelfth century,
Cosmas elaborated on this cosmography. Since the Bible tells us
that man lives on the 'face of the earth' there can be no antipodes.
In fact, if there were antipodes the sky would have to surround
the Earth whereas the Bible says the Earth is firmly fixed on its
foundations.

Later thinkers made some very important improvements on
the cosmology of Cosmas. At the centre of the universe was, of
course, the fixed Earth. Above the Earth were the moon, planets,
and sun each attached to one sphere. These eight spheres rotated
around the Earth in circular motion, the only kind possible for

heavenly objects. The spheres and heavenly bodies were made of tangible but incorruptible matter which was not subject to the physical laws of matter on Earth. Moreover, these heavenly bodies remained at a fixed distance from Earth, for the matter of which they were composed found itself congenially located where it was. There were, however, two other spheres above these eight. The ninth one, which carried no planet or star, was the prime mover of itself and the other eight spheres. This ninth sphere revolved faster than the others to complete its 24-hour journey around the Earth, for the spirits that moved it, being next to heaven, the tenth sphere, were more ardent than those moving the other eight. The tenth sphere was at rest and was inhabited by the beings Cosmas had already described.

A little more was gleaned about the Earth itself. The land extended over what is roughly modern Europe and the Near East, and Jerusalem was the centre of the habitable world. Inside the Earth was hell whose shape was that of a funnel with sinners arranged in rows running around the sloping wall. Satan resided at the very bottom. On the Earth was the still flourishing Garden of Eden, but this area was surrounded, unfortunately, by a wall of fire and was therefore inaccessible. The Earth contained both wondrous and monstrous creatures besides man. Most important were the angels and the devil with his demon assistants. Among mortal creatures similar to man were the satyrs; these had crooked noses, horns in their foreheads, and goat-like feet. It was readily ascertained that there were many varieties of satyrs. Some were headless; others, one-eyed; others possessed enormous ears; and still others were one-footed. The seas rivalled the earth in their marvels, even infested as it was with dragons who habitually fought elephants.

Nature was indeed marvellous and wondrous to behold. Of course, no one beheld it for there was no need to. Did not Saint Augustine say: 'Whatever knowledge man has acquired outside Holy Writ, if it be harmful it is there condemned; if it be wholesome, it is there contained.' Hence man had only to read the Scriptures and the writings of the early fathers of the Church in order to secure all the knowledge he should have. Biblical affirmations similarly answered the fundamental question of why the physical world, the animals, and the plants existed. These were created to serve man. The plants and animals furnished him with food just as the rain nourished his crops. Man was the centre of the universe not merely geographically but also in terms of ultimate purpose and design. Though the world of nature existed to serve man, the study of nature was to be avoided and nature itself even dreaded because Satan ruled the Earth and his cohorts were omnipresent. Science was actually sinful, and the knowledge obtained thereby was purchased at the price of eternal damnation.

All of nature served man, but man existed only to die and be reunited with God. Life on this Earth was of no real importance; only the after-life of the spirit mattered. Man had therefore to escape from this foul Earth into the divine empyrean. He had to wrest his soul from a stubborn flesh guilty of original sin by divesting himself of all earthly concerns and attachments. Participation in the bounty of nature, food, clothing, and sex had to be severely restricted, for these tainted the soul. Through these measures medieval man, certain of his sins and doubtful of salvation, prepared for an after-life, and might perhaps succeed in winning divine grace. This need to purify the soul by blotting out nature from man's thoughts and senses introduced a new dichotomy, an unending warfare between flesh and spirit, the world and God.

This account of the nature of man and his universe is a sample of the learning that was widespread towards the end of the Alexandrian period and during a good part of the ensuing medieval period. For reasons mentioned in the preceding chapter, the Alexandrian Greek civilization had degenerated rapidly. The late Alexandrian thinkers corrupted rather than improved their intellectual heritage. They neglected the sciences and mathematics, exhausted themselves in metaphysical disputes, and sought to reconcile Plato and Aristotle on subjects about which both philosophers were ignorant. Under the rising influence of Christianity, the Alexandrians deemed it important to explore the invisible world, to seek methods of freeing the soul from the body, and to discourse with demons and spirits. Their accomplishment was to convert philosophy into magic.

The decline in Greek and Roman learning was accelerated by the Church's battle against paganism. The Greek and Latin masterpieces contained a mythology that had to be erased from people's minds, and a morality opposed to Christian ethics. Also, the Greek and Roman emphasis on life in this world was regarded as misguided at the very least. What do a physical life, health, science, literature, and philosophy amount to compared with salvation of the soul? Why read the poets when the precepts of the Gospel should be pondered? Why make life on Earth agreeable and comfortable when it is but an insignificant prelude to eternal life elsewhere? Why seek to answer questions about natural phenomena when the nature of God and the relation of the human soul to God are yet to be explored and understood? It was but a step to the conclusion that all Greek and Roman learning was impious and heretical. Thus the opposition of the Church to paganism and its ideals fostered an anti-classical attitude throughout the Christian domains and absorbed all intellectual interests and energies in theological questions.

The region in which knowledge and learning had reached their highest peak and then dropped to their nadir embraced only those countries surrounding the Mediterranean Sea. Thus far, we should notice, central and northern Europe has played no role. What was the state of affairs in England, France, Germany, and other countries? How were they linked to the civilizations of Greece and Rome and how did they come to inherit the riches of Greek thought?

The Germanic tribes that inhabited Europe in the early centuries of our era, the forefathers of most Americans, were still barbarians. They lived in ignorance and poverty, sometimes euphemistically called virtuous simplicity. Industry was unknown. Trade was effected through barter and supplemented by plundering other tribes and more civilized regions. The political organization of each tribe was primitive and was headed by a valorous warrior. Political bonds were supplemented by religious ones. All these tribes worshipped the Sun, Moon, Fire, the Earth, and special deities who governed the daily affairs of life. Like most primitive peoples the Germanic tribes believed in divination and in human sacrifices to the gods.

An account of the learning, arts, and sciences of the Germanic tribes is readily given. There was nothing. These peoples were entirely unacquainted with the use of letters. Because writing was unknown it was impossible for any generation to pass on to its successors an adequate record of its discoveries, creations, or experiences. The oral transmission of knowledge may magnify trivial exploits into legends and cast a shadow of truth over fancies and superstitions but it does not promote the arts and sciences.

The barbarians were gradually civilized. The first major influence was exerted by the Romans who conquered portions of Europe and imposed some of their customs and institutions on the conquered areas. When the Empire collapsed the Church, the only remaining powerful organization in Europe, took control. In order to Christianize the heathens the Church introduced and supported schools, organized parishes, and supplied able leaders. By these means the barbarians became acquainted with writing, political institutions, law, ethics, and, of course, the Christian religion. Thus Europe received the legacy of Rome.

A people unacquainted with the rudiments of arithmetic could hardly be expected to advance mathematics. Actually history has no surprise for us in this instance. In no one of the civilizations that have contributed to our modern one did mathematical learning exist on as low a level as it did in medieval Europe. From the years 500 to 1400 there was no mathematician of note in the whole Christian world.

The progress that was made during this period was contributed by the Hindus and Arabs. We have already had occasion to see how the Hindus applied the Babylonian principle of place value to base ten and converted the Babylonian separation symbol into a full-fledged zero. In so far as history can ascertain the Hindus were entirely original in creating one other idea which proved immensely important later on. This was the concept of negative numbers. Corresponding to each number such as 5, they introduced a new number − 5 and called the old numbers positive to distinguish them from the new, negative ones. The Hindus showed, too, that these new numbers could be as useful as positive numbers by employing them to represent debts. In fact, they formulated the arithmetic operations on negative numbers with this application in mind.

These and other Hindu contributions were acquired by the Arabs who transmitted them to Europeans. The ideas were not absorbed into the body of mathematical learning, however, until well into the seventeenth century. The European universities of the medieval period offered only arithmetic and geometry, theoretical arithmetic consisting largely of simple calculations and complex superstitions. The geometry was confined pretty much to the first three books of Euclid, candidates for the master's degree being required to know no more. The farthest stage that was reached in some universities was the very elementary theorem that the base angles of an isosceles triangle are equal. A modicum of mathematics was also involved in the other two subjects of the quadrivium, these being music and astronomy. All in all, a learned European mathematician of a thousand years ago knew far less than any elementary-school graduate does today.

Yet even in this low state of civilization mathematics played a part. One role of mathematics, though not one always in favour with the Church, was to make astrological forecasts. In fact, early in the medieval period the word mathematics as distinguished from geometry meant astrology and professors of astrology were called *mathematicii*. At that time, astrology was in disfavour with the Roman emperors also, and so we find laws such as the Roman Code of Mathematics and Evil Deeds, which damned and forbade the art of mathematics. Later Roman and Christian emperors, though banishing astrologers from their kingdoms, employed them in their own courts as influential advisers. If there was anything in this business of foretelling the future, the rulers were not going to overlook it; neither were they going to let anyone else get hold of that knowledge.

Despite the moral and legal condemnation of astrology the subject flourished because, from Alexandrian times on, the great physicians, including Galen, believed that they could decide the proper medical treatments by consulting the stars. On the basis of true and spurious Arabic translations of Aristotle it was believed that the regular, circular motions of the stars controlled the ordered course of nature such as the seasons, night and day, and growth and decay. The planets, on the other hand, errant in their movements when viewed from the Earth, governed the variable and indeterminate life of man. Each planet influenced a particular organ of the body. Mars presided over the bile, blood, and kidneys; Mercury ruled the liver; and Venus ruled the genital organs. Each zodiacal sign governed some one region of the body such as the head, neck, shoulders, and arms. The appearances of the planets among the constellations of stars controlled human fortunes. And so mathematicians and physicians earnestly studied the motions of the stars and the planets and attempted to correlate their positions in the sky with the behaviour of the human body and with human events.

So much knowledge of mathematics was required for this purpose that physicians had to become learned in the field. In fact, they were astrologers and mathematicians far more than they were students of the human body. For several hundred years the words physician and algebraist were practically synonymous. When, for example, Samson Carrasco is thrown from his horse in *Don Quixote*, an *algebrista* is summoned to bind up his wounds. The medieval universities actually taught medical students the use of mathematics in astrology, the most famous of such places being Bologna, which had a school of Mathematics and Medicine as early as the twelfth century. Even Galileo lectured to medical students on astronomy so that they might apply it to astrology.

It is clear that in the medieval period astrology was not regarded as a superstition to be indulged in by the stupid or the naïve. It was a science whose principles were as seriously accepted as were Copernican astronomy and the law of gravitation in the nineteenth century. Roger Bacon, Cardan, and Kepler subscribed to it and used their scientific and mathematical knowledge in its service. The science of astrology has today degenerated to syndicated 'Lucky Day' columns in newspapers, to birthday-month horoscope reports dispensed in the five-and-ten-cent stores, and to the bargain subway scales which offer a person his exact weight and future for the same penny. The science of yesterday is the superstition of today.

To a modern man the Church's interest in mathematics is far

more understandable. In the first place, astronomy, geometry, and arithmetic were needed to keep the calendar. It was especially important to know the date of the Easter holiday. Every monastery in Europe had at least one monk who could do this work.

Mathematics was equally valuable to the Church as a preparation for theology. All that Plato and other Greeks of the Classical Age had found in mathematics as a preparation for philosophy, the Church accepted, merely substituting theology for philosophy. It was careful to state that not too much mathematics was needed for this purpose. Just so much was good for the mind, and no more. This interest in reasoning on the part of theologians who nevertheless accepted so much on faith and who appealed to the Scriptures or to the fathers of the Church as the arbiters of truth warrants a little examination.

The Greeks had many gods and no theology. The medieval period had one God and a great deal of theology. Early in the medieval period faith was indeed almost the sole support of that theology. Saint Augustine said: let us believe in order that we may know. However, as more and more Church fathers propounded doctrines and as scholars seeking to understand these doctrines faced the problem of reconciling opposing or conflicting statements, reason was employed to effect the necessary reconciliations. Reasoning also fortified faith with true argument, dialectic, and explanation. Reason similarly demonstrated the accord between systems of philosophy and Christian doctrines and between observed facts and Christian renditions.

Rather late in the medieval period, reason began to supplant faith as the chief support of Christian theology. This movement was stimulated by the translation into Latin from Arabic of numerous Greek manuscripts. In particular, Aristotle's immense learning and his logic became known. Because Christian theology had already incorporated some Aristotelianism, the Church scholars could not afford to ignore the vast body of knowledge now made available. The Church thereupon faced the task of reconciling Aristotelianism and Catholic theology and of harmonizing metaphysics with revelation. The possibilities of a thoroughly rational defence of Christianity appealed to the

Scholastics, of whom Saint Thomas Aquinas is the most noted. Aquinas undertook to provide a firm logical structure for theology and to combine Catholic doctrine and Aristotelian philosophy in one rational system. The result of his efforts, the *Summa Theologiae*, affords the most comprehensive and thorough exposition of Catholic philosophy ever constructed, while the organization of his material earned for his work the title of the 'spiritual Euclid'.

This brief glance at the rational interests of Catholic theologians, though it does not do justice to the high intellectual calibre of their works, may make clear why the Church kept alive at least a modicum of mathematics during the Middle Ages. There is, however, a more vital relation between mathematics and medieval theology. We have already indicated that the Church had a philosophy of nature. This philosophy asserted, first of all, that nature was designed by God, designed in fact to serve man. Every event, every being, served a purpose. It was a second tenet of this philosophy that nature was intelligible to man. God's ways and God's purposes could be understood if man but searched hard enough. The understanding would come not from the observation of nature, but from a proper study of the Scriptures, the word of God. Moreover, the Church urged its subjects to seek this understanding of God's purposes; the proper study of mankind was God. The average man might not attain complete comprehension – the ways of God are mysterious to some mortals – but meaning, reason, and purpose were there. 'Just are the ways of God and justifiable to men.'

Thus the late medieval scholars, the Scholastics in particular, not only provided the rational atmosphere in which modern mathematics and science were born, but they infused the great thinkers of the Renaissance with the belief that nature was the creation of God and that God's ways could be understood. It was this fundamental article of faith that dominated and inspired the Renaissance mathematicians and scientists. It was this faith that sustained the patient, tireless, arduous, and difficult researches of Copernicus, Brahe, Kepler, Galileo, Huygens, and Newton. True, these men abandoned the Scriptures and returned to Euclid for their premises and to the observation of nature for their purely scientific data, but they sought, for the most part, to do no more than to understand God's marvellous design. They were, and remained, orthodox adherents of the Christian religions. It is an irony of history that their researches produced laws which clashed with Church doctrines and that these researches ultimately undermined the Church's domination of thought.

We cannot dismiss the medieval period without asking why it failed to advance mathematics at least during the later years. In answering this question we are inevitably drawn to link and contrast the medieval civilization with the equally barren Roman era. It would appear that the Roman civilization was unproductive in mathematics because it was too much concerned with practical results to see farther than its nose. The medieval period, on the other hand, was unproductive because it was not concerned with the *civitas mundi* but, rather, with the *civitas dei* and with preparation for the latter world. One civilization was earth-bound, the other, heaven-bound. The practicality of the Romans bred sterility whereas the mysticism of the Church argued, in effect, for the complete neglect of nature, and its dogmatism confined the intellect and impeded the creative spirit. There is sufficient historical evidence to show that mathematics cannot bloom in either of these climates. It was true in the Greek period, and we shall shortly see it to be true again, that mathematics can flourish best in a civilization willing to ally itself with the world of nature and, at the same time, to permit the mind unlimited freedom of thought whether or not it promises immediate solutions of the problems of man and his universe.

VIII

Renewal of the Mathematical Spirit

If you do not rest on the good foundation of nature,
you will labour with little honour and less profit.

Leonardo da Vinci

VERY much ignored as a fascinating and influential figure of the
Renaissance is Jerome Cardan. In his *Book of My Life,* which is
comparable to Cellini's *Autobiography* and which makes Cellini
appear as a saint and recluse by comparison, Cardan candidly
revealed the most intimate and glowing details of his life and of
his times.

The career of this extraordinary rogue and scholar started ad-
venturously, he said in his confessions, when his mother tried
and failed to induce an abortion. The illegitimate and sickly
child was born in Milan in the year 1501. He described himself as
endowed at birth with only misery and scorn. Rather early in life
he prepared himself for the numerous careers he was to follow
– those of mathematician, physician, metaphysician, cheat,
gambler, murderer, and adventurer. Despite a wretched boy-
hood, many maladies, chronic illnesses, and extreme poverty, he
was finally graduated in medicine from the University of Pavia.
During his first forty years, poverty continued to stalk him; a
physical disability prevented him for a long time from satisfying
his strong desires for the pleasures of love; and illnesses continued to rob him of vigour. As if to vent his wrath on life he
cheated at games, was vindictive and consciously cruel in his
speech, and boasted of his superiority over his contemporaries.

During most of his life he practised medicine and debauchery.
In his spare time he produced some of the best mathematics of
the Renaissance. His rascality expressed itself in this activity also,
for the most famous result which appeared in his *Ars Magna,* his
greatest mathematical work, was created by another math-

ematician and published by Cardan without permission. During many years of his life he held professorships of mathematics and medicine in several Italian universities. His last years were spent as an astrologer in the papal court. Towards the end of his life he found that despite his infamies, he had managed to acquire a grandson, fame, wealth, learning, powerful friends, and belief in God, to whose goodness he owed the fifteen teeth still remaining in his mouth. It is said that he prognosticated his own death and committed suicide on the date predicted in order to maintain his reputation as an astrologer.

Cardan bridged the gap between the Middle Ages and modern times. In his metaphysics he was still tied to medievalism and fantasy. He was the rational apologist for palmistry, ghosts, portents, and astrology. He was also a firm believer in natural magic, a 'science' somewhat broader in scope than astrology. Through natural magic man learned about human character, the ways and purposes of nature, knowledge of the future, how the incorruptible heavenly bodies influence the daily actions and fate of man, and the art of prolonging life.

In his lewdness and rejection of authoritarian doctrines, as well as in his searching mathematical, physical, and medical studies, Cardan symbolized the revolt from a thousand years of intellectual serfdom, and the revival of interest in the physical world. His properly scientific investigations were modern in spirit and entirely free of mysticism and occultism. Despite the generous use of other people's creations, Cardan's great works on algebra and arithmetic were the first major contributions to modern mathematics and were undoubtedly among the best of the sixteenth century.

Cardan's *Ars Magna*, Copernicus' *On the Revolutions of the Heavenly Spheres*, and Vesalius' *On the Structure of the Human Body*, which appeared between 1543 and 1545, mark as clearly as the written word ever can the dividing line between medieval and modern thought. These works were so revolutionary that it is natural to inquire what forces broke down the medieval civilization and coalesced to form a new one.

The earliest influence tending to transform thought and life in medieval Europe was the introduction of Greek works. The first significant contact with these works was made through the Arabs. Late in the medieval period some of the Greek scholars, who resided in Constantinople, the centre of the Byzantine or Eastern Roman Empire, became discouraged by the poverty there and migrated to Italy. Those who remained were driven from their homes when the Turks captured the city and these, too, sought refuge in Italy. By the fifteenth century, it became possible to make translations into Latin directly from the Greek manuscripts these scholars brought with them. From this time on the impact of Greek knowledge on European thought was boundless. All the great scientists of the Renaissance acknowledged the Greeks as their inspiration and gave credit to that people for specific ideas. The Polish Copernicus, the German Kepler, the Italian Galileo, the French Descartes, and the English Newton received light and warmth from the sun of Greece.

Equally important in the fashioning of modern civilization was the rise of towns, cities, and a merchant class. Mining, manufacturing, cattle-raising on a large scale, and huge farms, the forerunners of present-day big business, became an important part of European life. Wealth begets wealth and worldliness. The merchants sought to enjoy the material things they handled; they demanded, also, the freedom to plan and do business within the framework of a government favourable to their interests. The Church, on the other hand, denounced profits, sanctified poverty and the simple life, and stressed denial in this world for the sake of an after-life. Inevitably the townspeople resented and rebelled against the limitations imposed by the Church.

Because the merchants sought to expand their trading interests, they promoted the geographical explorations of the fifteenth and sixteenth centuries. The discoveries of America and of a route to China around Africa enlarged man's horizons and brought much knowledge to Europe of strange lands, beliefs, religions, and ways of life. This knowledge challenged medieval dicta and stimulated men's imaginations.

In contrast to the slave classes of Egypt, Greece, and Rome and the serfdom of medieval feudalism, the new society possessed an expanding class of free labourers and artisans. The stimulus to profit by their labour caused these men to think through ideas pertinent to their work. Labourers seeking to increase their effectiveness and wage-paying employers commenced an active search for labour-saving devices. As a result, there arose a growing interest in machinery, materials, and nature. This social and economic movement fostered the conversion of the European civilization from feudalism and indifference to natural phenomena to industrialism and the investigation of physical problems. The great practical inventions that resulted from the work of the artisans were more numerous than anyone might have anticipated. Cotton paper and later rag paper replaced costly parchment; movable type replaced manuscript copyists. These inventions gave wings to thought, enabling it to fly over boundaries of nations and religions.

Another event of the Renaissance, the introduction of gunpowder in the fourteenth century, suggested a host of scientific problems. Gunpowder made possible bullets and cannon balls which could be fired effectively from great distances and with high velocities. To develop these weapons and to learn how to employ them effectively the princes of states spent huge sums which were out of all proportion to the scientific importance of the phenomena involved. But the needs of war have always aroused nations to put forth money and efforts unimaginable during times of peace.

Doubts about the soundness of Church science and cosmology, objections to the Church's suppression of experimentation and thought on problems which the new economic order created, the degeneration of the papal court to a level of morality which Christians would normally describe as pagan, and, not the least, serious intellectual schisms all culminated in the Protestant revolution. The men who revolted were supported by the merchant class, which was anxious to break the power of the Church, and by many secular princes who wished to rule unfettered.

The Reformation as such did not unshackle men's minds; nevertheless, indirectly it served the cause of free thought. When religious leaders such as Luther, Calvin, and Zwingli dared to challenge the papacy and Catholic doctrines, the common person felt encouraged to do the same. The Protestants, as revolters, were necessarily more tolerant and so gave protection to thinkers whom the Catholic Church would have suppressed. Reluctantly, too, the Protestants made rational interpretations of the Scriptures to combat Catholic doctrines, though Reason was 'the Devil's harlot' to Luther. In fact, the Protestants were forced at times to maintain that variations of belief are the necessary consequences of free inquiry. Finally, inasmuch as people were called upon to choose between Catholic and Protestant claims, independent thinking was unintentionally encouraged. With a 'plague on both your houses', many turned from the two faiths to other sources of knowledge, such as nature and the ancient classics.

It should be apparent even from this hasty sketch of the new forces at work in Europe that fundamental changes were about to take place in that civilization. Though there may be some dispute about which century marks the turning point from medieval to modern times, there is no doubt that by the fifteenth century Europe became an arena of seething minds which disputed bitterly on religious beliefs, advocated reason against scholastic tyranny and obsolete authority, and pitted Greek worldliness against Catholic other-worldliness. Intellects forced to speculate endlessly on a small group of concepts and fettered by limited knowledge and dogmatism finally burst their bonds. Men intolerant of control, ready to criticize the established canons of conduct, and enthusiastic about the liberty of the ancients asserted themselves against irksome authorities. Out of the intellectual ferment, minds eager to apprehend and devour new ideas sought ground firmer than the disputed Catholic theology on which to stand and build, and attempted to erect new approaches to the problems of man, nature, and the social order.

The material with which to build was readily secured. From the rich stores of Greek learning which had lain almost untouched for a thousand years Europeans derived a new spirit, new ideals, and a new outlook on the universe. The Greek works restored confidence in the sovereign powers of human reason and encouraged the Renaissance man to apply that faculty to the problems besetting his age. The love of a dispassionate search for truth was reborn and the search itself redirected to nature's laws instead of divine pronouncements gleaned from the Scriptures, to the universe of God instead of God. As if awakened from a long slumber Europeans discovered a 'brave new world' teeming with life and wondrous creatures, among which man himself stood forth as a biological and physical phenomenon worthy of observation and study. Men looked with enlivened curiosity at the heavens and were enthralled by the strange stories of those who sailed the seas and explored new lands. Beauty so long condemned to hell as a pagan goddess of the flesh was rediscovered both in literature and in the physical world, and in place of sin, death, and judgement men sought beauty, pleasure, and joy. The dignity of man, who had theretofore been denounced as a worthless sinner, was reaffirmed. Above all, the human spirit was emancipated and encouraged to roam freely over the universe.

A major positive doctrine of the Renaissance proclaimed the idea of 'back to nature'. Every variety of scientist abandoned the endless rationalizing on the basis of dogmatic principles vague in meaning and unrelated to experience, and turned to nature herself as the true source of knowledge. This appeal to nature and to observation had been urged long before by Roger Bacon and had even been pursued by a few noble and broad-minded thinkers earlier than the fifteenth century, among whom we may name William of Ockham, Nicholas of Oresme, and John Buridan. These men, however, spoke too soon and could not be heard above the din of endless theological disputation. But the thin stream slowly broadened out and gained force.

The back-to-nature movement had hardly been launched when a few scientists who were ardently engaged in it conceived an even more revolutionary idea. Whereas the Greeks and early Renaissance scientists sought knowledge of nature, Francis Bacon and René Descartes dared to suggest mastery and dreamed of man's conquest of the whole natural world. To Bacon the aim of science was not just speculative satisfaction but the establishment of the reign of man over nature and the increase in man's comforts and happiness. Descartes wrote:

It is possible to attain knowledge which is very useful in life, and instead of that speculative philosophy which is taught in the Schools, we may find a practical philosophy by means of which, knowing the force and action of fire, water, air, the stars, heavens and all other bodies that environ us, as distinctly as we know the different crafts of our artisans, we can in the same way employ them in all those uses to which they are adapted, and thus render ourselves the masters and possessors of nature.

The challenge thrown out by Bacon and Descartes was quickly taken up, and scientists plunged optimistically into the task of mastering nature. Today, three hundred years later, the heirs of these Renaissance thinkers and scientists are still at work on the task, the vision of Bacon and Descartes steadfast in their minds and urging them ever onward.

The movement to reconstruct all knowledge by the application of reason and the return to nature as the source of truth naturally brought to the fore the subject which had in the past contributed pre-eminently to both of these goals. Keen minds seeking to establish new systems of thought on the basis of certain cogent knowledge were attracted by the certitude of mathematics, for the truths of mathematics, however much they may have been ignored in medieval times, had really never been challenged or been subject to the slightest doubt by the true scholars. Moreover, mathematical demonstrations carried with them a compulsion and an assurance that had not been equalled in science, philosophy, or religion. Said Descartes:

I was especially delighted with the mathematics on account of the certitude and evidence of their reasonings: ... I was astonished that foundations, so strong and solid, should have had no loftier superstructure reared on them.

Leonardo, too, said that only by holding fast to mathematics can the mind safely penetrate the labyrinth of intangible and insubstantial thought.

To the Renaissance scientist, as to the Greek, mathematics was more than a reliable approach to knowledge; it was the key to nature's behaviour. The conviction that nature is mathematical and that every natural process is subject to mathematical laws began to take hold in the twelfth century when Europeans first obtained it from the Arabs, who in turn were quoting the Greeks. Roger Bacon, for example, believed that the book of nature was written in the language of geometry. In his day this doctrine sometimes took a rather unusual form. It was believed, for example, that the divine light is the cause of all phenomena and is the form of all bodies. Hence, the mathematical laws of optics were the true laws of nature.

Kepler, too, affirmed that the reality of the world consists of its mathematical relations. Mathematical laws are the true cause of phenomena. Mathematical principles are the alphabet in which God wrote the world, said Galileo; without their help it is impossible to comprehend a single word, and man wanders in vain through dark labyrinths. In fact, only the mathematically expressible properties of the physical world are really knowable. The universe is mathematical in structure and behaviour, and nature acts in accordance with inexorable and immutable laws.

Descartes, father of modernism, related a mystic experience which revealed to him the secret of nature. In a dream which he recalled distinctly and which, he said, occurred on 10 November 1619, the truth stood forth clearly that mathematics is the 'Open Sesame'. He awoke convinced that all of nature is a vast geometrical system. Thereafter he 'neither admits nor hopes for any principles in Physics other than those which are in Geometry or in abstract Mathematics, because thus all the phenomena of nature are explained, and some demonstrations of them can be given'.

Nature, then, was to be analysed and reduced to mathematical laws. But how should this process begin? What phenomena should be selected for investigation? What concepts are fundamental and at the same time mathematically expressible? To these questions the Renaissance students of nature fashioned their own answers.

Unlike the Greeks to whom objects and their shapes were fundamental, space being of concern only as marking the end or boundary of an object, the new scientist chose space itself as one concept which underlies all phenomena, and in which objects exist or extend and move (though Descartes insisted that some kind of non-perceivable matter exists where the solid matter of experience does not). The essence of objects or matter is space, and objects are essentially chunks of space, space solidified, or geometry incarnate. Granted this principle, matter was mathematically describable through the geometry of space. Time was introduced as another fundamental concept. Objects exist and move in time just as they do in space. Galileo then pointed out that time could be expressed mathematically, for the instants of time are but numbers and just as numbers follow each other, so do the instants in time.

As for objects themselves, their basic properties are extension and motion. Differences among bodies are differences in shape, density, and motion of their component particles, and these properties are real and expressible in mathematical terms. On the other hand, such qualities as colour, taste, warmth, and pitch are not real but are reactions of minds to the real, primary qualities. These secondary qualities could be dismissed in an analysis of the real world because they are but illusions or mere appearances.

Thus extension, or shape in space, and motion in space and time are the source of all properties and are the fundamental realities. In Descartes' words, 'Give me extension and motion and I will construct the universe.' Mathematics, through geometry and number, expresses these essences of objects. The motion of objects, he continued, is due to the mechanical action of forces which obey necessary and exact laws. Life itself, human, animal, and plant, is subject to these laws. Only God and the human soul were exempted by Descartes. In brief, the real world is the totality of mathematically expressible motions of objects in space and time, and the entire universe is a great, harmonious, and mathematically designed machine.

The notion of causality, the link between two events one of which seems to follow necessarily from the other, received a new formulation. The fact that the effect appears to follow the cause in time is due to the limitations of human sense perceptions. *Causa sive ratio*; cause is nothing but reason. The meaning of this doctrine is best explained by an analogy. Given the axioms of Euclidean geometry, the properties of a circle, such as the length of the circumference, the area, and the properties of inscribed angles, are all immediately determined as necessary logical consequences. In fact, Newton is supposed to have asked why anyone bothered to write out the theorems of Euclidean geometry since they are obviously implied by the axioms. Most human beings, however, take a long time to discover each of these properties. But this discovery in time, which seems to relate axioms and theorems in the same temporal sequence as cause and effect, is illusory. So it is with physical phenomena. To the divine understanding all phenomena are co-existing and are comprehended in one mathematical structure. The senses, however, recognize events one by one and regard some as the causes of others. We can understand now, said Descartes, why mathematical prediction of the future is possible; it is because the mathematical relationships are pre-existing. The ultimate in physical explanation is the mathematical relationship. The mathematical interpretation of nature became so popular and fashionable in 1650 that it spread throughout Europe and dainty, expensively bound accounts by its chief expositor, Descartes, adorned ladies' dressing tables.

One other very vital element in Renaissance thought has not yet been mentioned. The scientists of the period were born and educated in a religious world that also had a philosophy of nature. This philosophy, we know, asserted that the universe was designed by God and was the product of His handiwork; also, its rationale was accessible to man. The Catholic emphasis on the reasonableness of nature and on the all-important presence of God was impressed on every intellectual of the fifteenth and sixteenth centuries. Hence, these men faced the task of reconciling and fusing Catholic teachings and the Greek mathematical view of nature. Their solution is perhaps obvious. The universe is designed; it is rational; it is understandable to man. This much was common to the two philosophies. It was merely necessary to add that God designed and created the universe in accordance with mathematical laws to effect the reconciliation. In other words, by making God a devoted and supreme mathematician, it became possible to regard the search for the mathematical laws of nature as a religious quest. The study of nature became the study of God's word, His ways, and His will. The harmony of the world was God's mathematical arrangement and, added Descartes, the laws of nature remained constant because of the eternal invariableness of God's will.

God put into the world that rigorous mathematical order

which men comprehend only laboriously. Mathematical knowledge is absolute truth and as sacrosanct as any line of Scripture – superior, in fact, for there is much disagreement about the Scriptures but there can be none about mathematical truths. 'Nor,' said Galileo, 'does God less admirably discover Himself to us in Nature's actions, than in the Scripture's sacred dictions.'

Thus Catholic emphasis on a universe rationally designed by God and the Pythagorean-Platonic insistence on mathematics as the fundamental reality of the physical world were fused in a programme for science which in essence amounted to this: science was to discover the mathematical relationships that underlie and explain all natural phenomena and thus reveal the grandeur and glory of God's handiwork.

We can see that modern science derived its inspiration and initiation from a philosophy that affirmed the mathematical design of nature. Moreover, the goal of science was, similarly, a mathematical one, namely, the disclosure of that design. As Randall says in the *Making of the Modern Mind*,* 'Science was born of a faith in the mathematical interpretation of Nature, held long before it had been empirically verified.'

The nature of scientific activity as envisaged by the Renaissance thinkers is often incorrectly understood. Many people credit the rise of modern science to the introduction of experimentation on a large scale and believe that mathematics served only occasionally as a handy tool. The true situation, as indicated above, was actually quite the reverse. The Renaissance scientist approached the study of nature as a mathematician; that is, he sought and expected to find broad, profound, immutable, rational principles either through intuition or immediate sense perception, in much the same way as Euclid presumably found his axioms. There was to be little or no assistance from experimentation. He then expected to deduce new laws from these principles. The Renaissance scientist was a theologian with nature instead of God as his subject. For Galileo, Descartes, Huygens, and Newton, the deductive, mathematical part of the scientific enterprise always loomed larger than experimentation. Galileo valued a scientific principle, even when obtained by ex-

* London: George Allen & Unwin Ltd.

perimentation, far more because of the abundance of theorems
which flowed deductively from it than because it afforded knowledge in itself. Moreover, he said that he experimented rarely and
then primarily to refute those who did not follow the mathematics.

It is true that some experimentation did take place; a great deal
of it, however, was performed by artisans and technicians who
did not seek ultimate meanings and laws but common, practical
knowledge. Moreover, the experiments that had been performed
by the middle of the seventeenth century were not decisive. Not
only did mathematical theory precede and dominate experimentation in the formative period of modern science but, peculiarly
enough, experimentation was regarded as anti-scientific. The
turn to experimentation was an anti-rationalist movement, a
movement away from the unending and hitherto profitless speculation of a waning religious spirit and away from religious dogmatism so often proved wrong. Long after the Renaissance, the
experimentalists and theoreticians realized that they were pursuing
the same objectives and united their efforts.

What the great thinkers of the Renaissance envisaged as the
proper procedure for science did indeed prove to be the more
profitable course. The rational search for laws of nature produced, by Newton's time, very valuable results on the basis of the
slimmest observational and experimental knowledge. The great
advances of the sixteenth and seventeenth centuries were in astronomy where observation offered little that was new and in
mechanics where the mathematical theory attained com-
prehensiveness and perfection on the basis of very limited experimentation. The scientist is usually pictured in a laboratory piled
up with impressive equipment and gadgets; actually, during the
Renaissance the major scientists were 'pencil-pushers'.

IX

The Harmony of the World

> . . . how build, unbuild, contrive
> To save appearances, how gird the Sphere
> With Centrick and Eccentrick scribl'd o'er
> Cycle and Epicycle, Orb in Orb.
>
> *John Milton*

ON the title page of Copernicus' *On the Revolutions of the Heavenly Spheres,* published in the year of his death, 1543, appeared the legend originally inscribed on the entrance to Plato's academy: 'Let no one ignorant of geometry enter here.' The Renaissance had borne its first fruits.

Perhaps the enterprising merchants of the Italian towns received more than they bargained for when they aided the revival of Greek culture. They sought merely to promote a freer atmosphere; they reaped a whirlwind. Instead of continuing to dwell and prosper on firm ground, the terra firma of an immovable Earth, they found themselves clinging precariously to a rapidly spinning globe that was speeding about the sun at an inconceivable rate. It was probably sorry recompense to these merchants that the very same theory which shook the Earth free and set it spinning also freed the mind of man.

The reviving Italian universities were the fertile soil for these new blossoms of thought. There Nicolaus Copernicus became imbued with the Greek conviction that nature is a harmonious medley of mathematical laws and there, too, he became acquainted with the hypothesis – Hellenic also in origin – of planetary motion about a stationary sun. In Copernicus' mind these two ideas coalesced. Harmony in the universe demanded a heliocentric theory and he became willing to move heaven and earth in order to establish it.

Copernicus was born in Poland. After studying mathematics and science at the University of Cracow, he decided to go to

Bologna, where learning was more widespread. There he studied astronomy under the influential teacher, Novara, a foremost Pythagorean. In 1512 he assumed the position of canon of the Cathedral of Frauenburg in East Prussia, his duties being that of steward of Church properties and justice of the peace. During the remaining thirty-one years of his life, however, he spent much time in a little tower on the wall of the Cathedral closely observing the planets with naked eyes and making untold measurements with crude home-made instruments. The rest of his spare time he devoted to improving his new theory of the motions of heavenly bodies.

After years of observation and mathematical reflection, Copernicus finally circulated a manuscript describing this theory and his work on it. The reigning pope, Clement vi, approved of the work and requested its publication. But Copernicus hesitated. The tenure of office of the Renaissance popes was rather brief and a liberal pope might readily be succeeded by a reactionary one. Ten years later, Copernicus' friend Rheticus persuaded him to allow the publication, which Rheticus himself then undertook. While lying paralysed from an apoplectic stroke, Copernicus received a copy of his book. It is unlikely that he was able to read it, for he never recovered. He died shortly afterwards, in the year 1543.

At the time that Copernicus delved into astronomy the science was in about the same state in which Ptolemy had left it. It had become increasingly difficult, however, to include under the Ptolemaic heavens the knowledge and observations of Earth and sky accumulated, largely by the Arabians, during the succeeding centuries. By Copernicus' time it was necessary to invoke a total of 77 mathematical circles in order to account for the motion of the sun, moon, and five planets under the epicyclic scheme discussed in Chapter vi. It is no wonder that Copernicus grasped at the possibilities in the Greek idea of planetary motion about a stationary sun.

Already incorporated in Ptolemaic theory were some other Greek ideas which Copernicus adopted. He, too, believed that circular motion was the natural motion of heavenly bodies and therefore used the circle as the basic curve on which to build his theory. Accordingly, he supposed that each body, that is, moon or planet, moves on a circle whose centre moves on another circle. For some of these bodies he assumed that the centre of the latter circle moved on still a third circle, and where necessary, he introduced even a fourth circle. The centre of the last circle he assumed to be the sun, whereas Hipparchus and Ptolemy had taken it to be the Earth. For a mystic reason similar to that held by the Greeks, he retained the notion that each body or point moves along its circle at a constant speed, though the apparent motion of the body is not constant. A change in speed, Copernicus reasoned, could be caused only by a change in motive power, and since God, the cause of the motion, was constant, the effect could not be otherwise.

Then Copernicus proceeded to do what no Greek is known to have attempted; he carried out the mathematical analysis required by the heliocentric hypothesis. Merely by using the sun where Hipparchus and Ptolemy had used the Earth, Copernicus found that he was able to reduce the total number of circles involved from 77 to 31. Later, to secure better accord with observations, he refined his theory somewhat by putting the sun near, but not quite at, the centre of some of these aggregations of circles.

When Copernicus surveyed the extraordinary mathematical simplification that the heliocentric hypothesis afforded, his satisfaction and enthusiasm were unbounded. He had found a simpler mathematical account of the motions of the heavens, and hence one which must be preferred, for Copernicus, like all scientists of the Renaissance, was convinced that 'Nature is pleased with simplicity, and affects not the pomp of superfluous causes.' Copernicus could pride himself, too, in the fact that he had dared to think through what others, including Archimedes, had rejected as absurd.

Copernicus did not finish the job he had set out to do. Though the hypothesis of a stationary sun considerably simplified astronomical theory and calculations, the epicyclic paths of the planets did not quite fit observations and Copernicus' few attempts to patch up his theory, always on the basis of circular motions, did not succeed.

It remained for the German, Johann Kepler, some fifty years later, to complete and extend the work of Copernicus. Like most youths of those days who showed some interest in learning, Kepler was headed for the ministry. While studying at the University of Tübingen he obtained private lessons in Copernican theory from a teacher with whom he had become friendly. The simplicity of this theory impressed Kepler very much. Perhaps this interest awakened suspicions in the minds of the superiors of the Lutheran Church, for they questioned Kepler's devoutness, cut short his ministerial career, and assigned him to the professorship of Mathematics and Morals at the University of Gratz. This position called for a knowledge of astrology, and so he set out to master the rules of that 'art'. By way of practice he checked the predictions he made about his own fortune.

As an extracurricular activity he applied mathematics to matrimony. He had married a wealthy heiress while he was at Gratz. When this wife died he listed the young ladies eligible for the vacancy, rated each on a series of qualities, and averaged the grades. Women being notoriously less rational than nature, the highest-ranking prospect refused to obey the dictates of mathematics and declined the honour of being Mrs Kepler. Only by substituting a smaller numerical value was he able to satisfy the equation of matrimony.

Kepler's interest in astronomy continued and he left Gratz to become an assistant to that most famous observer, Tycho Brahe. On Brahe's death Kepler succeeded him as official astronomer, part of his duties being once again of an astrological nature for he was required to cast horoscopes for worthies at the court of his employer, Rudolph II. He reconciled himself to his work with the philosophical view that nature provided all animals with a means of existence. He was wont to refer to astrology as the daughter of astronomy who nursed her own mother.

During the years he spent as astronomer to the Emperor Rudolph, Kepler did his most serious work. It is extremely interesting that neither he nor Copernicus ever succeeded in ridding himself of the scholasticism from which his age was emerging. Kepler, in particular, mingled science and mathematics with theology and mysticism in his approach to astronomy, just as he combined wonderful imaginative power with meticulous care and extra-ordinary patience.

Moved by the beauty and harmonious relations of the Copernican system, he decided to devote himself to the search for whatever additional geometrical harmonies the data supplied by Tycho Brahe's observations might suggest and, beyond that, to find the mathematical relations binding all the phenomena of nature to each other. His predilection for fitting the universe into a preconceived mathematical pattern, however, led him to spend years in following up false trails. In the preface to his *Mystery of the Cosmos* (1596) we find him writing:

I undertake to prove that God, in creating the universe and regulating the order of the cosmos, had in view the five regular bodies of geometry as known since the days of Pythagoras and Plato, and that he has fixed according to those dimensions, the number of heavens, their proportions, and the relations of their movements.

And so he postulated that the radii of the orbits of the planets were the radii of spheres related to the five regular solids in the following way. The largest radius was that of the orbit of Saturn. In a sphere of this radius he supposed a cube to be inscribed. In this cube a sphere was inscribed whose radius was that of the orbit of Jupiter. In this sphere he supposed a tetrahedron to be inscribed and in this another sphere whose radius was that of the orbit of Mars, and so on through the five regular solids (Plate v). The scheme called for six spheres, just enough for the number of planets known in his day. The beauty and neatness of the scheme overwhelmed him so completely that he insisted for some time on the existence of just six planets because there were only five regular solids to determine the distances between them.

Although publication of this 'scientific' hypothesis brought fame to Kepler and makes fascinating reading even today, the deductions from it were, unfortunately, not in accordance with observations. He reluctantly abandoned his idea, but not before he had made extraordinary efforts to apply it in a modified form.

If the attempt to use the five regular solids to ferret out nature's secrets did not succeed, Kepler was eminently successful in later efforts to find harmonious mathematical relations. His most famous and important results are known today as Kepler's three laws of planetary motion. These laws became so famous and so valuable to science that he earned for himself the title of 'legislator of the sky'.

The first of these laws says that the path of each planet is not a circle but an ellipse with the sun slightly off centre at a point known as a focus of the ellipse (fig. 18). Substituting the ellipse for the circle eliminated the need for the several circular motions superimposed on one another that the epicyclic theory employed to describe the motion of a planet. (It is worth noticing that Kepler put to use mathematical knowledge which had been developed by the Greeks almost two thousand years earlier.) The simplicity gained by the introduction of the ellipse convinced him that he must abandon attempts to use circular motions.

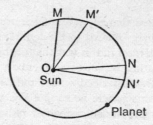

Figure 18. The elliptical and equal-area laws of planetary motion

Kepler's second law concerns the speed of the planets. Copernicus, we saw, insisted on the principle of constant speed, that is, each planet moved at a constant speed on its circle; the centre of this circle moved at a constant speed on another circle; and so on. Kepler at first held firmly to the doctrine that each planet moves along its ellipse at a constant speed, but observations finally compelled him to abandon this cherished belief. His joy was great when he discovered that he could replace it by an equally pleasing law, for his conviction that nature was mathematical was thereby reaffirmed.

If *MM'* and *NN'* (fig. 18) are distances traversed by a planet in equal intervals of *time* then, according to the principle of constant speed, *MM'* and *NN'* would have to be equal distances. According to Kepler's second law, however, *MM'* and *NN'* are generally not equal, but if *O* is the position of the sun, then the *areas OMM'* and *ONN'* are equal. Thus Kepler replaced equal distances by equal areas, and the mathematical design of the universe remained unshaken. To wrest such a secret from the heavens was indeed a triumph, for the relation described is by no means as easily discernible as it may appear to be here on paper. Kepler published this law and the law of elliptical motion in the year 1609 in a book entitled *On the Motions of the Planet Mars*.

Kepler's third law is as famous as his first two. It says that the square of the time of revolution of any planet is proportional to the cube of its average distance from the sun; that is, the ratio of the two quantities is the same for all planets. This formula can be used to compute the period of revolution of any planet from a knowledge of its average distance from the sun or, knowing the period of revolution, we can compute the average distance from the sun.

It is clear that mathematical concepts and mathematical laws are the essence of the new astronomical theory. But what is even more significant is the fact that its mathematical excellence endeared it to both Copernicus and Kepler despite many very weighty arguments against it. Indeed if either Copernicus or Kepler had been less the mathematician and more the scientist, or blind religionists, or even what the world calls sensible men, they would never have stood their ground. The scientific objections to a moving Earth were numerous. Neither man could explain how the heavy stuff of the Earth could be started and kept in motion, a major question asked by people who believed that only the heavenly bodies were light and could therefore be easily moved. About the best answer that Copernicus could give was that it is natural for any sphere to move. Equally troublesome was the objection: why doesn't the Earth's rotation cause objects on it to fly off into space just as an object whirled at the end of a string tends to fly off? In particular, why doesn't the Earth itself fly apart in pieces? The first question was not answered at all. To the latter question Copernicus replied that since the motion was natural, it could not have the effect of destroying the body. He also countered by asking why the skies did not fall apart under the motion presumed by the geocentric hypothesis. Entirely un-answered was the objection, related to the first question, that if the Earth rotates from west to east, an object thrown up into the air should fall back to the west of its original position. Again, if, as practically all scientists since Greek times believed, the motion of an object is proportional to its weight, why doesn't the Earth, in its motion around the sun, leave behind objects of lesser weight? Even the air surrounding the Earth should be left behind. Though Copernicus could not account for the fact that all objects on the Earth move with it, he 'disposed' of the motion of the air by arguing that the air is earthy and so rotates in sympathy with the Earth. All of the scientific objections to the new heliocentric theory given above were genuine and arose out of the fact that the age still accepted Aristotelian physics. The objections could not be and were not satisfactorily answered until Newtonian physics was created.

No less a personage than Francis Bacon, the father of em-pirical science, summed up in 1622 the scientific arguments against Copernicanism:

In the system of Copernicus there are found many and great incon-veniences; for both the loading of the earth with a triple motion is very incommodius and the separation of the sun from the company of the planets with which it has so many passions in common is likewise a difficulty and the introduction of so much immobility into nature by representing the sun and the stars are immovable . . . all these are the speculations of one, who cares not what fictions he introduces into nature, provided his calculations answer.

Although the clarity of Bacon's arguments could be surpassed, the opposition of a man of his reputation and ability could not be lightly brushed aside. Bacon's conservatism was due, inciden-tally, to his persistent inability to appreciate the importance of exact measurement with all his insistence on observation.

If Copernicus and Kepler had been more 'sensible', 'practical' men, they would never have defied their senses. We do not feel either the rotation or the revolution of the Earth despite the fact that Copernican theory has us rotating through space about three-tenths of a mile per second and revolving around the sun at the rate of about eighteen miles per second. On the other hand, we apparently do see the motion of the sun. To the famous astronomical observer, Tycho Brahe, these and other arguments were conclusive proof that the Earth must be stationary. In the words of Henry More 'sense pleads for Ptolemee'.

Were Copernicus and Kepler orthodox religionists, they would not have been willing even to investigate the possibilities of a heliocentric hypothesis. Medieval theology, buttressed by the Ptolemaic system, held that man was at the centre of the universe and that he was the apple of God's eye for whom God had specially created the sun, moon, and stars. By putting the sun at the centre of the universe, the heliocentric theory denied this comforting dogma. It made man appear to be one of a possible host of wanderers on many planets which, in turn, were drifting through a cold sky. He was an insignificant speck of dust on a whirling globe instead of chief actor on the central stage. It was unlikely, therefore, that he was born to live gloriously and to attain paradise upon his death, or that he was the object of God's ministrations. The sacrifice of Christ for insignificant man appeared pointless. The sky as the seat of God, the destination of the saints and of a Deity ascended from the Earth, and the paradise to which good people could aspire, was shattered by the passage of a speeding Earth. In short, the undermining of the Ptolemaic order of the universe removed cornerstones of the Christian edifice and threatened to topple the whole structure.

Copernicus' willingness to battle entrenched religious thinking is well evidenced by a passage in a letter he addressed to Pope Paul III:

If perhaps there are babblers who, although completely ignorant of mathematics, nevertheless take it upon themselves to pass judgement on mathematical questions and, improperly distorting some passages of the Scriptures to their purpose, dare to find fault with my system and censure it, I disregard them even to the extent of despising their judgement as uninformed.

Religion, physical science, common sense, and even astronomy bowed to mathematics at the behest of Copernicus and Kepler. Copernicus and Kepler had to combat many astronomical doctrines established either in Ptolemaic theory or in medieval embellishments of Aristotle. For example, the planets, sun, and moon were believed to be perfect, unalterable, and incorruptible, whereas the Earth had the contrary properties. The new theory classed the Earth with the other planets. Furthermore, the hypothesis of a moving Earth calls for motion of the stars relative to the Earth. But observations by men of the sixteenth and seventeenth centuries failed to detect this relative motion. Now no scientific hypothesis that is inconsistent with even one fact is really tenable. Nevertheless, Corpernicus and Kepler held to their heliocentric view. These sun-struck lovers of mathematics were designing a beautiful theory. If the theory did not fit all the facts, it was too bad for the facts.

Copernicus, though deliberately vague on the question of the motion of the Earth relative to the stars, at first disposed of the problem by stating that the stars were at an infinite distance. Apparently, he himself was not satisfied with this statement and so he assigned the problem to the philosophers. The true explanation, namely, that the stars were very far from the Earth, so far as to render their relative motion undetectable, was not acceptable to the Renaissance 'Greeks', who still believed in a closed and limited universe. The true distances involved were utterly beyond any figure they would have thought reasonable. Actually, the problem of accounting for the motion of the stars relative to the Earth was not solved until 1838 when the mathematician Bessel finally measured the parallax of the nearest star and found it to be 0·76″.

In view of all these arguments and forces working against the new theory, why did Copernicus and Kepler advocate it? Knowing that the great explorations of their age demanded a more accurate astronomy, one is tempted to ascribe the motivation for their work to the need for more reliable geographical information and improved techniques in navigation. But Copernicus and Kepler were not at all concerned with these pressing, practical problems. What these men did owe to their times was the opportunity to come into contact with Greek thought, an opportunity furnished by the revival of learning in Italy. Copernicus, we saw, studied there and Kepler benefited by Copernicus' work. Also both men owed to their times an atmosphere certainly more favourable to the acceptance of new ideas than the one that prevailed two centuries earlier. The geographical explorations, the Protestant Revolution, and so many other exciting movements were challenging conservatism and complacency, that one new theory did not have to bear the brunt of the natural opposition to change.

Actually, Copernicus and Kepler developed their most revolutionary theory to satisfy certain philosophical and religious interests. Having become convinced of the Pythagorean doctrine that the universe is a systematic, harmonious structure whose essence is mathematical law, they set about discovering this essence. Copernicus' published works give unmistakable, if indirect, indications of his reasons for devoting himself to astronomy. He valued his theory of planetary motion not because it improves navigational procedures but because it reveals the true harmony, symmetry, and design in the divine workshop. It is wonderful and overpowering evidence of God's presence. Writing of his achievement, which was thirty years in the making, Copernicus expressed his gratification:

We find, therefore, under this orderly arrangement, a wonderful symmetry in the universe, and a definite relation of harmony in the motion and magnitude of the orbs, of a kind that it is not possible to obtain in any other way.

He did mention in the preface to his major work, *De Revolutionibus*, that he was asked by the Lateran Council to help in reforming the calendar which had become deranged over a period of many centuries. Though he wrote that he kept this problem in mind it is quite apparent that it never dominated his thinking.

Kepler, too, made clear his dearest interests. His published work, the fruit of his labours, attest to the sincerity of his search for harmony and law in the creations of the divine power. In the preface to his *Mystery of the Cosmos* he said:

Happy the man who devotes himself to the study of the heavens; he learns to set less value on what the world admires the most; the works of God are for him above all else, and their study will furnish him with the purest of enjoyments.

A major treatise entitled *The Harmony of the World*, which Kepler published in 1619, actually expounded a system of heavenly harmonies, a new 'music of the spheres', which made use of the varying velocities of the six planets. These harmonies were enjoyed by the sun which Kepler endowed with a soul specifically for this purpose. Lest it be supposed that this treatise was just a lapse into poetic mysticism, we should realize that it also announced his celebrated third law of motion.

The work of Copernicus and Kepler was the work of men searching the universe for the harmony which their commingled religious and scientific beliefs assured them must exist, and exist in aesthetically satisfying mathematical form. It is true that Ptolemaic theory also offered mathematical laws of the universe and Copernicus and Kepler did agree that since astronomy was just geometry and geometry was truth, either theory could be true because both were good geometry. But the new theory was mathematically simpler and more harmonious.

To men who were convinced that an omnipotent being designing a mathematical universe would certainly prefer these superior features, the new theory was necessarily right. Indeed, only a mathematician who was assured that the universe was rationally and simply ordered would have had the mental fortitude to buck the prevailing philosophical, religious, and scientific beliefs, and the perseverance to work out the mathematics of such a revolutionary astronomy. Only men possessed of unshakable convictions in regard to the importance of mathematics in the design of the universe would have dared to uphold the new theory against the powerful opposition it was sure to encounter. It is a historical fact that Copernicus did address himself only to mathematicians because he expected that these alone would understand him, and in this respect he was not disappointed.

Granted that it was the superior mathematics of the new theory which inspired Copernicus and Kepler, and later Galileo, to repudiate religious convictions, scientific arguments, common sense, and well-entrenched habits of thought, how did the theory help to shape modern times?

First, Copernican theory has done more to determine the content of modern science than is generally recognized. The most powerful and most useful single law of science is Newton's law of gravitation. Without anticipating here the discussion reserved for a more appropriate place in this book we can say that the best experimental evidence for this law, the evidence which established it, depends entirely on the heliocentric theory.

Second, this theory is responsible for a new trend in science and human thought, barely perceptible at the time but all-important today. Since our eyes do not see, nor our bodies feel, the rotation and revolution of the Earth, the new theory rejected the evidence of the senses. Things were not what they seemed to be. Sense data could be misleading and reason was the reliable guide. Copernicus and Kepler thereby set the precedent that guides modern science, namely, that reason and mathematics are more important in understanding and interpreting the universe than the evidence of the senses. Vast portions of electrical and atomic theory and the whole theory of relativity would never have been conceived if scientists had not come to accept the reliance upon reason first exemplified by Copernican theory. In this very significant sense Copernicus and Kepler began the Age of Reason, in addition to fulfilling the cardinal function of scientists and mathematicians, that is, to provide a rational comprehension of the universe.

By deflating the stock of Homo sapiens, Copernican theory reopened questions that the guardians of Western civilization had been answering dogmatically upon the basis of Christian theology. Once there had been only one answer; now there are ten or twenty to such basic questions as: *Why does man desire to live and for what purpose? Why should he be moral and principled? Why seek to preserve the race?* It is one thing for man to answer such questions in the belief that he is the child and ward of a generous, powerful, and provident God. It is another to answer them knowing that he is a speck of dust in a cyclone.

Copernican theory flung such questions in the faces of all thinking men and women, and, as thinking beings, they could not reject the challenge. Their struggles to recover mental equilibrium, which was even further upset by the mathematical and scientific work following Copernicus and Kepler, provide the key to the history of thought of the last few centuries.

Much evidence can be found in the literature since Kepler's times that points to this agitation aroused by the new and disturbing thoughts. The metaphysical John Donne, though trained in and content with the encyclopedic and systematic scholasticism, was compelled to acknowledge the undesirable complexity to which Ptolemaic theory had led:

> We think the heavens enjoy their spherical
> Their round proportion, embracing all;
> But yet their various and perplexed course,
> Observed in divers ages, doth enforce
> Men to find out so many eccentric parts,
> Such diverse downright lines, such overthwarts,
> As disproportion that pure form.

Though the argument for Copernicanism was clear to Donne he could only deplore the fact that the sun and planets no longer ran in circles around the Earth.

Milton, also, pondered the challenge to Ptolemaic theory but made no decisive choice. Both theories are described in *Paradise Lost*. Unable to meet the new mathematics on its own ground he turned instead to rebuking its creators. Man should admire, not question, the works of God.

> From Man or Angel the great Architect
> Did wisely to conceal, and not divulge
> His secrets to be scann'd by them who ought
> Rather admire; . . .

> *

> Solicit not thy thoughts with matters hid,
> Leave them to God . . .
> . . . be lowly wise;
> Think only what concerns thee and thy being.

Yet even Milton was unconsciously moved to accept a more mysterious and a vaster space than the compact, thoroughly defined space of Dante, for example.

The gentle remonstrations of the milder poets, Ben Jonson's satire, Bacon's scientific arguments as well as personal jealousy, the ridicule of professors, the mathematical refutations of the brilliant Cardan, the resentment of astrologers who feared for their livelihood, Montaigne's scepticism, complete rejection from Shakespeare, and condescending mention from John Milton earned for Copernicus a reputation as a new Duns Scotus, the learned crazy one. In 1597 Galileo wrote to Kepler describing Copernicus as one 'who though he has obtained immortal fame among the few, is, nevertheless, ridiculed and hissed by the many, who are fools'.

Nevertheless the opinion of the few prevailed. The cultural revolution gained momentum; people were compelled to think, to challenge existing dogmas, and to re-examine long-accepted beliefs. And from this criticism and re-examination there emerged many of the philosophical, religious, and ethical principles now accepted without question in Western civilization.

By far the greatest value of the heliocentric theory to modern times is the contribution it made to the battle for freedom of thought and expression. The treatment that the heliocentric hypothesis received at the outset illustrates one fairly safe generalization: the reaction to change is reaction. Because man is conservative, a creature of habit, and convinced of his own importance, the new theory was decidedly unwelcome. Moreover, the vested interests of well-entrenched scholars and religious leaders caused them to oppose it. The most momentous battle in history, the battle for the freedom of the human mind, was joined on the issue of the right to advocate heliocentrism, and among the most violent of the anti-Copernicans were the Protestants who had so recently broken from traditionalism themselves.

The self-appointed representatives of God began the battle with vicious attacks. Martin Luther called Copernicus an 'upstart astrologer' and a 'fool who wishes to reverse the entire science of astronomy', Calvin thundered: 'Who will venture to place the authority of Copernicus above that of the Holy Spirit?' Do not Scriptures say that Joshua commanded the sun and not the Earth to stand still? That the sun runs from one end of the heavens to the other? That the foundations of the Earth are fixed and cannot be moved? Let us learn how to go to heaven and not how the heavens go, protested a Cardinal. The Inquisition condemned the new theory as 'that false Pythagorean doctrine utterly contrary to the Holy Scriptures', and in 1616 the Index of Prohibited Books banned all publications dealing with Copernicanism. Indeed, if the fury and high office of the opposition are a good indication of the importance of an idea no more valuable one was ever advanced.

The spirit of inquiry became so shackled in that age that when Galileo discovered the four satellites of Jupiter with his small telescope, some scientists and religionists refused to look through his instrument to see those bodies for themselves. And many who did tempt the devil by looking refused to believe their own eyes. It was this bigoted attitude that made it dangerous to advocate the new theory. One risked the fate of Giordano Bruno, who was put to death by the Inquisition 'as mercifully as possible and without the shedding of blood', the horrible formula for burning a prisoner at the stake.

Despite the earlier ecclesiastic prohibition of works on Copernicanism, Pope Urban VIII did give Galileo permission to publish a book on the subject, for the Pope believed there was no danger that anyone would ever prove the new theory necessarily true. Accordingly in 1632 Galileo published his *Dialogue on the Two Chief Systems of the World* in which he compared the geocentric and heliocentric doctrines. In order to please the Church and so pass the censors he incorporated a preface to the effect that the latter theory was only a product of the imagination. Unfortunately, Galileo wrote too well and the Pope began to fear that the argument for heliocentrism, like a live bomb wrapped in silver foil, could still do a great deal of damage to the Catholic faith. The Church roused itself once more to do battle against a heresy 'more scandalous, more detestable, and more pernicious to Christianity than any contained in the books of Calvin, of Luther, and of all other heretics put together'. Galileo was again called by the Roman Inquisition and compelled on the threat of torture to declare: 'The falsity of the Copernican system cannot be doubted, especially by us Catholics . . .'

The threat of burning faggots, the wheel, the rack, the gallows, and other ingenious refinements of torture were definitely more conductive to orthodoxy than to scientific progress. When he heard of Galileo's persecution, Descartes, who was a nervous and timid individual, refrained from advocating the new theory and actually destroyed one of his own works on it.

The heliocentric theory became, however, a powerful weapon with which to fight the suppression of free thought. The truth (at least to the seventeenth and eighteenth centuries) of the new theory and its incomparable simplicity attracted more and more adherents as people gradually realized that the teachings of religious leaders could be fallible. It soon became impossible for these leaders to retain their authority over all Europe and the way was prepared for freer thought in all spheres. Certainly the emancipation of science from theology dates from this controversy.

The import of this battle and its favourable outcome should not be lost to us. Those who still enjoy and those who have lost the freedoms so recently acquired in Western civilization cannot fail to appreciate how much was at stake in the battle to advance the heliocentric theory and how much we owe to the men of gigantic intellect and extraordinary courage who waged the fight. Fortunately for us the very fires that consumed the martyrs to free inquiry dispelled the darkness of the Middle Ages. The fight to establish the heliocentric theory weakened the stranglehold of ecclesiasticism on the minds of men. The mathematical argument proved more compelling than the theological one and the battle for the freedom to think, speak, and write was finally won. The scientific Declaration of Independence is a collection of mathematical theorems.

X

Painting and Perspective

The world's the book where the eternal sense
  Wrote his own thoughts; the living temple where,
  Painting his very self, with figures fair
  He filled the whole immense circumference.

T. Campanella

DURING the Middle Ages painting, serving somewhat as the handmaiden of the Church, concentrated on embellishing the thoughts and doctrines of Christianity. Towards the end of this period, the painters, along with other thinkers in Europe, began to be interested in the natural world. Inspired by the new emphasis on man and the universe about him the Renaissance artist dared to confront nature, to study her deeply and searchingly, and to depict her realistically. The painters revived the glory and gladness of an alive world and reproduced beautiful forms which attested to the delightfulness of physical existence, the inalienable right to satisfy natural wants, and the pleasures afforded by earth, sea, and air.

For several reasons the problem of depicting the real world led the Renaissance painters to mathematics. The first reason was one that could be operative in any age in which the artist seeks to paint realistically. Stripped of colour and substance the objects that painters put on canvas are geometrical bodies located in space. The language for dealing with these idealized objects, the properties they possess as idealizations, and the exact relationships that describe their relative locations in space are all incorporated in Euclidean geometry. The artists need only avail themselves of it.

The Renaissance artist turned to mathematics not only because he sought to reproduce nature but also because he was influenced by the revived philosophy of the Greeks. He became thoroughly familiar and imbued with the doctrine that mathematics is the essence of the real world, that the universe is ordered and explicable rationally in terms of geometry. Hence, like the Greek philosopher, he believed that to penetrate to the underlying significance, that is, the reality of the theme that he sought to display on canvas, he must reduce it to its mathematical content. Very interesting evidence of the artist's attempt to discover the mathematical essence of his subject is found in one of Leonardo's studies in proportion. In it he tried to fit the structure of the ideal man to the ideal figures, the square and circle (Plate VI).

The sheer utility of mathematics for accurate description and the philosophy that mathematics is the essence of reality are only two of the reasons why the Renaissance artist sought to use mathematics. There was another reason. The artist of the late medieval period and the Renaissance was, also, the architect and engineer of his day and so was necessarily mathematically inclined. Businessmen, secular princes, and ecclesiastical officials assigned all construction problems to the artist. He designed and built churches, hospitals, palaces, cloisters, bridges, fortresses, dams, canals, town walls, and instruments of warfare. Numerous drawings of such engineering projects are in da Vinci's notebooks and he, himself, in offering his services to Lodovico Sforza, ruler of Milan, promised to serve as an engineer, constructor of military works, and designer of war machines, as well as architect, sculptor, and painter. The artist was even expected to solve problems involving the motion of cannon balls in artillery fire, a task which in those times called for profound mathematical knowledge. It is no exaggeration to state that the Renaissance artist was the best practising mathematician and that in the fifteenth century he was also the most learned and accomplished theoretical mathematician.

The specific problem which engaged the mathematical talents of the Renaissance painters and with which we shall be concerned here was that of depicting realistically three-dimensional scenes on canvas. The artists solved this problem by creating a totally new system of mathematical perspective and consequently refashioned the entire style of painting.

The various schemes employed throughout the history of

painting for organizing subjects on plaster and canvas, that is, the various systems of perspective, can be divided into two major classes, conceptual and optical. A conceptual system undertakes to organize the persons and objects in accordance with some doctrine or principle that has little or nothing to do with the actual appearance of the scene itself. For example, Egyptian painting and relief work were largely conceptual. The sizes of people were often ordered in relation to their importance in the politico-religious hierarchy. Pharaoh was usually the most important person and so was the largest. His wife would be next in size and his servants even smaller. Profile views and frontal views were used simultaneously even for different parts of the same figure. In order to indicate a series of people or animals one behind the other, the same figure was repeated slightly displaced. Modern painting, as well as most Japanese and Chinese painting, is also conceptual (Plate xxvii).

An optical system of perspective, on the other hand, attempts to convey the same impression to the eye as would the scene itself. Although Greek and Roman painting was primarily optical, the influence of Christian mysticism turned artists back to a conceptual system, which prevailed throughout the Middle Ages. The early Christian and medieval artists were content to paint in symbolic terms, that is, their settings and subjects were intended to illustrate religious themes and induce religious feelings rather than to represent real people in the actual and present world. The people and objects were highly stylized and drawn as though they existed in a flat, two-dimensional vacuum. Figures that should be behind one another were usually alongside or above. Stiff draperies and angular attitudes were characteristic. The backgrounds of the paintings were almost always of a solid colour, usually gold, as if to emphasize that the subjects had no connection with the real world.

The early Christian mosaic 'Abraham with Angels' (Plate vii), a typical example of the Byzantine influence, illustrates the disintegration of ancient perspective. The background is essentially neutral. The earth, tree, and bushes are artificial and lifeless, the tree being shaped peculiarly to fit the border of the picture. There is no foreground or base on which the figures and objects stand. The figures are not related to each other and, of course, spatial relations are ignored because measures and sizes were deemed unimportant. The little unity there is in the picture is supplied by the gold background and the colour of the objects.

Though remnants of an optical system used by the Romans were sometimes present in medieval painting, this Byzantine style predominated. An excellent example, indeed one that is regarded as the flower of medieval painting, is 'The Annunciation' (Plate VIII) by Simone Martini (1285–1344). The background is gold. There is no indication of visual perception. The movement in the painting is from the angel to the Virgin and then back to the angel. Though there is loveliness of colour, surface, and sinuous line, the figures themselves are unemotional and arouse no emotional response in the onlooker. The effect of the whole is mosaic-like. Perhaps the only respect in which this painting makes any advance towards realism is in its use of a ground plane or floor on which objects and figures rest and which is distinct from the gilt background.

Characteristic Renaissance influences which steered the artists towards realism and mathematics began to be felt near the end of the thirteenth century, the century in which Aristotle became widely known by means of translations from the Arabic and the Greek. The painters became aware of the lifelessness and unreality of medieval painting and consciously sought to modify it. Efforts towards naturalism appeared in the use of real people as subjects of religious themes, in the deliberate use of straight lines, multiple surfaces, and simple forms of geometry, in experiments with unorthodox positions of the figures, in attempts to render emotions, and in the depiction of drapery falling and folding around parts of bodies as it actually does rather than in the flat folds of the conventional medieval style.

The essential difference between medieval and Renaissance art is the introduction of the third dimension, that is, the rendering of space, distance, volume, mass, and visual effects. The incorporation of three-dimensionality could be achieved only by an optical system of representation, and conscious efforts in this direction were made by Duccio (1255–1319) and Giotto

(1276–1336), at the beginning of the fourteenth century. Several devices appeared in their works that are at least worth noticing as stages in the development of a mathematical system. Duccio's 'Madonna in Majesty' (Plate IX) has several interesting features. The composition, first of all, is severely simple and symmetrical. The lines of the throne are made to converge in pairs and thus suggest depth. The figures on either side of the throne are presumably standing on one level but they are painted one above the other in several layers. This manner of depicting depth is known as terraced perspective, a device very common in the fourteenth century. The drapery is somewhat natural as exemplified by the folds over the Madonna's knee. Also there is some feeling for solidity and space and some emotion in the faces. The picture as a whole still contains much of the Byzantine tradition. There is a liberal use of gold in the background and in the details. The pattern is still mosaic-like. Because the throne is not properly foreshortened to suggest depth, the Madonna does not appear to be sitting on it.

Even more significant is Duccio's 'The Last Supper' (Plate X). The scene is a partially boxed-in room, a background very commonly used during the fourteenth century and one that marks the transition from interior to exterior scenes. The receding walls and receding ceiling lines, somewhat foreshortened, suggest depth. The parts of the room fit together. Several details about the treatment of the ceiling are important. The lines of the middle portion come together in one area, which is called the vanishing area for a reason that will be made clear later. This technique was consciously used by many painters of the period as a device to portray depth. Second, lines from each of the two end-sections of the ceiling, which are symmetrically located with respect to the centre, meet in pairs at points which lie on one vertical line. This scheme, too, known as vertical or axial perspective, was widely used to achieve depth. Neither scheme was used systematically by Duccio but both were developed and applied by later painters of the fourteenth century. Suggestions of the real world, such as the bushes on the left side of the painting, should be noticed.

Unfortunately, Duccio did not treat the whole scene in 'The

Last Supper' from a single point of view. The lines of the table's edges approach the spectator, contrary to the way in which the eye would see them. The table appears to be higher in the back than in the front and the objects on the table do not seem to be lying flat on it. In fact they project too far into the foreground. Nevertheless, there is a sense of realism particularly in regard to the larger features of the painting.

It can be said that three-dimensionality is definitely present in Duccio's work. The figures have mass and volume and are related to each other and to the composition as a whole. Lines are used in accordance with some particular schemes, and planes are foreshortened. Light and shadow are also used to suggest volume.

The father of modern painting was Giotto. He painted with direct reference to visual perceptions and spatial relations and his results tended towards a photographic copy. His figures possessed mass, volume, and vitality. He chose homelike scenes, distributed his figures in a balanced arrangement, and grouped them in a manner agreeable to the eye.

One of Giotto's best paintings, 'The Death of St Francis' (Plate XI), like Duccio's 'The Last Supper', employs the popular transitional device, a partially boxed-in room. The room does suggest a localized three-dimensional scene as opposed to a flat two-dimensional scene existing nowhere. The careful balance of the component objects and figures is clearly intended to appeal to the eye. Equally obvious are the relations of the figures to each other though none is related to the background. In this painting and in others by Giotto, the portions of the rooms or buildings shown seem to stand on the ground. Foreshortening is employed to suggest depth.

Giotto is not usually consistent in his point of view. In his 'Salome's Dance' (Plate XII), the two walls of the alcove on the right do not quite jibe with each other, nor do the table and ceiling of the dining-room. Nevertheless, the three-dimensionality of this painting can no longer be overlooked. Rather interesting and significant is the bit of architecture at the left. The real world is introduced even at the expense of irrelevance.

Giotto was a key figure in the development of optical perspective. Though his paintings are not visually correct and though he did not introduce any new principle, his work on the whole shows great improvement over that of his predecessors. He himself was aware of the advances he had made, for he often went to unnecessary lengths in order to display his skill. This is almost certainly the reason for the inclusion of the tower in his 'Salome's Dance'.

Advances in technique and principles may be credited to Ambrogio Lorenzetti (active 1323–48). He is noteworthy for the organization of his themes in realistic, localized areas; his lines are vigorous and his figures robust and humanized. Progress is evident in the 'Annunciation' (Plate XIII). The ground plane on which the figures rest is now definite and clearly distinguished from the rear wall. The ground also serves as a measure of the sizes of the objects and suggests space extending back to the rear. A second major advance is that the lines of the floor which recede from the spectator meet at one point. Finally, the blocks are foreshortened more and more the farther they are in the background. On the whole Lorenzetti handled space and three-dimensionality as well as anyone in the fourteenth century. Like Duccio and Giotto he failed to unite all the elements in his paintings. In the 'Annunciation' the wall and floor are not related. Nevertheless, there is good intuitive, though not mathematical, handling of space and depth.

With Lorenzetti we reach the highest level attained by the Renaissance artists before the introduction of a mathematical system of perspective. The steps made thus far towards the development of a satisfactory optical system show how much the artists struggled with the problem. It is evident that these innovators were groping for an effective technique.

In the fifteenth century the artists finally realized that the problem of perspective must be studied scientifically and that geometry was the key to the problem. This realization may have been hastened by the study of ancient writings on perspective which had recently been exhumed along with Greek and Roman art. The new approach was, of course, motivated by far more than the desire to attain verisimilitude. The greater goal was understanding of the structure of space and discovery of some of the secrets of nature. This was an expression of the Renaissance philosophy that mathematics was the most effective means of probing nature and the form in which the ultimate truths would be phrased. These men who explored nature with techniques peculiar to their art had precisely the spirit and attitude of those other investigators of nature who founded modern science by means of their mathematics and experiments. In fact, during the Renaissance, art was regarded as a form of knowledge and a science. It aspired to the status of the four Platonic 'arts': arithmetic, geometry, harmony (music), and astronomy. Geometry was expected to supply the badge of respectability. Equally enticing as a goal in the development of a scientific system of perspective was the possibility of achieving unity of design.

The science of painting was founded by Brunelleschi, who worked out a system of perspective by 1425. He taught Donatello, Masaccio, Fra Filippo, and others. The first written account, the *della Pittura* of Leone Battista Alberti, was published in 1435. Alberti said in this treatise on painting that the first requirement of the painter is to know geometry. The arts are learned by reason and method; they are mastered by practice. In so far as painting is concerned, Alberti believed that nature could be improved on with the aid of mathematics, and towards this end he advocated the use of the mathematical system of perspective known as the focused system.

The great master of perspective and, incidentally, one of the best mathematicians of the fifteenth century, was Piero della Francesca. His text *De Prospettiva Pingendi* added considerably to Alberti's material, though he took a slightly different approach. In this book Piero came close to identifying painting with perspective. During the last twenty years of his life he wrote three treatises to show how the visible world could be reduced to mathematical order by the principles of perspective and solid geometry.

The most famous of the artists who contributed to the science of perspective was Leonardo da Vinci. This striking figure of incredible physical strength and unparalleled mental endowment prepared for painting by deep and extensive studies in anatomy, perspective, geometry, physics, and chemistry. His attitude towards perspective was part and parcel of his philosophy of art. He opened his *Trattato della Pittura* with the words, 'Let no one who is not a mathematician read my works.' The object of painting, he insisted, is to reproduce nature and the merit of a painting lies in the exactness of the reproduction. Even a purely imaginative creation must appear as if it could exist in nature. Painting, then, is a science and like all sciences must be based on mathematics, 'for no human inquiry can be called science unless it pursues its path through mathematical exposition and demonstration'. Again, 'The man who discredits the supreme certainty of mathematics is feeding on confusion, and can never silence the contradictions of sophistical sciences, which lead to eternal quackery.' Leonardo scorned those who thought they could ignore theory and produce art by mere practice: rather, 'Practice must always be founded on sound theory.' Perspective he described as the 'rudder and guide rope' of painting.

The most influential of the artists who wrote on perspective was Albrecht Dürer. Dürer learned the principles of perspective from the Italian masters and returned to Germany to continue his studies. His popular and widely read treatise *Underweysung der Messung mit dem Zyrkel und Rychtscheyd* (1528) affirmed that the perspective basis of a picture should not be drawn free-hand but constructed according to mathematical principles. Actually, the Renaissance painters were incomplete in their treatment of the principles of perspective. Mathematicians of a later period, notably Brook Taylor and J. H. Lambert, wrote definitive works.

It is fair to state that almost all the great artists of the fifteenth and early sixteenth centuries sought to incorporate mathematical principles and mathematical harmonies in their paintings, with realistic perspective a specific and major goal. Signorelli, Bramante, Michelangelo, and Raphael, among others, were deeply interested in mathematics and in its application to art. They deliberately executed difficult postures, developed and handled foreshortening with amazing facility, and at times even suppressed passion and feeling, all in order to display the scientific elements in their work. These masters were aware that art, with all its use of individual imagination, is subject to laws.

The basic principle of the mathematical system which these artists developed may be explained in terms used by Alberti, Leonardo, and Dürer. These men imagined that the artist's canvas

Figure 19. Dürer: The Designer of the Sitting Man

is a glass screen through which he looks at the scene to be painted, just as we might look through a window to a scene outside. From one eye, which is held fixed, lines of light are imagined to go to each point of the scene. This set of lines is called a *projection*. Where each of these lines pierces the glass screen a point is marked on the screen. This set of points, called a *section* creates the same impression on the eye as does the scene itself. These artists then decided that realistic painting must produce on canvas the location, size, and relative positions of objects exactly as they would appear on a glass screen interposed between the eye and the scene. In fact, Alberti proclaimed that the picture is a section of the projection.

Figure 20. Dürer: The Designer of the Lying Woman

This principle is illustrated in several woodcuts executed by Dürer. The first two of these (figs. 19 and 20) show the artist holding one eye at a fixed point while he traces on a glass screen, or on paper which is ruled in squares corresponding to squares on the glass screen, the points in which lines of light from the eye to the scene cut the screen. The third of these woodcuts (fig. 21) shows how the artist can trace the correct pattern on the glass screen even though he is supposedly far from the screen. In this woodcut the eye viewing the scene is effectively at the point where the rope is knotted to the wall. The fourth woodcut (fig. 22) shows a pattern traced out on a screen.

Figure 21: Dürer: The Designer of the Can

Since canvas is not transparent and since an artist may wish to paint a scene that exists only in his imagination, he cannot paint a Dürer 'section' simply by tracing points. He must have rules to guide him. And so the writers on perspective derived from the principle of projection and section a set of theorems that com-

Figure 22: Dürer: The Designer of the Lute

prise the system of focused perspective. This is the system that has been adopted by nearly all artists since the Renaissance.

What are the principal theorems or rules of the mathematical science of perspective? Suppose the canvas is held in the normal vertical position. The perpendicular from the eye to the canvas, or an extension of it, strikes the canvas at a point called the principal vanishing point (the reason for the term will be apparent shortly). The horizontal line through the principal vanishing point is called the horizon line because, if the spectator were looking through the canvas to open space, the horizon line would correspond to the actual horizon. These concepts are illustrated in fig. 23. This figure shows a hallway viewed by a person whose eye is at point *O* (not shown) which lies on a line perpendicular

T-F

to the page and through the point P. P is the principal vanishing point and the line $D_2 PD_1$ is the horizon line.

The first essential theorem is that all horizontal lines in the scene that are perpendicular to the plane of the canvas must be drawn on the canvas so as to meet at the principal vanishing point. Thus lines such as AA', EE', DD', and others (fig. 23) meet at P. It may seem incorrect that lines which are actually parallel should be drawn to meet. But this is precisely how the eye sees parallel lines, as the familiar example of the apparently converging railroad tracks illustrates. It is perhaps clear now why the point P is called a vanishing point. There is no point corresponding to it in the actual scene, since the parallel lines of the scene itself do not meet.

Another theorem to be deduced from the general principle that the picture should be a section of the projection is that any set of parallel horizontal lines which are not perpendicular to the plane

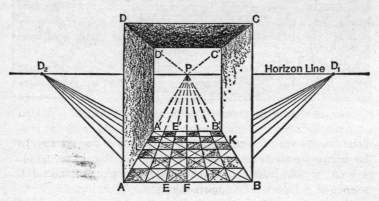

Figure 23. Sketch of a hallway according to the focused system of perspective

of the canvas but meet it at some angle must be drawn so as to converge to a point which lies somewhere on the horizon line depending on the angle which these lines make with the plane of the canvas. Among such sets of horizontal parallel lines there are two very important ones. Lines such as AB' and EK of figure 23, which in the actual scene are parallel and make a 45° angle with the plane of the canvas meet at a point D_1, which is called a diagonal vanishing point. The distance PD_1 must equal the distance OP, that is, the distance from the eye to the principal vanishing point. Similarly parallel horizontal lines such as BA' and FL, which in the actual scene make a 135° angle with the canvas, must be drawn so as to meet at a second diagonal point, D_2 in figure 23, and PD_2 must also equal OP. Parallel lines of the actual scene that rise or fall as they recede from the spectator must also meet in one point, which will be above or below the horizon line. This point would be the one in which a line from the eye parallel to the lines in question pierces the canvas.

The third theorem that follows from the general principle of projection and section is that parallel horizontal lines of the scene which are parallel to the plane of the canvas are to be drawn horizontal and parallel, and that vertical parallel lines are to be drawn vertical and parallel. Since to the eye *all* sets of parallel lines appear to converge, this third theorem is not in harmony with visual perception. This inconsistency will be discussed later.

Long before the creation of the system of focused perspective artists had realized that distant objects should be drawn foreshortened. They had great difficulty, however, in determining the proper amount of foreshortening. The new system provided the requisite theorems which may also be deduced from the general principle that the painting is a section of the projection. In the case of the square floor blocks in figure 23, the proper handling of the diagonal lines such as AB', BA', EK, and FL determines the correct foreshortening.

There are many other theorems for the trained artist to use if he wishes to achieve the realism the focused system permits. Pursuit of these specialized results, however, would carry us too far afield. There is one point that is implicit in what has been discussed and that is of importance to the layman viewing a painting designed in accordance with the focused system. The position of the artist's eye is inseparable from the design of the painting. To obtain the correct effect the spectator should view the painting from this position, that is, the spectator's eye should be at the level of the principal vanishing point and directly in front of it at a distance equal to the distance from the principal vanishing point to either diagonal vanishing point. Actually it would be well if paintings were hung so that they might be raised or lowered to suit the viewer's height.

Before we examine some great paintings designed according to the system of focused perspective we should point out that the system does not furnish a faithful reproduction of what the eye sees. The principle that a painting must be a section of a projection requires, as already stated, that horizontal parallel lines which are parallel to the plane of the canvas as well as vertical parallel lines, are to be drawn parallel. But the eye viewing such lines finds that they appear to meet just as other sets of parallel lines do. Hence in this respect at least the focused system is not visually correct. A more fundamental criticism is the fact that the eye does not see straight lines at all. The reader may convince himself of this fact if he will imagine himself in an airplane looking down on two perfectly parallel, horizontal railroad tracks. In each direction the tracks appear to meet on the horizon. Two straight lines, however, can meet in only one point. Obviously, then, since the tracks meet at the two horizon points, to the eye they must be curves. The Greeks and Romans had recognized that straight lines appear curved to the eye. Indeed, Euclid said so in his *Optics*. But the focused system ignores this fact of perception. Neither does the system take into account the fact that we actually see with two eyes, each of which receives a slightly different impression. Moreover, these eyes are not rigid but move as the spectator surveys a scene. Finally, the focused system ignores the fact that the retina of the eye on which the light rays impinge is a curved surface, not a photographic plate, and that seeing is as much a reaction of the brain as it is a purely physiological process.

In view of these deficiencies in the system, why did the artists adopt it? It was, of course, a considerable improvement over the inadequate systems known to the fourteenth century. More important to the fifteenth- and sixteenth-century artists was the fact that the system was a thoroughly mathematical one. To people already impressed with the importance of mathematics in understanding nature, the attainment of a satisfactory mathematical system of perspective pleased them so much that they were blind to all its deficiencies. In fact, the artists believed it to be as true as Euclidean geometry itself.

Let us now examine the progeny of the wedding of geometry and painting. One of the first painters to apply the science of perspective initiated by Brunelleschi was Masaccio (1401–28). Although later paintings will show more clearly the influence of the new science, Masaccio's 'The Tribute Money (Plate xiv) is far more realistic than anything done earlier. Vasari said that Masaccio was the first artist to attain the imitation of things as they really are. This particular painting shows great depth, spaciousness, and naturalism. The individual figures are massive; they exist in space and their bodies are more real than Giotto's. The figures stand on their own feet. Masaccio was also the first to use a technique which supplements geometry, namely, aerial perspective. By diminishing the intensity of the colour as well as the size of objects farther in the background, distance is suggested. Masaccio was, in fact, a master at handling light and shade.

One of the major contributors to the science of perspective was Uccello (1397–1475). His interest in the subject was so intense that Vasari said Uccello 'would remain the long night in his study to work out the vanishing points of his perspective' and when summoned to bed by his wife replied, 'How sweet a thing is this perspective.' He took pleasure in investigating difficult problems, and he was so distracted by his passion for exact perspective that he failed to apply his full powers to painting. Painting was an occasion for solving problems and displaying his mastery of perspective. Actually his success was not complete. His figures are generally crowded on one another and his mastery of depth was imperfect.

Unfortunately, the best examples of Uccello's perspective have been so much damaged by time that they cannot be reproduced. One scene from the sequence entitled 'Desecration of the Host' does give some indication of his work (Plate xv). His 'Perspective Study of a Chalice' (Plate xvi) shows the complexity of surfaces, lines, and curves involved in an accurate perspective drawing.

The artist who perfected the science of perspective was Piero della Francesca (1416–92). This highly intellectual painter had a passion for geometry, and planned all his works mathematically to the last detail. The placement of each figure was calculated so as to be correct in relation to other figures and to the organization of the painting as a whole. He even used geometrical forms for parts of the body and objects of dress and he loved smooth curved surfaces and solidity.

Piero's 'The Flagellation' (Plate XVII) is a masterpiece of perspective. The choice of principal vanishing point and the accurate use of the principles of the focused system tie the characters in the rear of the courtyard to those in front, while the objects are all accommodated to the clearly delimited space. The diminution of the black inlays on the marble floor is also precisely calculated. A drawing in Piero's book on perspective shows the immense labour which went into this painting. Here as well as in other paintings Piero used aerial perspective to enhance the impression of depth. The whole painting is so carefully planned that movement is sacrificed to unity of design.

Piero's 'Resurrection' (Plate XVIII) is judged by some critics to be one of the supreme works of painting in the entire world. It is almost architectural in design. The perspective is unusual: there are two points of vision and therefore two principal vanishing points. As is evident from the fact that we see the necks of two of the sleeping soldiers from below, one principal vanishing point is in the middle of the sarcophagus. Then unconsciously the eye is carried up to the second principal vanishing point which is in the face of Christ. The two pictures, that is the lower and upper parts, are separated by a natural boundary, the upper edge of the sarcophagus, so that the change in point of view is not disturbing. By making the hills rise rather sharply Piero unified the two parts at the same time that he supplied a natural-appearing background for the upper one. It has sometimes been said that Piero's intense love for perspective made his pictures too mathematical and therefore cool and impersonal. However, a look at the sad, haunting, and forgiving countenance of Christ shows that Piero was capable of expressing delicate shades of emotion.

Leonardo da Vinci (1452–1519) produced many excellent examples of perfect perspective. This truly scientific mind and subtle aesthetic genius made numerous detailed studies for each painting (Plate XIX). His best-known work and perhaps the most famous of all paintings is an excellent example of perfect perspective. The 'Last Supper' (Plate XX) is designed to give exactly the impression that would be made on the eye in real life. The viewer feels that he is in the room. The receding lines on the walls, floor, and ceiling not only convey depth clearly but converge to one point deliberately chosen to be in the head of Christ so that attention focuses on Him. It should be noticed, incidentally, that the twelve apostles are arranged in four groups of three each and are symmetrically disposed on each side of Christ. The figure of Christ Himself forms an equilateral triangle; this element of the design was intended to express the balance of sense, reason, and body. Leonardo's painting should be compared with Duccio's 'The Last Supper' (Plate X).

A few more examples of paintings that incorporate excellent perspective will indicate perhaps the widespread appeal and application of the new science. Though Botticelli (1444–1510) is most widely known for such paintings as 'Spring' and the 'Birth of Venus' where the artist expresses himself in pattern, lines, and curves and where realism is not an objective, he was capable of excellent perspective. One of the finest of his numerous works, 'The Calumny of Apelles' (Plate XXI), shows his mastery of the science. Each object is sharply drawn. The various parts of the throne and of the buildings are well executed and the foreshortening of all the objects is correct.

A painter who exhibited great skill in perspective was Mantegna (1431–1506). Anatomy and perspective were ideals with him. He chose difficult problems and used perspective to achieve harsh realism and boldness. In his 'St James Led to Execution' (Plate XXII) he deliberately chose an eccentric point of view. The principal vanishing point is just below the bottom of the painting and to the right of centre. The whole scene is successfully treated from this unusual point of view.

The sixteenth century witnessed the culmination of the great Renaissance developments in realistic painting. The masters displayed perfect perspective and form, and emphasized space and colour. The ideal of form was loved so much that artists were indifferent to content. The distinguished pupil of Leonardo and Michelangelo, Raphael (1483–1520), supplied many excellent examples of the ideals, standards, and accomplishments towards which the preceding centuries had been striving. His 'School of Athens' (Plate XXIII) portrays a dignified architectural setting in which harmonious arrangement, mastery of perspective, and exactness of proportions are clear. This painting is of interest not merely because of its superb treatment of space and depth, but because it evidences the veneration that the Renaissance intellectuals had for the Greek masters. Plato and Aristotle, left and right, are the central figures. At Plato's left is Socrates. In the left foreground Pythagoras writes in a book. In the right foreground Euclid or Archimedes stoops to demonstrate some theorem. To the right of this figure Ptolemy holds a sphere. Musicians, arithmeticians, and grammarians complete the assemblage.

The Venetian masters of the sixteenth century subordinated line to colour and light and shade. Nevertheless they too were masters of perspective. The expression of space is fully three-dimensional, and organization and perspective are clearly felt. Tintoretto (1518–94) is representative of this school. His 'Transfer of the Body of St Mark' (Plate XXIV) shows perfect treatment of depth; the foreshortening of the figures in the foreground should be noticed.

We shall take time for just one more example. We have already mentioned Dürer (1472–1528) as one of the writers on the subject of perspective who greatly influenced painters north of the Alps. His 'St Jerome in his Study' (Plate XXV), an engraving on copper, shows what Dürer himself could do in practice. The principle vanishing point is at the right centre of the picture. The effect of the design is to make the spectator feel that he is in the room just a few feet away from St Jerome.

The reader may now test his acuteness on the subject of perspective by seeing how many absurdities he can detect in William Hogarth's steel engraving entitled 'False Perspective' (Plate XXVI).

The examples given above of paintings which use the focused system of perspective could be multiplied a thousandfold. These few are sufficient, however, to illustrate how the use of mathematical perspective emancipated figures from the gold background of medieval painting and set them free to roam the streets and hills of the natural world. The examples also illustrate a secondary value in the use of focused perspective, namely, that of promoting the unity of composition of the painting. Our account of the rise of this system may have shown, too, how the theorems of mathematics proper and a philosophy of nature in which mathematics was dominant determined the course of Western painting. Though modern painting has departed sharply from a veridical description of nature, the focused system is still taught in the art schools and is applied wherever it seems important to achieve a realistic effect.

XI

Science Born of Art: Projective Geometry

The moving power of mathematical invention is not
reasoning but imagination.

A. de Morgan

THE most original mathematical creation of the seventeenth century, a century in which science provided the dominant motivation for mathematical activity, was inspired by the art of
painting. In the course of their development of the system of
focused perspective the painters introduced new geometrical
ideas and raised several questions that suggested an entirely new
direction for research. In this way the artists repaid their debt to
mathematics.

The first of the ideas arising out of the work on perspective is
that there is a distinction between the world accessible to man's
sense of touch and the world he sees. Correspondingly, there
should be two geometries, a tactile geometry and a visual geometry. Euclidean geometry is tactile because its assertions agree
with our sense of touch but not always with our sense of sight.
For example, Euclid deals with lines that never meet. The existence of such lines can be vouched for by the hands but not by the
eye. We never *see* parallel lines. The rails do appear to meet off in
the distance.

There are many other reasons for characterizing Euclidean
geometry as a tactile geometry. For example, it treats congruent
figures, or figures that can be superposed one on the other.
Superposition is an act performed by the hands. Also, the
theorems of Euclidean geometry frequently deal with measurement, another act performed by the hand. Finally, Euclid's world
was finite, a world virtually accessible to the sense of touch. Thus
he did not consider a straight line in its entirety but rather regarded it as a segment that can be extended as far as is necessary

in either direction. There was no attempt to consider what happens at great distances from a given figure.

Since Euclidean geometry could reasonably be regarded as disposing of problems created by the sense of touch, it remained to investigate the geometry of the sense of sight. Towards this end the work on perspective offered a second major suggestion. The basic idea in the system of focused perspective is that of pro-

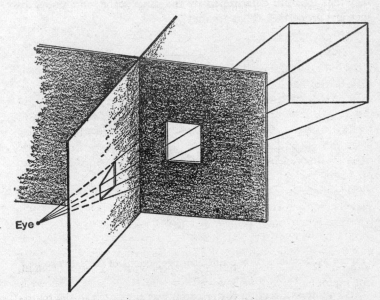

Figure 24. Two different sections of the same projection

jection and section. A projection is a set of lines of light from the eye to the points of an object or scene; a section is the pattern formed by the intersection of these lines with a glass sheet placed between the eye and the object viewed. Though the section on a glass sheet will vary in size and shape with the position and angle at which the sheet is held, each of these sections (fig. 24) creates the same impression on the eye as does the object itself.

This fact suggests several large mathematical questions. Suppose we consider two different sections of the same projection.

Since they create the same impression on the eye they should have many geometrical properties in common. Just what properties do the sections have in common? Also, what properties do the object and a section determined by it have in common? Finally, if two different observers view the same scene, two different projections are formed (fig. 25). If a section of each of these projections is made, these two sections should possess, in view of the fact that they are determined by the same scene, common geometrical properties. What are they?

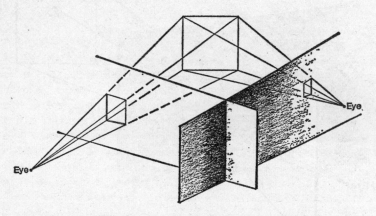

Figure 25. Sections of two different projections of the same scene

Still another direction for research was suggested to the mathematician by the work on perspective. The artist, we saw, cannot paint objects as they are. Instead he must draw parallel lines so that they converge on the canvas; he must also introduce foreshortening and other devices in order to give the eye the illusion of reality. To execute this plan the artist needs theorems that give him the location of lines and tell him what other lines any given line must intersect. Mathematicians were thereby motivated to search for theorems on the intersection of lines and of curves.

The first major mathematician to explore the suggestions arising out of the work on perspective was the self-educated architect and engineer, Girard Desargues (1593–1662). His motive in undertaking these studies was to help his colleagues in engineering, painting, and architecture. 'I freely confess,' he wrote, 'that I never had taste for study or research either in physics or geometry except in so far as they could serve as a means of arriving at some sort of knowledge of the proximate causes ... for the good and convenience of life, in maintaining health, in the practice of some art ... having observed that a good part of the arts is based on geometry, among others the cutting of stones in architecture, that of sun-dials, that of perspective in particular.' He began by organizing numerous useful theorems and disseminated these findings through lectures and handbills. Later he wrote a pamphlet on perspective which attracted very little attention.

Desargues advanced from this first work to highly original mathematical creation. His chief contribution, the foundation of projective geometry, appeared in 1639 but, like his services to artists, was hardly noticed. All the printed copies of this book were lost. Though a few of his contemporaries appreciated his work, most either ignored or mocked it. After devoting a few more years to architectural and engineering problems Desargues retired to his estate. Two of his contemporaries, Philippe de la Hire and Blaise Pascal, did study and advance Desargues's brain child before the subject passed into a long period of oblivion. Fortunately La Hire made a manuscript copy of Desargues' book and this record, discovered by chance two hundred years later, tells us what Desargues contributed.

The most startling, though not the most significant, fact about the new geometry of Desargues is that it contains no parallel lines. Just as the representation of parallel lines on canvas requires their meeting at a point, so parallel lines in space (in Euclid's sense) are required by Desargues to meet in a point which may be infinitely distant but which is nevertheless assumed to exist. This point is the counterpart in real space to the point where the parallel lines, if drawn on canvas, intersect. The addition of this 'point at infinity' represents no contradiction of Euclid's geometry but rather an extension, one that conforms to what the eye sees.

The basic theorem of projective geometry, a theorem now fundamental in all of mathematics, comes from Desargues and is named after him. It illustrates how mathematicians responded to the questions raised by perspective.

Suppose the eye at point *O* looks at a triangle *ABC* (fig. 26). The lines from *O* to the various points on the sides of the triangle constitute, as we know, a projection. A section of this projection

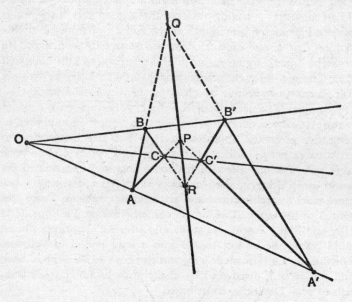

Figure 26. Desargues's theorem

will then contain a triangle *A'B'C'*, where *A'* corresponds to *A*, *B'* to *B*, and *C'* to *C*. The two triangles, *ABC* and *A'B'C'* are said to be perspective from the point *O*. Desargues states his theorem as follows

The pairs of corresponding sides, *AB* and *A'B'*, *BC* and *B'C'*, and *AC* and *A'C'* of two triangles perspective from a point meet, respectively, in three points that lie on one straight line.

With specific reference to our figure the theorem says that if we prolong sides *AC* and *A'C'*, they will meet in a point *P*; sides *AB* and *A'B'* prolonged will meet in a point *Q*; and sides *BC* and $B'C'$ prolonged will meet in a point R. And P, Q, and R will lie on a straight line. The theorem holds whether the triangles lie in the same or in different planes.

Equally typical of theorems in projective geometry is one proved, at the age of sixteen, by the precocious French thinker, Pascal, with whom we shall deal more fully later. This theorem was incorporated by Pascal in an essay on conics, an essay so brilliant that Descartes could not believe it was written by one so young. Pascal's theorem, like Desargues', states a property of a geometrical figure that is common to all sections of any projection of that figure. In more mathematical language, it states a

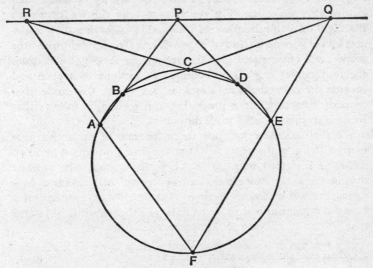

Figure 27. Pascal's theorem

property of a geometrical figure that is invariant under projection and section.

Pascal had this to say: Draw any six-sided polygon (hexagon) with vertices on a circle and letter the vertices A, B, C, D, E, F (fig. 27). Prolong a pair of *opposite* sides, AB and DE for example, until they meet in a point P. Prolong another pair of opposite sides until they meet in a point Q. Finally, prolong the third pair until they meet in a point R. Then, Pascal asserts, P,

Q, and R will always lie on a straight line. In other words,

If a hexagon is inscribed in a circle, the pairs of opposite sides intersect, respectively, in three points which lie on one straight line.

The concepts of projective geometry illuminate even familiar mathematics. As we saw in Chapter IV, the Greeks knew that the circle, parabola, ellipse, and hyperbola are sections of a cone (fig. 7 in Chapter IV). If we think of an eye placed at O, the vertex of the cone, and if we think of lines such as OA on the surface of the cone as lines of light from O to points on the circle ABC, then the lines form a projection and the circle, parabola, ellipse, and hyperbola appear as sections made by various planes cutting this projection. The reader can verify this by focusing a flashlight on a circular piece of wire and by observing the shadow cast by the wire on a sheet of paper. When the paper is turned the section will change and give the various conic sections. Because the four curves can all be obtained as sections of a cone and because Pascal's theorem states a fact about the circle that remains invariant under projection and section, it follows that Pascal's theorem applies to all the conics.

We shall consider just one more theorem of projective geometry. Pascal's theorem tells us something about a hexagon which is inscribed in a circle. C. J. Brianchon, who worked during the early nineteenth-century revival of projective geometry, created a famous theorem that describes a property of a hexagon circumscribed about a circle. His theorem (fig. 28) states that

If a hexagon is circumscribed about a circle the lines joining opposite vertices meet in one point.

As we might expect, Brianchon's theorem applies not only to the circle but to any conic section.

The theorems of Desargues, Pascal, and Brianchon are indications of the type of theorem proved in projective geometry and they must suffice as illustrations. We may characterize all of the theorems in this field by saying that they centre about the ideas of projection and section and state properties of geometric figures that are common to sections of the same projection or different projections of the same object.

Whereas the patronage of artists by princes, secular and clerical, made possible the extraordinary activity in painting and subsequently led to projective geometry, it was the expanding needs of the rapidly rising middle class of the period that prompted an interest in map-making. The search for trade routes in the sixteenth century involved extensive geographical explorations, and maps were needed to assist in the explorations and to keep pace with the discoveries.

It must not be inferred from this that preceding civilizations had not made maps. Indeed the Greeks, Romans, and Arabians made maps that were accepted for centuries. The explorations of the fifteenth and sixteenth centuries, however, revealed the inaccuracies and inadequacies of the existing maps and created a

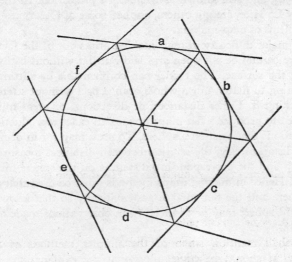

Figure 28. Brianchon's theorem

demand for better and more up-to-date ones. Moreover, the revival of the idea that the Earth is a sphere called for maps drawn on that basis. It raised such questions as how a course should be set out on a plane map so that it corresponds to the shortest distance on the sphere. The printing of maps was begun in the second half of the fifteenth century, and the great commercial centres, Antwerp and Amsterdam, soon became centres for the art of map-making.

Though the practical interests of map-makers are quite remote from the aesthetic interests of painters, both activities are intimately related through mathematics. Mathematically, the problem of making a map is that of somehow projecting figures from a sphere on to a flat sheet, the latter being but the section of the projection. Hence the principles involved here are the same as those in the sciences of perspective and projective geometry. In the sixteenth century, map-makers employed these and related ideas to develop new methods, the most famous of which is the one developed by the Flemish cartographer, Gerard Mercator (1512–94), and still known as Mercator's projection. In the next century La Hire, among others, applied some of Desargues' ideas to problems of map-making.

The major difficulty in map-making arises out of the fact that a sphere cannot be slit open and laid out flat without badly distorting the surface. The reader can confirm this by slitting and attempting to flatten out a whole orange peel without stretching or cracking it. Either distances, or directions, or areas must be distorted to produce a flat map; none is an exact reproduction of the relations that exist on a sphere. To use a map for information about distances, say, the relation between distances measured on the map and the corresponding distances on the sphere must be known. Hence in making maps methods must be used that relate the sphere and the flat surface systematically so that knowledge about the sphere may be deduced from observations made on the flat map.

We shall mention some of the simpler methods of map-making. It should be understood that the explanations given below cover only the geometrical principles involved in these methods. To show how measurements made on a particular map may be converted into corresponding information about the sphere would require the introduction of much more mathematics.

A simple scheme of map-making is known as the Gnomic projection. We imagine that an eye is placed at the centre of the

Earth and that it is looking at the Western Hemisphere. Each line of sight is continued past the Earth until it reaches a point on a plane that is tangent to the Earth's surface at some convenient point in the Western Hemisphere (fig. 29). If this point is on the equator we obtain a map such as that shown in figure 30.

It will be noticed that the meridians of longitude appear as straight lines. In fact any great circle on the Earth, that is any

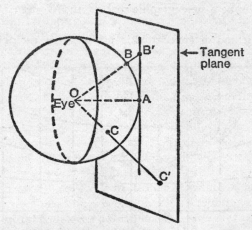

Figure 29. The principle of the gnomonic projection

circle whose centre is the centre of the Earth, such as the equator or a longitude circle, will project into a straight line under this scheme. This property is quite important. The shortest distance along the surface of the Earth between two points on the surface is given by the arc of the great circle joining these points. This arc will project into a straight-line segment joining the projections of the two points. Since ships and planes generally follow great circle routes, these routes are readily plotted as straight-line paths on the map. In addition, all points on the map have the correct directions from the centre and the correct directions from each other. A bad feature of this method of map projection is that the regions along the edges of the hemisphere being portrayed are projected very far out on the map with great distortion in the distances, angles, and areas involved. For this reason the map in figure 30 cannot show the entire hemisphere.

Projection and section are used in a different way in a second method of map-making known as stereographic polar projection. Suppose an eye is located on the equator in the middle of the Eastern Hemisphere and looks at points in the Western Hemisphere (fig. 31). Let a plane cut through the Earth between the two hemispheres. A section of the lines of sight made by the plane gives us a stereographic map of the Western Hemisphere (fig. 32).

The method of stereographic projection is useful because it

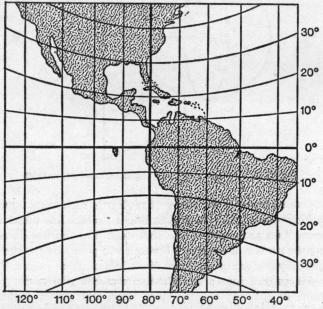

Figure 30. Gnomonic map of the Western Hemisphere

preserves angles. That is, if two curves meet at angle C on the sphere, the images of these curves on the map will meet at an angle C' which equals angle C. For example, the circles of latitude cross the meridians at right angles on the sphere. The projections of these curves meet at right angles on the map. Unfortunately the stereographic projection does not preserve area. The region near the centre of the map is reduced to about one-fourth of its actual size on the sphere. Near the edges of the map, however, the areas are almost correct.

The most widely known method of map-making is the Mercator projection. The principle involved in this method cannot be presented in terms of projection and section but it can be described approximately by a related projection. The latter method, known as the perspective cylindrical projection, employs a cylinder which surrounds the Earth and is tangent to it along some great circle. In figure 33 this circle is the equator. The lines that constitute the projection emanate from the centre of the Earth, point O in figure 33, and extend to the cylinder. Thus the point P on the Earth's surface is projected on to P′

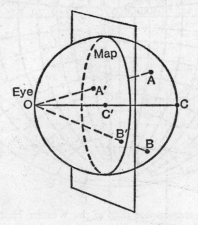

Figure 31. The principle of the stereographic projection

on the cylinder. The cylinder is now slit along a vertical line and laid flat. On the flat map the parallels of latitude appear as horizontal lines and the meridians as vertical lines. No points on the map correspond to the North and South Poles.

The essential difference between the perspective cylindrical projection and the Mercator projection is in the spacing of the parallels of latitude especially in the extreme northern and southern regions. Figure 34 illustrates the Mercator projection. The importance of this scheme is twofold. In the first place, as in the case of stereographic projection, it preserves angles. Second, in

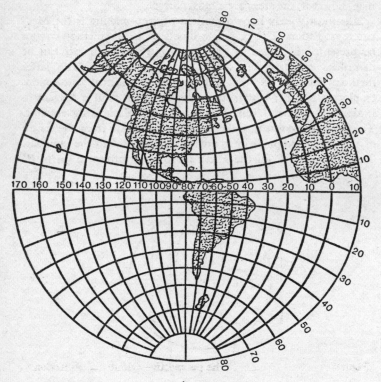

Figure 32. Stereographic map of the Western Hemisphere

steering a ship it is convenient to follow a course with constant compass bearings; this means a course which crosses the successive meridians on the sphere at the same angle. Such a course is known as a rhumb line or loxodromic curve. This course appears as a straight line on a map made according to the Mercator projection. Hence it is especially easy to lay out a ship's course and follow it on such a map.

A great-circle route, it should be noticed, does not imply constant compass bearing except when the great circle is the equator or a meridian of longitude. Hence on a Mercator map the great-circle route appears as a curve. It is the practice in navigation to approximate this curve by several short rhumb lines, thus permitting the ship to keep constant compass bearing along each rhumb and at the same time to take some advantage of the shortest distance afforded by the great-circle route.

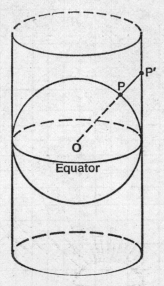

Figure 33. The principle of the perspective cylindrical projection

The Mercator method of map projection is so common that most people hardly realize the distortion it introduces. Greenland appears almost as large as South America though actually it is one-ninth as large. Canada appears twice as large as the United States; it is one and one-sixth as large. Despite such distortions the map is so useful in navigation for the reason given above that it is the one most widely used.

These brief descriptions of the geometrical principles underlying several methods of map-making do not exhaust the variety of methods nor do they give any indication of the mathematics that must be used to interpret measurements made on the map in terms of what is actually the case on the sphere. It should be

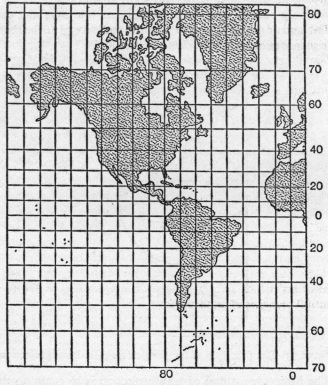

Figure 34. Mercator projection of the Western Hemisphere

clear, however, that mathematics is essential to map-making and, in particular, that projection and section are as extensively employed as in the study of perspective. Also, just as the use of projection and section in perspective gave rise to mathematical questions, so did it happen in map-making. In connection with maps it is important for practical reasons to know the properties common to a region on the sphere and the corresponding region of the map. For example, the fact that the sizes of angles are preserved in a particular method of map projection is very useful. Hence map-making, like perspective, has been the source of many new mathematical problems.

The ideas discussed in this chapter have centred about the notion of projection and section. The painters were led to this notion in their efforts to construct a satisfactory optical system of perspective. The mathematicians derived from the notion a totally new subject of investigation – projective geometry. And the map-makers employed the notion to design new map projections. All three fields, therefore, are intimately related by one basic mathematical concept.

Projective geometry proper can be applied to some practical problems; however, it has been cultivated primarily for the intrinsic interest men have found in it, for its beauty, its elegance, the latitude it affords to intuition in the discovery of theorems, and the strict deductive reasoning it demands for the proofs. After being temporarily neglected in favour of applied mathematics this subject was actively investigated during the nineteenth century and proved to be the mother of many new geometries. Perhaps because painting coloured his thoughts, the 'science born of art' which Desargues created is today one of the most beautiful branches of mathematics.

XII

A Discourse on Method

As long as algebra and geometry proceeded along separate paths, their advance was slow and their applications limited.

But when these sciences joined company, they drew from each other fresh vitality and thenceforward marched on at a rapid pace towards perfection.

Joseph Louis Lagrange

APPLIED mathematics in the modern sense of the term was not the creation of the engineer or the engineering-minded mathematician. Of the two great thinkers who founded this subject one was a profound philosopher, the other a gamester in the realm of ideas. The former devoted himself to critical and profound thinking about the nature of truth, the existence of God, and the physical structure of the universe. The latter lived an ordinary life as a lawyer and civil servant; at night, he indulged himself in mental sprees by creating and lavishly offering to the world million-dollar theorems. The work of both men in many fields will be immortal.

The profound philosopher was René Descartes (1596–1650), who was born to moderately wealthy parents in La Haye, France. At the age of eight he was sent to be educated at the Jesuit College of La Flèche where he became interested in mathematics. In his seventeenth year, at the conclusion of his study of the usual school subjects, he decided to learn more about himself and the world from first-hand experience. He began these studies by living a gay life in Paris after which he retired to a quiet corner of the city for a period of reflection. This was followed by participation in military campaigns, travel, more life in Paris, more war, more Paris, and finally by a decision to settle down.

Perhaps because Descartes thought he could attain complete seclusion in Holland he secured a house in Amsterdam. He lived in solitude, except for the company of his mistress and a child, and spent the major part of the following twenty years in writing. There he penned his best works and acquired fame almost as soon as his first book was published. As he continued to write, his audience and he himself became more and more impressed with the greatness of his work. Profound thoughts set forth in literary classics which revealed the clarity, precision, and effectiveness of the French language made both Descartes and philosophy popular.

After twenty years of retirement he was persuaded to tutor Queen Christina of Sweden, and so he moved to Stockholm. The queen preferred to begin her day at five in the morning by studying in an ice-cold library and Descartes was obliged to meet her at that hour. These demands, however, were too much for frail René. His flesh was weak and his spirit unwilling. He caught cold and died in the year 1650.

When Descartes was still in school at La Flèche he began to wonder how it was that man professed to know so many truths. It was partly because he had a critical mind and partly because he lived at a time when the world outlook that had dominated Europe for a thousand years was being vigorously challenged, that Descartes could not be satisfied with the tenets so forcibly and so dogmatically pronounced by his teachers and by leaders of sects other than his own. He felt all the more justified in his doubts when he realized that he was in one of the most celebrated schools of Europe and that he was not an inferior student. At the end of his course of study he concluded that there was no sure body of knowledge anywhere. All his education had advanced him only to the point of discovering man's ignorance.

To be sure, he did recognize some values in the usual type of studies. He agreed that, 'Eloquence has incomparable force and beauty; that Poesy has its ravishing graces and delights'; however, he judged these to be gifts of nature rather than fruits of study. He revered Theology because it pointed out the path to heaven and he, too, aspired to heaven but 'being given assuredly to understand that the way is not less open to the most ignorant than to the most learned, and that the revealed truths which lead to heaven are above our comprehension', he did not presume to subject them to the impotence of his reason. Philosophy, he granted, 'affords the means of discoursing with an appearance of truth on all matters, and commands the admiration of the more simple'. It had produced nothing, however, which was beyond dispute or above all doubt, though it had been cultivated for ages by the most distinguished men. He, therefore, did not presume that his success with traditional philosophy would be any greater. 'Jurisprudence, Medicine, and the other Sciences, secure for their cultivators honours and riches. . . .' Nevertheless, inasmuch as they borrow their principles from Philosophy, he judged that no solid superstructures could be reared on foundations so infirm and he was, thank Heaven, not compelled to make merchandise of Science for the betterment of his fortune. 'As for Logic, its syllogisms and the majority of its other precepts are of avail rather in the communication of what we already know, or . . . even in speaking without judgement of things of which we are ignorant, than in the investigation of the unknown. . . .' Numerous 'highly useful precepts and exhortations to virtue are contained in treatises on Morals'; but the disquisitions of the ancient moralists were towering and magnificent palaces with no better foundation than sand and mud. In all these fields, real or verifiable truth was noticeable by its absence.

During his years of soldiering, travelling, and living in Paris he pondered the question of how one can obtain truths. Gradually a programme for securing them became clear to him. He began by discarding all the opinions, prejudices, and so-called knowledge that he had thus far acquired. In addition, he rejected all knowledge based on authority and divested himself of all preconceived notions.

To reject falsity, however, did not in itself produce truth. The problem he then set for himself was to find the method of establishing new truths. The answer, he said, came to him in a dream while he was on one of his military campaigns.

The 'long chains of simple and easy reasonings by means of which geometers are accustomed to reach the conclusions of their most difficult demonstrations' led him to believe that 'all things to the knowledge of which man is competent are mutually connected in the same way. . . .' He decided, then, that a sound body

of philosophy could be deduced only by the methods of the geometers, for only they had been able to reason clearly and unim-
peachably and to arrive at indubitable truths. Having concluded
that mathematics 'is a more powerful instrument of knowledge
than any other that has been bequeathed to us by human agency',
he sought to distil from a study of the subject some general
principles that would provide the method of obtaining exact
knowledge in all fields, or, as he called it, a 'universal mathematics'. That is, he proposed to generalize and extend the
methods used by mathematicians in order to make them applicable to all investigations. In essence, the method would be an
axiomatic, deductive construction for all thought. The conclusions would be theorems derived from the axioms.

Guided by the methods of the geometers Descartes carefully
formulated the rules that would direct him in his search for
truth. He decided, first, that he would accept nothing as true
which was not so clear and distinct to his mind that all doubt was
excluded. Thus, he rejected sense data and, accordingly, all qualities of objects, such as taste and colour, which might be the
individual reactions of the perceiver rather than the intrinsic
characteristics of the objects themselves. The second principle of
his method was to divide large problems into smaller ones. The
third one stated that he would proceed from the simple to the
complex; and fourth, he would enumerate and review the steps
of his deductive reasoning so thoroughly that nothing would be
inadvertently omitted. These principles are the core of his
method.

He had, however, to find the simple, clear, and distinct truths
that would play the same part in his philosophy that axioms play
in mathematics proper. The results of his search are famous.
From the one reliable source that his doubts left unscathed – his
consciousness of self – he extracted the building blocks of his
philosophy: (*a*) I think, therefore I am; (*b*) each phenomenon
must have a cause; (*c*) an effect cannot be greater than the cause;
and (*d*) the ideas of perfection, space, time, and motion are
innate to the mind.

Since man doubts so much and knows so little he is not a
perfect being. Yet, according to axiom (*d*), his mind does possess

the idea of perfection and, in particular, of an omniscient, omnipotent, eternal, and perfect being. How do these ideas come about? In view of axiom (c) the idea of a perfect being could not be derived from or created by the imperfect mind of man. Hence it could be obtained only from the existence of a perfect being, who is God. Therefore God exists.

A perfect God would not deceive us and so our intuition can be trusted to furnish some truths. Hence the axioms of mathematics, for example, our clearest intuitions, must be truths. The theorems of mathematics, however, do not possess the simplicity and obviousness of the axioms. How can we be sure of their truth? Man does reason in a manner he believes to be infallible, but what guarantee has he that the methods of reasoning necessarily lead to truths? Again falling back on a God who would not deceive man, Descartes argued that the conclusions, too, must be truths and therefore must be correct assertions about the real world. From such foundations Descartes proceeded to build his philosophy of man and the universe.

His story of his search for method and of the application of the method to problems of philosophy was presented in his famous *Discourse on Method*. The supremacy of human reason, the invariability of natural laws, the doctrine of extension and motion as the essence of physical objects, the distinction between body and mind, and the distinction between qualities that are real and inherent in objects and qualities that are only apparently present but are actually due to the reaction of the mind to sense data are elaborated in these writings and have been influential in shaping modern thought.

It is not our purpose here to elaborate on the philosophical paths that Descartes followed, however worthy of study they are in their own right. What is relevant to our story is that the truths of mathematics and mathematical method served as a beacon light to a great thinker groping his way through the intellectual storms of the seventeenth century. His philosophy may indeed be characterized as mathematized philosophy. It is far less mystical, metaphysical, and theological, and far more rational than those of his medieval and Renaissance predecessors. He examined carefully the meaning and reasoning involved in all of his steps; he taught men to look within themselves for truths; and he cast off pupilage to antiquity and authority. With Descartes theology and philosophy parted company.

The method Descartes abstracted from mathematics and generalized he then reapplied to mathematics; with it he succeeded in creating a brand new way of representing and analysing curves. This creation, now known as co-ordinate geometry, is the basis of all modern applied mathematics. It will be valuable as far into the future as man can see whereas Descartes' philosophy, like most philosophies, is tied to a particular time. Before we examine Descartes' thinking in mathematics proper we must pause to acknowledge the equally worthy and independent efforts of his countryman and co-discoverer of co-ordinate geometry, Pierre Fermat.

In contrast to Descartes' adventurous, romantic, and purposive life Fermat's was dull, highly conventional, and matter of fact. He was born in 1601 into the family of a French leather merchant. After studying law at Toulouse he spent most of his life as a civil servant. Fermat's home life, too, was quite ordinary. He was married at the age of thirty and was devoted to his wife and five children. He lived quietly, ignored problems involving God, man, and the nature of the universe, and relaxed at night with his favourite pastime, mathematics. Whereas to Descartes mathematics served to solve philosophical and scientific problems and to master nature, to Fermat the subject offered beauty, harmony, and the pleasures of contemplation. Despite the brief amount of time he was able to spend on the subject and the pleasure-seeking attitude with which he approached it, he established himself after sixty-four years of life as one of the truly great mathematicians of all times.

His contributions to the calculus were first rate though somewhat overshadowed by those of Newton and Leibniz. He shared with Pascal the honour of creating the mathematical theory of probability, and shared with Descartes the creation of co-ordinate geometry. He founded single-handed one major branch of mathematics, the theory of numbers. In all these fields this 'amateur' produced brilliant results and left his impress. Though not concerned with a universal method in philosophy Fermat did seek a general method of working with curves. And here his thoughts joined company with those of Descartes.

We must digress briefly to understand why it was that the great mathematicians of the time were so much concerned with the study of curves. In the early part of the seventeenth century, mathematics was still essentially a body of geometry with algebraic appendages, and the heart of this body was Euclid's contribution. Euclidean geometry confines itself to figures formed by straight lines and circles but by the seventeenth century the advances of science and technology had produced a need to work with many new configurations. Ellipses, parabolas, and hyperbolas became important because they described the paths of the planets and comets. Parabolas were, also, the paths of projectiles such as cannon balls. The motion of the moon was intensively studied to help locate ships at sea. The curved path of light rays through the atmosphere was of interest to astronomers and artists, while the curvature of lenses was studied for use in spectacles, the telescope, and microscope, and for an understanding of the operation of the human eye. Actually, both Descartes and Fermat were very much interested in optics. Descartes published an essay on the passage of light through lenses, and Fermat contributed several fundamental laws, one of which will concern us in a later chapter. Unfortunately, Euclid offered no information on the curves involved in these and numerous other practical problems, and extant Greek works on the conic sections were inadequate.

Not only did the Greek works fail to supply the desired knowledge about important curves but they also failed to supply broadly applicable mathematical methods of obtaining that knowledge. Every proof in Euclid called for some new, often ingenious, approach. The Greek mathematicians with ample time at their disposal and no concern for immediate application apparently did not miss this lack of a general procedure. But the multifarious practical and scientific needs of the seventeenth century put pressure on the mathematicians to solve difficult problems in short order.

At this juncture Descartes and Fermat stepped into the picture. They were definitely dissatisfied with the limited methods used in Euclidean geometry. Descartes explicitly criticized the geometry of the ancients as being too abstract and so much tied to figures 'that it can exercise the understanding only on condition of greatly fatiguing the imagination'. Algebra, too, was criticized because it was so completely subject to rules and formulas 'that there results an art full of confusion and obscurity calculated to embarrass, instead of a science fitted to cultivate the mind'. On the other hand, both men recognized that geometry supplied information and truth about the real world. They also appreciated the fact that algebra could be employed to reason about abstract and unknown quantities; and it could be used to mechanize the reasoning process and minimize the effort needed to solve problems. It is potentially a universal science of method. Descartes and Fermat therefore proposed to borrow all that was best in geometry and algebra and correct the defects of one with the help of the other.

We can best understand what these men accomplished in the task they set for themselves by following the reasoning of Descartes, though our account may differ from his in details. We saw that in his general study of method he had decided to solve all problems by proceeding from the simple to the complex. Now the simplest figure in geometry is the straight line. He therefore sought to approach the study of curves through straight lines and he found the way to do this.

Let there be given, said Descartes, any curve such as the one shown in figure 35. This curve can be thought of as being generated by a point P which lies on a vertical line PQ. As the line moves to the right P itself moves up or down in accordance with the shape of the curve. Thus any curve can be studied by studying the motion of a point P which moves up or down on a straight line as the line itself moves parallel to its former positions. So far, so good. But how can one characterize any curve by the behaviour of P?

For this purpose Descartes used algebra, for he knew that algebraic language is a simple device to aid the memory and that it embraces many facts in a short space. As the vertical line moves to the right (fig. 35) its distance from a fixed position at O, say, can be used to denote its position. This distance is denoted by *x*. The position of *P* on the moving line can be specified by stating its distance above the fixed horizontal line *OQ*. This distance can be denoted by *y*. Thus for each position of *P* there will be a value of *x* and a value of *y*. For the same *x* two different curves will differ in the *y*-values. Hence what characterizes a curve is some

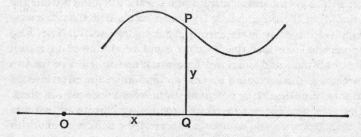

Figure 35. A curve generated by the motion of a straight-line segment of varying length

relation between *x* and *y* that holds for points *P* on this curve and that would be different for a different curve.

Let us see how this idea applies to a simple curve such as a straight line passing through the point *O* and making an angle of 45° with the horizontal (fig. 36). If the moving line *QP* moves any distance *x* to the right, the point *P* has to rise a distance *y* equal to *x* to reach the straight line, for Euclidean geometry tells us the *OQP* is an isosceles right triangle and the *OQ* must therefore equal *QP*. Hence

(1) $y = x$

is the relation that characterizes the points of the straight line concerned. Thus the point *P* for which the distance *OQ* is 3 and the distance *PQ* is 3 is a point on the line because its *x*-value of 3 and its *y*-value of 3 satisfy the equation $y = x$.

In order to include points such as *P′* on the straight line and at the same time distinguish *P′* from *P*, it is agreed to use negative numbers to represent the distances moved by *PQ* to the left of *O* and distances *QP* below the horizontal line *OQ*. Thus the *x*- and *y*-values of *P′* are both negative and equal, and it is still true that

$y = x$. On the other hand, for the point R which is not on the line $P'OP$, the y-value or the distance QR is not equal to x; therefore for points off the line it is not true that $y = x$.

We may systematize the thoughts contained in the discussion above as follows. To discuss the equation of a curve we introduce a horizontal line which will be called the X-axis (fig. 37), a point O on this line which is called the origin, and a vertical line

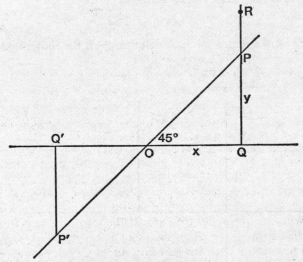

Figure 36. A straight line making a 45° angle with the horizontal

through O which is called the Y-axis. If P is any point on a curve, there are two numbers that describe its position. The first is the distance from O to the foot, Q, of the perpendicular from P to the X-axis. This number is called the x-value or abscissa of P. The second number is the distance PQ and this number is called the y-value or ordinate of P. The two numbers are called the co-ordinates of P and are generally written thus: (x, y). It is agreed that if P is to the right of the Y-axis, its x-value is taken to be positive, and if to the left, negative. In like manner, if P is above the X-axis its y-value is taken to be positive, whereas if it is below, its y-value is negative. The curve itself is then described

algebraically by stating some equation which holds for the *x*- and
y-values of points on that curve and only for those points.

Just to illustrate Descartes' idea once more we shall apply it to
the circle in figure 38. Suppose the circle has radius 5. Let *P* be
any point on the curve and let *x* and *y* be its co-ordinates. Then
the Pythagorean theorem of Euclidean geometry, which says that
the sum of the squares of the arms of a right triangle equals the
square of the hypotenuse, tells us that

(2) $$x^2 + y^2 = 25.$$

This relation holds for *each* point on *P* on the circle; that is, the

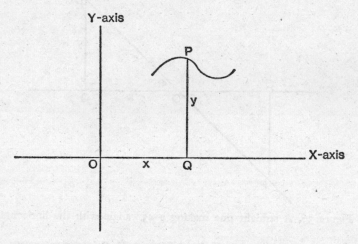

Figure 37. The rectangular co-ordinate system

x and *y* of each point are such that $x^2 + y^2 = 25$. For example,
the point which has the co-ordinates (3, 4) lies on the circle because $3^2 + 4^2 = 25$. However, (3, 2) are not the co-ordinates of
a point on the circle because $3^2 + 2^2$ does not equal 25. We say
that equation (2) is *satisfied* by the co-ordinates of a point if the
substitution of its abscissa for *x* and its ordinate for *y* makes
the left side equal the right side. The co-ordinates of a point on
the circle satisfy the equation; the co-ordinates of a point not on
the circle do not satisfy the equation.

We have thus far illustrated how a curve can be represented by an equation that characterizes this curve in a unique way. Descartes' idea also permits us to reverse the process above. Suppose we start with an equation, such as

(3) $$y = x^2.$$

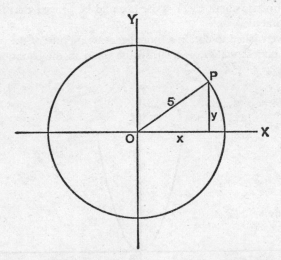

Figure 38. A circle placed on a rectangular co-ordinate system

What curve may be associated with this equation? Let us think once more in terms of the behaviour of the point P on the moving line PQ. As PQ moves to the right of O, the distance OQ, which is the x-value of P, is positive. Now equation (3) says that the y-value of P, or the distance PQ, must always equal x^2. When x is positive, so is x^2. Hence P must lie above the X-axis. Moreover, when x is small so is x^2, whereas when x gets larger, x^2 increases very rapidly. Therefore we know, at least roughly, what the curve looks like for positive x (fig. 39). Now as PQ moves to the left of O, the x-value of P is negative. But x^2 is still positive because the square of a negative number is positive. Hence P will be above the X-axis. Moreover, the value of x^2 is the same for a given negative value of x as it is for the corresponding positive value of x. For example, x^2 is 9 when $x = -3$ as well as when $x = +3$. Hence the point P will move in the same way to the left of the Y-axis as it does to the right. The complete curve is shown in figure 39, wherein it is understood that the curve continues indefinitely upward to the right and to the left. Our analysis of the equation $y = x^2$ shows that the curve is symmetric about the Y-axis. It could be proved that the curve is a parabola.

If we wished to obtain a more accurate picture of the curve, we could choose values of x, substitute them in the equation $y = x^2$,

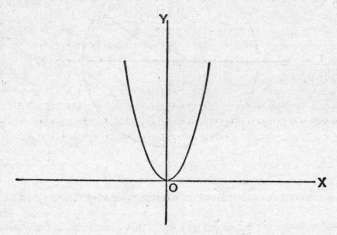

Figure 39. The curve of $y = x^2$

and calculate the corresponding y values. Thus when $x = 1$, $y = 1$; when $x = 2$, $y = 4$; when $x = 2\frac{1}{2}$, $y = {}^{25}\!/_{4}$; and so forth. Since each of these pairs of co-ordinates, for example $(2, 4)$, represents a point on the curve we could plot these points and join them by a smooth curve. The more co-ordinates we calculate, the more points can be plotted and the more accurately the curve can be drawn.

The heart of Descartes' and Fermat's idea is now before us. To each curve there belongs an equation that uniquely describes the points of that curve and no other points. Conversely, each

equation involving x and y can be pictured as a curve by interpreting x and y as co-ordinates of points. Formally stated, *the
equation of any curve is an algebraic equality which is satisfied
by the co-ordinates of all points on the curve but not by the co-
ordinates of any other point.* The association of equation and
curve, then, is the brand new thought. By combining the best of
algebra and the best of geometry Descartes and Fermat had a new
and immensely valuable method for studying geometric figures.
This is the essence of the idea Descartes embodied in an appendix to his *Discourse on Method* as proof of what his general
method in philosophy could accomplish when applied to mathematics. And indeed, in two or three months, Descartes succeeded in solving many difficult problems by using his new
method.

Beyond the analysis of properties of individual curves, the
association of equation and curve makes possible a host of

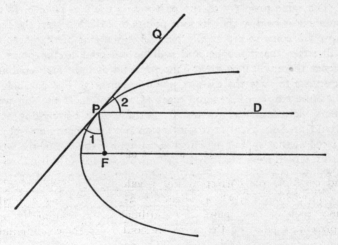

Figure 40. The focusing property of a parabola

scientific applications of mathematics. In this connection we
shall examine an application of the parabola wherein the
equation of the curve has proved invaluable. A parabola is
always symmetric with respect to one line which is called its

axis. In figure 39 this axis of symmetry is the Y-axis. In figure 40, the axis is the horizontal line shown. On this axis there is a point F, called the focus, such that if P is *any* point on the parabola, the line PF and the line through P parallel to the axis, PD in figure 40, make equal angles with the tangent PQ at P. That is, angle 1 equals angle 2.

Now suppose that the parabola is a cross-section of a reflecting surface and that a small source of light is placed at F. The light rays emanating from F will strike the parabola and, very fortunately for us, will be reflected along lines parallel to the axis. Thus a typical light ray originating at F will take the path FPD. The effect is that all the light will be concentrated in the direction of the axis and will produce a strong beam of light. In practical applications of this principle we use a surface that is obtained by rotating the parabola about its axis; a familiar example is furnished by an automobile headlight (see also fig. 44).

The same property of the parabola is used in reverse. If the parabola is held so that its axis points to a distant star, the light rays will come in practically parallel to the axis of the parabola, will strike the parabola, and will be reflected to the point F. Hence there will be a great concentration of light at F enabling scientists to view the distant star more clearly. The parabola is therefore employed in some types of telescopes. If the sun were being viewed instead of a star, the light rays converging at F would produce great heat and set on fire an inflammable object placed at that spot. This effect accounts for the use of the word *focus*, which means 'a hearth' or 'burning-place' in Latin.

Since practical applications of mathematics are not our primary concern in this book we shall merely mention in passing that all the conic sections possess properties similar to the one just described for the parabola. Therefore, these curves are effectively employed in lenses, telescopes, microscopes, X-ray machines, auditoriums, radio antennas, searchlights, and hundreds of other major devices. When Kepler introduced the conic sections in astronomy they became basic in all astronomical calculations including those of eclipses and the paths of comets. The conic sections are used also in the design of cables and roadways for bridges. In all these applications the equations of these curves have rendered possible, or at the very least expedited, calculations. Where the methods of Euclidean geometry would have required elaborate and complicated constructions and would have furnished lengths which could be measured only approximately, Descartes' algebraic equation is a much simpler tool and furnishes answers to as many decimal places as individual cases require. Co-ordinate geometry may not have lived up to Descartes' expectation that it would solve all geometric problems, but it solves many more than he could have envisioned in the seventeenth century.

Really important ideas are usually germinal in that they, in turn, suggest unsuspected notions and relationships. Descartes' association of equation and curve automatically uncovered a new world of curves. For each algebraic equation in x and y a curve exists that is described by that equation. Since the number and variety of equations that can be written down is unlimited, so is the range of curves. And these numerous curves, discovered only through their equations, in turn have proved useful in new and manifold applications.

The association of equation and curve did more than open up a new world of curves; it held forth prospects of new spaces. The extension of the idea to three-dimensional space suggested itself immediately. Beyond that was the provocative challenge to extend the idea to still higher dimensions. We must look into these more recent ramifications of co-ordinate geometry, for these extensions are at the basis of the most complicated and sophisticated of modern scientific developments, including the theory of relativity.

We shall consider first the extension of co-ordinate geometry to three-dimensional space. Earlier we saw that the position of a point in the plane can be described by a pair of numbers or co-ordinates. It will be apparent in a moment that the position of a point in space can be specified by a triplet of numbers. Let A be any plane, as the plane of this page, and let it be held horizontally. Suppose that in this plane the direction in which positive x-values are measured is indicated by OX (fig. 41) and the direction in which positive y-values are measured is indicated by OY.

Now, every point P in space is above or below the plane A a distance we shall represent by z; z is positive for points above A and negative for points below it. For example, if P is 4 units above A its z-value is 4. The position of P in space is completely described by noticing first that it is directly above the point R in the horizontal plane. R has x- and y-co-ordinates because it is in the plane A. Suppose these co-ordinates are (3, 2). Then the numbers 3, 2, and 4 completely determine the position of P, and no other point in space answers to this description. We therefore call 3, 2, and 4 the co-ordinates of P and write them in the form

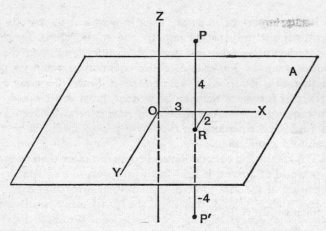

Figure 41. The three-dimensional rectangular co-ordinate system

(3, 2, 4). To a point in the plane A, such as R, a third co-ordinate, 0, is assigned so that the co-ordinates of R in the three-dimensional co-ordinate system we are now erecting are (3, 2, 0). The point P' with co-ordinates (3, 2, −4) is also shown in figure 41. The intersection O of the three axes OX, OY, and OZ is called the origin of the three-dimensional system and has the co-ordinates (0, 0, 0).

By means of our three-dimensional co-ordinate system it is possible to relate algebraic equations and geometrical figures in space. To illustrate such a relation let us consider the sphere. By definition the sphere is the set of all points in space at a given distance from a fixed point called the centre of the sphere. Suppose that all points of our sphere are 5 units from the centre and that the sphere is located so that its centre is at the origin of a

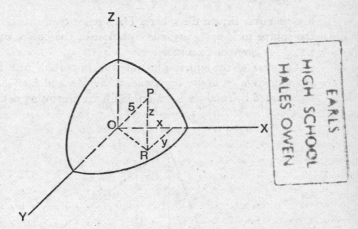

Figure 42. A sphere placed on a three-dimensional rectangular co-ordinate system

three-dimensional co-ordinate system (fig. 42). Let (x, y, z) be the co-ordinates of *any* point P on the sphere. Then x and y are the arms of a right triangle (lying in the horizontal plane) whose hypotenuse is OR. By the Pythagorean theorem

$$x^2 + y^2 = OR^2$$

OR and z, however, are the arms of the right triangle ORP, whose hypotenuse is OP or 5 units. Then

$$OR^2 + z^2 = 25.$$

But OR^2 has a value from the preceding equation. If we substitute this value we obtain the equation

$$x^2 + y^2 + z^2 = 25.$$

This is the equation of a sphere in the sense that the left side equals the right side when and only when the co-ordinates of a point on the sphere are substituted for x, y, and z. The point $(0, 3, 4)$, for example, satisfies the equation because

$$0^2 + 3^2 + 4^2 = 25.$$

and therefore lies on the sphere. The similarity of the equation of the sphere to $x^2 + y^2 = 25$, the equation of the circle, should be noted for later consideration.

The case of the sphere illustrates an important new fact. An equation in x, y, and z represents a *surface*, and each surface is represented by such an equation. Without presenting details here

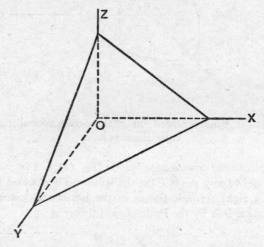

Figure 43. The plane corresponding to $3x + 4y + 5z = 6$

we shall mention some equations and their surfaces because it may help the reader to follow our discussion of four-dimensional geometry.

An equation of the form

$$3x + 4y + 5z = 6,$$

(the numbers are arbitrary) represents the set of points on a plane (fig. 43). The similarity of this equation to the equation of a straight line, which in a two-dimensional co-ordinate system is exemplified by $3x + 4y = 6$, is obvious.

An equation of the form

$$x^2 + y^2 = z$$

represents a paraboloid (fig. 44). The paraboloid shown here has roughly the shape of a mixing bowl or an automobile headlight. This equation is very much like the equation $y = x^2$ which represents a parabola.

The sphere, plane, and paraboloid are the analogues in three-dimensional space of the circle, line, and parabola, and this relation reveals itself when their equations are compared. If we could take the time to examine the equations of other surfaces we would find that they are the natural extensions of the equations of curves that have similar geometrical properties.

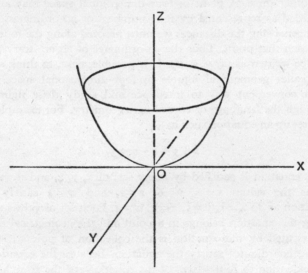

Figure 44. The paraboloidal surface corresponding to $x^2 + y^2 = z$

Language conveys ideas; a rich language may also suggest new ideas. At least this is the case in mathematics where the language often proves cleverer than the people who invent it. The algebraic language of co-ordinate geometry proved to have unexpected power, for it dispenses with the need for geometrical thinking.

Consider the equation $x^2 + y^2 = 25$, which we know represents a circle. Where is the rounded figure, the path that knows no end, the beauty of shape? All in the formula. Algebra has replaced geometry; the mind has replaced the 'eye'. We can find all the properties of the geometrical circle in algebraic properties of this equation. This fact suggested to the mathematicians that through the algebraic representation of geometric figures they could explore a concept advanced even before the time of Descartes and Fermat but theretofore unapproachable – namely, four-dimensional geometry.

What is a four-dimensional geometry? Approached pictorially the concept has no meaning. But we can *think* about *four* mutually perpendicular lines, that is four lines each perpendicular to the other three. A point in four-dimensional space may also be regarded as represented by four numbers or co-ordinates, these numbers being the distances we must proceed along the four axes to reach that point. Thus the co-ordinates of an arbitrary point can be written as (x, y, z, w). It is possible, next, to think about particular geometrical figures in four-dimensional space. The most convenient way to introduce and study these figures is through the language of co-ordinate geometry. For example, we can set up an equation such as

$$x + y + z - w = 5.$$

This equation is satisfied by many sets of x, y, z, and w values. Thus the values $x = 1$, $y = 6$, $z = 2$, and $w = 4$ satisfy this equation as do $x = 1$, $y = 5$, $z = 3$, $w = 4$. Each set of values satisfying the equation belongs to a point and the geometrical figure represented by the equation is the collection of points each of whose coordinates satisfy the equation. Because the equation is the extension to four letters of the equations of the straight line and plane, we call this figure a hyperplane. Similarly, we can speak of the figure belonging to the equation

$$x^2 + y^2 + z^2 + w^2 = 25$$

as a hypersphere because this equation is an extension to four letters of the equations of circle and sphere. The equations in four letters are the algebraic description of figures in four-dimensional space.

The figures of four-dimensional geometry exist in the same sense as do the figures in two and three dimensions. The hyperplane is as 'real' as the circle and sphere. The same applies to all other objects of higher dimensional geometry. The difficulty most people experience in accepting a four-dimensional geometry and the corresponding equations is due to the fact that they confuse mental constructions and visualization. All of geometry, including two- and three-dimensional Euclidean geometry, deals as Plato emphasized, with ideas that exist in the mind only. Fortunately we can visualize or picture the two- and three-dimensional ideas by means of drawing on paper, and these drawings help us to remember and to organize our thoughts. But the pictures are not the subject matter of geometry and we are not permitted to reason from them. It is true that most people including mathematicians, lean upon these pictures as a crutch and find themselves unable to walk when the crutch is removed. For a tour of the domains of higher dimensional geometry, however, the crutch is not available. No one, not even the most gifted mathematician, can visualize four-dimensional structures; he must rely on his mind alone. The structures themselves are then treated by means of their equations.

As a matter of fact, it is possible to visualize *sections* of figures in four-dimensional space. The meaning of this statement can be explained by reference to a three-dimensional situation. Suppose we wanted to study the ellipsoid (for example, the surface of a football) in detail. To circumvent the difficulty in visualizing the whole figure, a difficulty not too great in this case, a favourite mathematical device is to take plane sections of the ellipsoid and study these. From these sections – ellipses such as A and B in figure 45 – we can obtain knowledge of the entire ellipsoid. Thus the problem of studying a figure in three-dimensional space is reduced to that of studying figures in two-dimensional space.

In a similar manner, we can examine two- and three-dimensional sections of four-dimensional geometrical figures and deduce knowledge about them from the sections. 'But,' the reader might object, 'we know what the plane sections of the ellipsoid are because we can visualize the whole figure. How can we do that for our four-dimensional world? The answer is, *by means of the algebraic equations.* We find the *equation* of the section first and obtain its shape with our knowledge of ordinary, two-dimensional and three-dimensional co-ordinate geometry.

In still another way we are able to visualize figures of a four-dimensional space. To study an elliptical section of the ellipsoid we can confine ourselves to the plane in which the ellipse lies, that is, we need to consider a two-dimensional world only. Now let us consider a curve of a four-dimensional world. If this curve happens to lie in a plane it can be completely visualized despite the fact that it is part of a four-dimensional world.

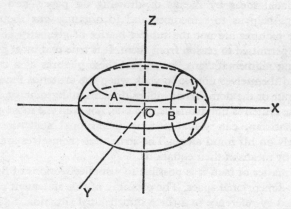

Figure 45. Two-dimensional sections of the ellipsoid

If we can study four-dimensional figures in terms of two- and three-dimensional sections, why admit the four-dimensional world in the first place? The answer is that the proper relationships of these various sections to each other can exist only in such a world, just as the proper relationship of sections *A* and *B* of the ellipsoid in figure 45 can exist only in three-dimensional space.

The notion of a four-dimensional geometry is actually a very helpful one in studying physical phenomena. There is a point of view from which the physical world can and should be regarded as four-dimensional. Any event occurs in a certain place and at a certain time. To describe this event as distinguished from other events we should give the position and time of its occurrence. Its position in space can be specified by three numbers, that is, its co-ordinates in a three-dimensional co-ordinate system: the time of occurrence can be specified by a fourth number. The four numbers x, y, z, and t, and no fewer, thus serve to specify the event unmistakably. The four numbers are the co-ordinates of a point in a four-dimensional space-time world. Hence it is natural to think of the world of events as a four-dimensional world and to study physical events in that light.

Let us consider, as a specific example, the motion of a planet. To locate a planet properly we must specify not only the position of the planet but the time when the planet occupies that position. Hence four numbers are actually required to describe the location of the planet, and these four numbers may be regarded as a point in a four-dimensional geometry. The successive locations of the planet may also be described as points of a four-dimensional world and the entire motion of the planet in space-time is described by a hypercurve. We cannot visualize or draw such a curve but we can represent it by an equation or, more accurately, by a set of equations, in four letters. If the equations are correctly chosen then they embody a complete description of the motion just as $x^2 + y^2 = 25$ is a complete description of the circle. And just as we can deduce facts about the circle by studying its equation so we can deduce facts about the motion of a planet by studying the representative equations.

Perhaps we should take this occasion to point out that a great deal of nonsense has been written about what it would be like if we lived in a world of four spatial dimensions. Many writers have declared that in a world of four spatial dimensions people could eat an egg without breaking the shell, or leave a room without passing through the walls, floor, or ceiling. These writers are reasoning by analogy with comparable situations in lower dimensions. To pass from a point A inside a square to a point B outside (fig. 46) and remain in the plane of the paper, the boundary C must be crossed. But we may avoid C if we are allowed to employ a third dimension and go out of the plane of the paper. Similarly, to pass from a point A inside a cube to a point B

outside (fig. 47), the surface of the cube must be crossed – as long as we are confined to three dimensions. However, and here the reasoning by analogy enters, if we could employ a fourth dimension the surface of the cube could be avoided.

Now such speculations would be harmless if they did not give the impression that mathematicians actually believe in the real existence of a world of four spatial dimensions and hope some day to train our visual apparatus to perceive this world. No such belief is held nor is such a project contemplated.

The notions of dimension and of higher dimensional geometry are fascinating branches of mathematics. But these topics take us far beyond the time and work of Descartes and Fermat. It is their work and the lesson to be derived from it that concern us in this chapter. What then is that lesson? First, mathematics was an inspiration and a guiding light in Descartes' philosophical thinking. Second, a philosophical interest in method and an intellectual delight in mathematical activity produced the co-ordinate geometry on which practically all applications of math-

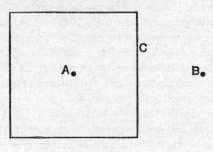

Figure 46.

ematics to the physical world depend. The line of development from Descartes through Newton to Einstein is as straight as any mathematical idealization could conceive.

Through Descartes' work the importance of mathematics was considerably increased, for he was the first influential thinker to demonstrate to the world the nature and value of mathematical method in man's search for truth. He offered a plan of attack on problems to a world lost in the morass of confusion that characterizes the end of an era. Just how much the world profited by Descartes' proselytizing for the cause of mathematical method will be apparent within the space of a few chapters.

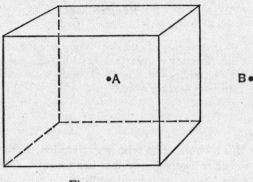

Figure 47

XIII

The Quantitative Approach
to Nature

So that we may say the door is now opened, for the first
time, to a new method fraught with numerous and wonderful results which in future years will command the
attention of other minds.

Galileo

ONE day a young man who was a student at the University of
Pisa visited the cathedral of that famous town. The services must
have been boring for instead of listening attentively he watched
the swinging of a great hanging lamp. He soon noticed that when
the lamp swung through a wide arc the time it took to perform
one swing seemed to be the same as when it swung through a
narrow arc. He did not use his pocket watch to check this observation for the simple reason that such a timepiece had not as yet
been invented; but he did think to use his pulse beat. The observation turned out to be correct, and a mere youth had discovered
a scientific law governing all pendulum motion: the time required by a pendulum to perform a swing is independent of the
amplitude of the swing. Not long afterward this law was used to
design the serviceable clock which the young man lacked. More
important, the discovery suggested a new concept of scientific
activity which defines modern science and at the same time
endows it with its 'magical' power. This is the concept we intend
to examine.

The young man who daydreamed in church, Galileo Galilei,
son of a musician, was born in Florence in 1564, the year of
Shakespeare's birth. At the age of seventeen he entered the University of Pisa to study medicine and, while there, learned mathematics in private lessons from a practical engineer. His reading
of Euclid and Archimedes fired a natural genius for mathematics

and science and so, with his father's consent, he turned his attention to those fields.

The scope of Galileo's interests and activities was unbelievably broad even for a great intellect of the Age of Genius. He was always keenly interested in mechanical devices and was mechanically dexterous himself. At home he kept a workshop in which he spent a great deal of his time. There he produced so many new and ingenious devices that he can be called the father of modern invention. The telescope, or 'perplexive glasses' as Ben Jonson called them, with which Galileo discovered the moons of both Jupiter and Saturn, the star composition of the Milky Way, the phases of Venus, and the mountains and valleys on the moon, was of his own design. These observations, incidentally, showed that the heavenly bodies possessed the same properties as did the Earth and therefore constituted additional weighty evidence for the heliocentric theory. Another of Galileo's inventions was a pulsilogium, a device that utilized its own law of pendulum motion to record pulse rates mechanically.

Though his scientific research overshadows his other activities, Galileo was also a major literary figure, and it is acknowledged that he wrote the best Italian prose of the seventeenth century. He experimented with literary forms, criticized and wrote poetry, and lectured for a while on Dante. Even his scientific writings are famous, not merely because they present his astronomical and physical researches but because they are literary classics as well. Galileo's interest in the art of writing was supplemented by devotion to painting and by skilful musicianship, both of which often consoled him in his years of trouble.

The most artistic and the most fruitful creation of the myriad-minded Galileo was a grand plan for reading the book of nature. In essence it offered a totally new concept of scientific goals and of the role of mathematics in achieving them. Though earlier, less comprehensive, and generally abortive efforts of forerunners should be acknowledged, Galileo explicitly formulated the plan and put it into effect by establishing a number of fundamental laws. When Galileo died in 1642, full of fame and years, modern science was already well started on its successful career, an accomplishment that must be credited to his almost single-handed efforts. It is Galileo's plan for studying and mastering nature that will concern us in this chapter.

Doubtless almost every twentieth-century person is aware that something revolutionary occurred in the field of science about the year 1600. Why did the scientific activity that was initiated in the seventeenth century prove so effective? Were the contributors such as Descartes, Galileo, Newton, Huygens, and Leibniz greater intellects than those found in earlier civilizations? Hardly. The profoundly learned Aristotle and the brilliant Archimedes both possessed intellects as fine as those of any of the seventeenth-century scientists. Was it because of the increased use of observation, experiment, and induction, methods urged by Roger Bacon and Francis Bacon? Apparently not. The turn to observation and experimentation may have been an innovation in the Renaissance but it was a method of approach at least familiar to Greek scientists. Nor does the mere use of mathematics in scientific studies explain the amazing accomplishments of modern science, for though the seventeenth-century scientist knew that the goal of his work should be to ferret out the mathematical relationships behind various phenomena, the search for such relationships in nature was not new to science. The belief in the mathematical design of nature had been put to the test even in Greek times.

The secret of the success of modern science was the selection of a new goal for scientific activity. This new goal, set by Galileo and pursued by his successors, is that of obtaining *quantitative descriptions* of scientific phenomena *independently of any physical explanations*. The revolutionary character of this new concept of science will be appreciated more if it is compared with the scientific activity of preceding ages.

Greek scientists concentrated on *explaining why* phenomena occur as they do. Aristotle, for example, spent much time in trying to explain why bodies thrown up into the air fall to the Earth. The Greek mathematician and engineer, Heron, used the principle that nature abhors a vacuum to explain other phenomena. Similarly, Greek physics accounted for the absence of apparent forces causing the circular motion of the heavenly bodies by arguing that circular motion was natural and hence needed no forces to start it or keep it going. Still other 'explanations' hardly seem to penetrate the phenomena they dealt with. For example, according to Plato, the Earth maintains its fixed position in the centre of the universe 'for a thing in equilibrium in the middle of any uniform substance will not have cause to incline more or less in any direction'.

Medieval Europe was also concerned with *why* things happen, only the explanations were always in terms of the purpose of a phenomenon. An 'explanation' of rain was that it watered man's crops. The crops grew to feed man, and man lived to serve God and worship Him. St Thomas, following Aristotle, discussed motion from the standpoint of why it happens and said that it is the act of that which is in potentiality and seeks to actualize itself. Whether or not these explanations appear satisfactory to us, they were nevertheless the answers given to questions asked in earlier scientific activity.

Galileo was the first man to realize that such speculations in regard to the causes and reasons for events had not advanced scientific knowledge very far and that they had not given man much power to predict and control the course of nature. For these reasons he proposed to replace them by a quantitative description of phenomena.

His proposal may be clarified by an example. In the simple situation in which a ball is dropped from a person's hand we might speculate endlessly as to *why* the ball falls. Galileo advised us to do otherwise. The distance the ball falls from its starting point increases as time elapses from the instant it is dropped. In mathematical language the distance the ball falls and the time that elapses as it falls are called *variables* since both change as the ball falls. Let us seek, said Galileo, some mathematical relation between these variables. The answer that Galileo sought is written nowadays in that scientific shorthand known as a formula; for the situation under discussion, this formula is $d = 16t^2$. This formula says that the number of feet, d, which the ball falls in t seconds is 16 times the square of the number of seconds. For example, in 3 seconds the ball falls 16 times 3^2 or 144 feet;

in 4 seconds, the ball falls 16 times 4^2 or 256 feet; and so forth.

It should be noticed, first, that the formula is compact, precise, and quantitatively complete. For each value of one variable, *time* in this case, the corresponding value of the other, *distance*, may be calculated exactly. This calculation can be performed for millions of values of the time variable, actually an infinite number of values, so that the simple formula $d = 16t^2$ contains an infinite amount of information.

The formula is a way of representing a relation between variables. The relationship itself, which may be known to exist on physical grounds, is called today a *function* or *functional relation*. Such relations hold in practically every sphere. Since the pressure of the atmosphere varies with the elevation above the surface of the Earth, there is a functional relation between the pressure and the altitude. Similarly, the cost of a manufactured article depends on, or is a function of, the cost of raw materials, labour costs, and overheads. In this last-mentioned example four variabes are involved, one of which, the cost of the article, depends on the other three.

It is very important to realize that the mathematical formula is a description of what occurs and not an explanation of a causal relationship. The formula $d = 16t^2$ says nothing about *why* a ball falls or whether balls have fallen in the past or will continue to fall in the future. It merely gives quantitative information on *how* a ball falls. And even though such formulas are used to relate variables which the scientist suspects are causally related, it is nevertheless true that he does not have to investigate, nor understand, the causal connection in order to treat the situation successfully. It is this fact that Galileo saw clearly when he emphasized mathematical description against the less successful qualitative and causal inquiries into nature.

It was Galileo's decision, then, to seek the mathematical formulas that describe nature's behaviour. This thought, like most thoughts of genius, may leave the reader unimpressed on first contact. There seems to be no real value in these bare mathematical formulas. They explain nothing. They simply *describe* in precise language. Yet such formulas have proved to be the most valuable knowledge man has ever acquired about nature.

We shall find that the amazing practical as well as the theoretical accomplishments of modern science have been achieved mainly through the quantitative, descriptive knowledge that has been amassed and manipulated rather than through metaphysical, theological, and even mechanical explanations of the causes of phenomena. The history of modern science is the history of the gradual elimination of gods and demons and the reduction of vague notions about light, sound, force, chemical processes, and other concepts to number and quantitative relationships.

The decision to seek the formulas that describe phenomena leads in turn to the question: what quantities should be related by formulas? A formula relates the numerical values of varying physical entities such as pressure and temperature. Hence these entities must be measurable. The principle Galileo followed next was to measure what is measurable and to render measurable what is not yet so. His problem then became that of isolating those aspects of natural phenomena which are basic and capable of measurement.

In pursuit of this objective he had to break new ground. His predecessors of the medieval period, following Aristotle, had approached nature in terms of concepts such as origins, essences, form, quality, causality, and ends. These categories do not lend themselves to quantitization. Instead Galileo proceeded to exploit a philosophy of nature founded by both himself and Descartes. The latter had already fixed on matter moving in space and time as the fundamental phenomenon of nature. All effects were explainable in terms of the mechanical effects of such motions. Matter itself was, in fact, a collection of atoms whose motions determined not only the behaviour of an object but also the sensations produced by that object.

Galileo therefore sought to isolate the characteristics of matter in motion that could be measured and then be related by mathematical laws. By analyzing and reflecting on natural phenomena he decided to concentrate on such concepts as space, time, weight, velocity, acceleration, inertia, force, and momentum. Later scientists added power, energy, and other concepts. In the selection of these particular properties and concepts Galileo again showed genius, for the ones he chose are not immediately discernible as the most important nor are they readily measurable. Some, such as inertia, are not even obviously possessed by matter; their existence had to be inferred from observations. Others, such as momentum, had to be created. Yet these concepts did prove to be most significant in the rationalization and conquest of nature.

There is another element in the Galilean approach to science that proved equally important in the sequel. Science was to be patterned on the mathematical model. Galileo and his immediate successors felt sure that they could find some laws of the physical world which would appear to be as unquestionably true as the axiom of Euclid that a straight line may be drawn through any two points. Perhaps contemplation, experimentation, or observation would suggest these axioms of physics; at any rate, once they were discovered their truth would be intuitively evident. With such fundamental intuitions these seventeenth-century scientists hoped to deduce a number of other truths in precisely the manner in which Euclid's theorems followed from his axioms.

In order to appreciate the significance of Galileo's plan it is necessary to realize that science is not a series of experiments regardless of how intelligently or skilfully they are executed; nor is it a series of facts experimentally or theoretically deduced. The positive content of a science is a body of theory which encompasses, organizes, relates, and illuminates a multitude of seemingly disconnected facts in a coherent and consistent fashion and which is capable of leading to new conclusions about the physical world. The individual facts or experiments are of little value in themselves. The value lies in the theory that unites them. The distances of the planets from the sun are details. The heliocentric theory is knowledge of the first magnitude. Thus another of Galileo's innovations was to make the scientific theory, the connective tissue among facts, a body of mathematical laws deducible from a set of axiomatic ones.

The Galilean plan contained, then, three main features. The first was to seek quantitative descriptions of physical phenomena and embody these in mathematical formulas. The second was to isolate and measure the most fundamental properties of phenomena. These would be the variables in the formulas. The third was to build up science deductively on the basis of fundamental physical principles.

To put this plan into execution Galileo had to find fundamental laws. We might obtain a mathematical formula relating the number of marriages in Siam and the price of horseshoes in New York City as these quantities vary from year to year. Such a formula is of no value to science, however, for it does not encompass, either directly or by implication, any useful information. The search for fundamental laws was another immense task because once again Galileo had to break with his predecessors. His approach to the study of matter in motion had to take into account an Earth moving through space and rotating on its axis, and these facts in themselves invalidated much of the only significant system of mechanics which the Renaissance world possessed, namely, the mechanics of Aristotle.

In treating the behaviour of objects on the Earth this ancient sage had taught that each has a natural place and that the natural state of a body is one of rest in that natural place. Heavy objects have their natural place at the centre of the Earth, which of course is the centre of the universe. Light objects, such as gases, have their natural place in the sky. Objects not in their natural place but otherwise undisturbed by external forces will seek that place. Thus arises natural motion. For example, an object released from the hand will seek the centre of the Earth and move towards it. When, however, an object is thrown or pulled, the resulting motion is violent in nature.

Since rest is the natural state, both natural and violent motion must be due to some force that acts continually; otherwise the motion would cease. Also, all motion is continually subject to resistance. In any case, the velocity of the motion can be expressed by the formula (using modern notation) $V = F/R$; or, stated in words, the velocity depends directly on the force and inversely on the resistance. In the case of natural motion the force is the weight of the object and the resistance comes from the medium in which the object moves. Hence heavier bodies must fall faster in a given medium because F in the formula $V = F/R$ is larger and so V must be. In violent motion the force is applied by human hands or some man-made mechanism and the resistance is due to the weight. Then for lighter bodies the resistance R is less and therefore the velocity V is greater. Hence lighter bodies move faster when a given force is applied.

Special theory was required to explain some phenomena. For example, a body that is dropped always gains speed. Now the force in this natural motion is supplied by the weight and this quantity, as well as the resistance of the medium, is constant. Hence, by the formula $V = F/R$, the velocity should be constant. The acceleration, or increase in velocity, was accounted for by the rush of air from the front to the back of the body. This air supposedly exerted force on the back and thus increased the velocity. Less scientifically minded people explained that a body moved more jubilantly as it neared home.

These laws of Aristotle are compounded of two parts of observation and eight parts of aesthetic and philosophical principles. Nevertheless, they served as the foundation for untold volumes of religion, philosophy, and science written over many centuries. We may be sure that Galileo's task in unearthing fundamental laws of nature, like Copernicus' advocacy of the heliocentric theory, was infinitely harder because he had to break with two thousand years of established thought.

According to Aristotle a force is required to keep a body in motion. Hence to keep an automobile or a ball moving, even on a very smooth surface, some propelling force should be present. But Galileo had greater insight into this phenomenon than did Aristotle. Actually a rolling ball or moving automobile is hindered somewhat by the resistance of air and retarded by friction between it and the surface on which it rolls. If these hindering actions were not present *no* propelling force would be needed to keep the automobile rolling. It would continue at the same speed *indefinitely*; moreover, it would follow a straight-line path. This fundamental law of motion, that *a body undisturbed by forces will continue indefinitely at a constant speed and in a straight line*, which was discovered by Galileo, is now known as Newton's first law of motion. It is obviously a more penetrating principle than the one Aristotle produced for the same situation. The law says that a body will change its speed only if it is acted upon by a force. Thus bodies possess the property of resisting change in speed. This property of matter, namely resistance to change in speed, is called its *inertial mass* or simply its mass.

It should be pointed out, before we immerse ourselves further in Galileo's ideas, that his very first principle is in contradiction with that of Aristotle. Does this mean that Aristotle made obvious blunders or that his observations were too crude or too few to yield the correct principle? Not at all. It is unlikely that mere observation would have led Aristotle to improve on himself, or others to improve on Aristotle. Aristotle was a realist and he taught what observations actually do suggest. Galileo's method, however, was more sophisticated and consequently more successful. Galileo approached the problem as a mathematician. He idealized the phenomenon by ignoring some facts to favour others, just as the mathematician idealizes the stretched string and the edge of a ruler by concentrating on some properties to the exclusion of others. By ignoring friction and air resistance and by imagining motion to take place in a pure Euclidean vacuum he discovered the correct fundamental principle. His trick was to geometrize the problem and then obtain the law.

We may, however, ask, are not friction and air resistance real effects? Do they not cause an object to lose speed and eventually to stop altogether? They do, sometimes; and when this occurs, friction and air resistance should be taken into account. They are, however, additional effects superimposed on the fundamental phenomenon, namely, that an object in motion continues at a constant speed indefinitely. Sometimes friction and air resistance are practically negligible, as when a one-pound piece of lead falls to the ground from a height of a few hundred feet. Also, recognition of the fact that these additional forces are present makes it possible to minimize their effect. Oil, ball bearings, and smooth surfaces reduce friction in moving machinery. Where the effect cannot be minimized, recognition of its existence allows us to take it into account explicitly and thereby predict the correct motion. Galileo's point here is precisely the point the mathematician makes when he treats ideal figures. The measurement of real triangles would produce angle sums varying from perhaps 160° to 200°. The basic fact is that the angle sum of an ideal triangle is 180° and in so far as a real triangle approximates the ideal triangle, its angle sum will approximate 180°. The paradox behind the achievements of modern science is that the scientist or mathematician appears to distort a problem by idealizing it so much that he affronts common sense, and then he proceeds to obtain the correct solution. An account of just how successfully Galileo's approach proved to be will appear shortly.

What can be said about the motion of a body if some force is applied to it? Here Galileo made a second fundamental discovery. The continuous application of a force causes a body to gain or lose velocity. Let us call the gain or loss in velocity per unit of time the acceleration of the body. Thus if a body gains velocity at the rate of 30 feet per second each second, its acceleration is 30 feet per second each second, or in abbreviated form, 30 ft/sec.². The second law of motion states that if a force causes a body to gain or lose velocity then the *force*, expressed in some suitable unit, *is equal to the product of the mass of the body and its acceleration.* Expressed as a formula this law says

(1) $$F = ma.$$

This formula is most significant. It implies that a constant force produces a constant acceleration on a constant mass, for if F and m are fixed, a must be also. For example, a constant air resistance causes a constant loss in velocity, and this accounts for the fact that an object rolling or sliding on a smooth floor will lose velocity continually until it has zero velocity.

Conversely, if a moving object does possess acceleration, that is, if a in formula (1) is not zero, then the force F cannot be zero. Now an object falling to the Earth from some height does possess acceleration. Hence some force must be acting. In Galileo's time the notion had already gained some acceptance that this force must be the pull of the Earth. Without, however, wasting much time on speculation about this notion, Galileo investigated the quantitative facts about falling bodies.

He discovered that if air resistance is neglected *all* bodies falling to the surface of the Earth have the same constant acceleration, that is, they gain velocity at the same rate, 32 feet per second each second. If the body is dropped, that is, merely allowed to fall from the hand, it will start with zero velocity. Hence at the end of one second its velocity is 32 feet per second; at the end of two seconds its velocity is 32 times 2 or 64 feet per second; and so forth. At the end of t seconds its velocity is $32t$ feet per second; in symbols,

$$(2) \qquad\qquad v = 32t.$$

This formula tells us exactly how the velocity of a falling body increases with time. It says, too, that a body which falls for a longer time will have a greater velocity. This is a familiar fact, for most people have observed that bodies dropped from high altitudes hit the ground at higher speeds than do bodies dropped from low altitudes.

We cannot multiply the velocity by the time in order to find the distance that a dropped body falls in a given amount of time. This would give the correct distance only if the velocity were constant. Galileo proves, however, that the correct formula for the distance the body falls in t seconds is

$$(3) \qquad\qquad d = 16t^2,$$

d being the number of feet the body falls in t seconds. For example, in three seconds, the body falls 16.3^2 or 144 feet.

By dividing both sides of formula (3) by 16 and then taking the square root of both sides, we obtain the result that the time required for an object to fall a given distance d is given by the formula $t = \sqrt{d/16}$. It will be noticed that the mass of the falling body does not appear in this formula. Hence all bodies take the same time to fall a given distance. This is the lesson Galileo is supposed to have learned by dropping objects from the tower of Pisa. People still find it difficult to believe, nevertheless, that a piece of lead and a feather when dropped from a height in a vacuum reach the ground in the same time.

Another useful formula can be derived by combining formulas (2) and (3). By dividing both sides of formula (2) by 32 we get

$$t = v/32.$$

If we substitute this value of t in formula (3) we obtain

$$d = 16(v/32)^2 = 16(v/32)(v/32),$$

or

(4) $d = v^2/64.$

Formula (4) tells us that if we know the velocity of a freely falling body then we can calculate the distance it has fallen to attain that velocity.

Multiplying both sides of this formula by 64 gives

$$v^2 = 64d,$$

or

(5) $v = \sqrt{64d}.$

Formula (5) gives the velocity acquired by an object in falling a distance d.

Let us take one more example of how the laws of motion can be used to derive a significant formula. Consider the phenomenon of a ball thrown straight up into the air. Of course, the height of the ball above the ground changes continually as does the elapsed time. Let t be the number of seconds the ball travels, counting from the instant it is thrown up, and let h be the height above the ground attained by the ball in t seconds. A useful formula to have in such a situation is the one relating the variables h and t.

Suppose the ball is thrown into the air with enough force to give it a speed of 100 feet per second as it leaves the hand. If no other forces were to act on the ball, then according to Newton's first law of motion this speed would remain constant. In t seconds the ball would travel upwards a distance equal to its speed multiplied by the number of seconds it travels, or, in this case, a distance of $100t$. At the same time that the ball travels upwards, however, it is pulled towards the Earth, as is any ball that is merely dropped. According to formula (3) the distance the ball is pulled towards the Earth in t seconds is $16t^2$ feet. Hence the motion of the ball is the result of two separate motions taking place simultaneously, a rise of $100t$ feet in t seconds and a fall of $16t^2$ feet in the same t seconds. The height h of the ball above the ground in t seconds is therefore

(6) $h = 100t - 16t^2.$

The derivation of formulas such as (4), (5), and (6) illustrates in a small way how Galileo hoped to carry out his programme of deriving the important laws of nature from a few basic ones. We can see that mathematical reasoning supported by the physical axioms permits deductive derivation of laws. These examples, as well as others we shall examine shortly, also illustrate how the mathematician can sit back in his armchair and obtain dozens of significant laws of nature. His tools, aside from paper and pencil, are the axioms and the theorems of mathematics and the axioms of physics such as the laws of motion. Mathematical deduction, the essence of his work, produces knowledge of the physical world.

From these proofs Galileo proceeded to an observation which he embodied in another law of motion. If one body is carried by another, as a passenger is carried by an aeroplane, the first shares the motion of the second. This seems obvious enough. But if the passenger should suddenly be ejected from the plane he would still have the horizontal motion of the plane; in fact, he would travel *right along with the plane* if it were not for air resistance and the downward pull of the Earth. This law explains why it is that objects on the Earth are not left behind by its rotation and its revolution around the sun.

The potential value of this law to the motion of projectiles is obvious enough, and Galileo soon capitalized on it. While study-ing the motion of projectiles he observed that an object's motion can result from two *independent* simultaneous motions. The meaning of this discovery can be clarified by an example. An object dropped from an aeroplane flying horizontally possesses two motions. In accordance with the law just described one is straight out in the same direction as the plane is going; this motion takes place at the velocity of the plane. The other motion is straight down. The combination of these two simultaneous motions causes the object to travel downwards along a curve which, as Galileo pointed out, is part of a parabola. However, the horizontal and vertical motions of the falling object are independ-ent of each other. If the plane were travelling faster, the horizon-tal motion of the object would be faster while the downward motion would be the same. Hence the object would take *the same time* to reach the ground as it did before, though it would travel farther horizontally before reaching the ground. Thus though the object might leave the plane at the point *O* in figure 48, it would hit the ground at the point *Q* when the plane's speed is greater rather than at the point *P* when the speed is less, but the time required to reach *P* or *Q* would be the same.

Galileo applied this principle of simultaneous independent motions to the motion of a cannon ball and proved that the path in this case too is part of a parabola and that the greatest range is obtained by firing the ball at an angle of 45° to the ground.

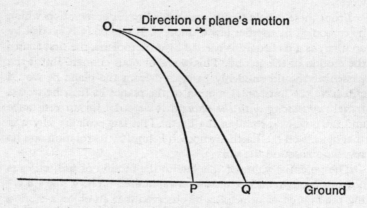

Figure 48. Two freely falling objects with different horizontal velocities reach the ground in the same time

All these results and many others were expounded by Galileo in his *Discourses and Mathematical Demonstrations Concerning Two New Sciences,* a masterpiece on which he laboured for more than thirty years. With this book Galileo launched modern physical science on its mathematical course, founded the science of mechanics, and set the pattern for modern scientific thought. Unfortunately, by the time the manuscript was ready to be published Galileo was out of favour with the Church and publication of any work by him was forbidden. He then had to arrange secretly for its publication in Holland and to pretend that he had had nothing to do with the printing. He maintained that a copy of the manuscript had fallen by chance into the hands of the Dutch publishers, who proceeded without his permission. Galileo died a few years after the publication in 1638, and with him the independent spirit of Italian thought died also.

XIV

The Deduction of Universal Laws

> I wish that old Copernicus could see
> How, through his truth, that once dispelled a dream,
> Broke the false axle-trees of heaven, destroyed
> All central certainty in the universe,
> And seemed to dwarf mankind, the spirit of man
> Laid hold on law, . . .
> And mounting, slowly, surely, step by step,
> Entered into its kingdom and its power.*
>
> *Alfred Noyes*

FORTUNATELY for science and mathematics, in a country with a freer intellectual atmosphere than Italy's a worthy successor to Galileo was born. In 1642, the very year of Galileo's death, on a farm located in a secluded English hamlet, a woman recently widowed gave birth to a frail and premature child. From such an insignificant origin and with a body so weak that his life was despaired of, Isaac Newton lived to be eighty-five and to acquire fame as great as any man's.

Except for a strong interest in mechanical contrivances Newton, like many geniuses, showed no special promise as a youth. For the negative reason that he showed no interest in farming his mother sent him to Cambridge. Despite several advantages of attendance there, such as the opportunity to study the works of Copernicus, Kepler, and Galileo and the opportunity to listen to the famous mathematician Isaac Barrow, Newton seemed to profit little. He was even found to be weak in geometry and at one time almost changed his course of study from natural philosophy to law. Four years of study ended as

* Reprinted from *Watchers of the Sky* by Alfred Noyes, copyright by Alfred Noyes, 1922, with the permission of the publisher J. B. Lippincott Company, N.Y., Mr Alfred Noyes and the publisher Wm Blackwood & Sons Ltd, Edinburgh and London.

unimpressively as they began and Newton returned home – to study.

This quiet and unobtrusive intellect burst forth brilliantly when, between the ages of twenty-three and twenty-five, Newton made three gigantic steps which secured his reputation and advanced modern science enormously. The first was the discovery of the secret of colour which he arrived at by decomposing white light; the second was the creation of the calculus, which we shall discuss later; and the third was his proof of the universality of the law of gravitation.

Had he announced any one of these achievements to the scientific world, he would have earned enduring fame at once; but Newton said nothing about them. When a plague which had been raging in London abated he returned to Cambridge to secure his master's degree and then became a fellow. When he was twenty-seven his teacher, Barrow, resigned and Newton, now recognized at least as a serious student of mathematics, was appointed in his place. His success as a lecturer did not parallel his success in research. At times no one attended. The original material he presented was not even noticed, much less acclaimed.

He finally published his work on the composite nature of white light, accompanying it with a presentation of his philosophy of science. Both the philosophy and the work on light were criticized and some scientists rejected both *in toto*. Newton was disgusted and resolved to refrain from further publication. When, several years later, he broke this resolution to announce further discoveries, he became embroiled in scientific controversies and arguments over priority of discovery, which confirmed his inclination to keep his research to himself. Were it not for the urging and financial assistance of the astronomer Edmond Halley, the *Mathematical Principles of Natural Philosophy* (1687), which embodied the fruit of Newton's work, would never have been published.

After the publication he finally did receive widespread acclaim. The *Principles* went through many editions, and popularizations became common. By 1789, forty editions had appeared in English, seventeen in French, eleven in Latin, three in German, and at least one in Portuguese and Italian. Among the popularizations was one entitled *Newtonianism for Ladies* which also went through many editions. Actually the *Principles* needed popularization, for the book is extremely difficult to read and is not at all clear to laymen, despite statements by educators to the contrary. The greatest mathematicians worked for a century to elucidate fully the material of the book.

Newton's fame spread until it became comparable to Einstein's today. Newton gave due credit to his predecessors: 'If I have seen a little farther than others it is because I have stood on the shoulders of giants.' Nor did he feel that his work was of incomparable importance: 'I do not know what I may appear to the world; but to myself I seem to have been only like a boy playing on the seashore, and diverting myself in now and then finding a smoother pebble or a prettier shell than ordinary, whilst the great ocean of truth lay all undiscovered before me.'

Of the great contributions of his youth, Newton's philosophy of science and his work on gravitation are most relevant to our present subject. The philosophy stated more explicitly the programme for science which Galileo had initiated: From clearly verifiable phenomena laws are to be framed that state nature's behaviour in the precise language of mathematics. By the application of mathematical reasoning to these laws new ones are to be deduced. Like Galileo, Newton wished to know *how* the Almighty had fashioned the universe but he was not presumptuous enough to inquire towards what end, nor did he hope to fathom the mechanism behind many phenomena. He said: 'To tell us that every species of things is endowed with an occult specific quality by which it acts and produces manifest effects, is to tell us nothing: *But to derive two or three general principles of motion from phenomena, and afterwards to tell us how the properties and actions of all corporeal things follow from those manifest principles, would be a very great step in philosophy* [science] *though the causes of those principles were not yet discovered:* and therefore I scruple not to propose the principles of motion above mentioned, they being of very general extent, and leave their causes to be found out.'

* The italics are Newton's.

In this task of describing nature Newton's most famous contribution was to unite heaven and Earth. Galileo had viewed the heavens as no man had previously been able to, but his successes in describing nature mathematically were limited to motions on or near the surface of the Earth. During Galileo's lifetime his contemporary, Kepler, had obtained his three famous mathematical laws on the motions of the heavenly bodies and had thereby clinched the argument for the heliocentric theory. Thus while one scientist was building the science of earthly motions, the other perfected the theory of the heavenly motions. The two branches of science seemed to be independent of each other. The challenge to find some relationship between them stirred the great scientists. It was met by the greatest one.

There was good reason to believe that some unifying principle did exist. Under Galileo's first law of motion bodies should continue to move in straight lines unless disturbed by forces. Hence the planets, set into motion somehow, should move in straight lines whereas, according to Kepler, they moved in ellipses around the sun. Some force must therefore be acting so as to deflect the planets continually from straight-line paths, just as a weight swung at the end of a string does not fly off in a straight line because the hand exerts a force pulling it in. Presumably the sun itself was acting as an attracting force on the planets. The scientists of Newton's day also appreciated the fact that the Earth attracts bodies to it. This attraction accounted for the fall to Earth of a body released from the hand; otherwise, since the body receives no force from the hand it would, according to the first law of motion, remain suspended in air. Since both the Earth and the sun attract bodies, the idea of unifying both actions under one theory was advanced and discussed even in Descartes' time.

Newton converted a common thought into a mathematical problem and, without determining the physical nature of the forces involved, solved this problem by brilliant mathematics. He was able to show that the very same mathematical formula describes the action of the sun on the planets as well as the action of the Earth on objects near it. Because the same formula described both classes of phenomena he concluded that the same force operated in both cases. Story has it that the identity of the Earth's pull on objects and the sun's pull on the Earth was brought to Newton's attention by the fall of an apple from a tree. The mathematician Gauss, however, said that Newton told this story to dispose of stupid persons who asked him how he discovered the law of gravitation. At any rate, this apple, unlike another that played a role in history, improved the status of man.

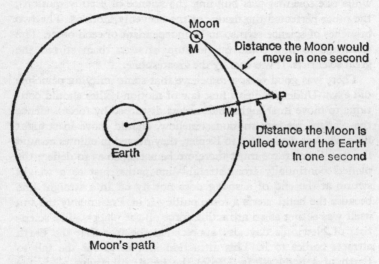

Figure 49. The gravitational effect of the Earth on the moon

Newton's reasoning in showing that the same formula applied to heavenly and earthly bodies is now classic. We shall consider a somewhat crude version of it which may nevertheless give the essence. The path of the moon around the Earth is roughly a circle. Since the Moon, M in fig. 49, does not follow a straight-line path such as MP, it evidently is pulled towards the Earth by some force. If MP were the distance the moon might have moved in one second with no gravitational force acting, then PM' is the distance the moon is pulled towards the Earth during that second. Newton used PM' as a measure of the

Earth's attractive force on the moon. The corresponding quantity in the case of a body near the surface of the Earth is 16 feet, for a body dropped from the hand is pulled 16 feet towards the Earth in the first second. Newton wished to show that the same force accounted for both PM' and the 16 feet.

Rough calculation had led him to believe that the force attracting one body to another depends on the square of the distance between the centres of the two bodies involved, and that this force decreases as the distance increases. The distance between the centre of the moon and the centre of the Earth is about 60 times the radius of the Earth. Hence the effect of the Earth on the moon should be $1/(60)^2$ of its effect on a body near the surface of the Earth, that is, the moon should be pulled towards the Earth $1/(60)^2$ of 16 or ·0044 feet each second. By using some numerical results obtained by means of the trigonometric ratios, Newton found that the moon is pulled towards the Earth by just that amount each second. Thus he had obtained a most important piece of evidence to the effect that *all* bodies in the universe attracted each other in accordance with the same law.

More extensive investigations showed Newton that the precise formula for the force of attraction between *any* two bodies is given by the formula

$$(1) \qquad\qquad F = kMm/r^2,$$

where F is the force of attraction, M and m are the masses of the two bodies, r is the distance between them, and k is the same for *all* bodies. For example, M could be the mass of the Earth and m, the mass of an object near or on the surface of the Earth. In this case r is the distance from the centre of the Earth to the object. Formula (1) is, of course, the law of gravitation.

Having obtained the correct form of this law by studying the motion of the moon, Newton showed next that the law could be applied to motions on or near the Earth. According to this law the Earth attracts each body. We feel this pull of the Earth on a body when we hold it. When M is the mass of the Earth and m is the mass of the body, then F in formula (1) measures the pull of the Earth on the body or the *weight* of the body. We should note, then, that *weight* is a force, whereas *mass* is a quality of objects that concerns resistance to change in motion.

Newton was careful to distinguish between these two related properties of matter, that is, mass and weight. While the mass of a body is constant, its weight may vary. For example, if the distance of a body from the centre of the Earth is altered, the weight of the body is altered. Specifically, if a body of mass m is taken up 4000 miles above the Earth, its distance from the centre of the Earth is doubled. Now, if r in formula (1) represents its original distance from the centre, $2r$ represents its new distance. To calculate the weight of the mass at the new position we replace r by $2r$. The denominator in formula (1) becomes $(2r)^2$ or $4r^2$. Hence F would be only a fourth as much as when the body is on the Earth. That is, a body of mass m would weigh only a fourth as much 4000 miles above the Earth as it does on the surface of the Earth. To summarize, we have been demonstrating that though the mass of an object remains constant, its weight can be varied by altering its distance from the centre of the Earth.

Consider another consequence of formula (1). Let M be the mass of the Earth and m the mass of an object near the surface. If we rewrite formula (1) as

$$F = \frac{kM}{r^2}\, m,$$

and divide both sides of this equation by m we obtain

(2) $$\frac{F}{m} = \frac{kM}{r^2}.$$

Now regardless of what object near the Earth's surface we consider, the quantities on the right side of formula (2) are the same, because r is about 4000 miles, M is the mass of the Earth, and k is the same for all bodies. Hence for *any* object near the surface of the Earth, the ratio F/m, that is, the ratio of weight to mass, is constant. Thus the two distinct properties of matter are related quantitatively in a very simple way. An explanation of this surprising relationship was not known until the theory of relativity was created. Since we almost always deal with objects near the surface of the Earth we are misled by this constant relationship between mass and weight and often confuse the two. For example, if we try to start an automobile by pushing it, we are inclined to attribute the need for force to the weight of the automobile. Actually, it is the mass exhibiting its resistance to change in motion.

From the second law of motion and the law of gravitation we can deduce still another consequence. The second law of motion says that any force acting on a body of mass m gives the body an acceleration. In particular, the force of gravity exerted by the Earth on a body should give it an acceleration. But the force of gravity is

$$(3) \qquad F = \frac{kMm}{r^2},$$

whereas the relation of any force to the acceleration it causes is

$$(4) \qquad F = ma.$$

When the force F in formula (4) is that of gravity we can equate the right sides of formulas (3) and (4) because the left sides are then equal; that is,

$$ma = \frac{kMm}{r^2}.$$

We may divide both sides of this last equation by m and obtain

$$(5) \qquad a = \frac{kM}{r^2}.$$

This result says that the acceleration imparted to an object by the force of gravity of the Earth is always kM/r^2. Since k is a constant, M is the mass of the Earth, and r is the distance of an object from the centre of the Earth, the quantity kM/r^2 is the same for *all* bodies near the surface of the Earth. Hence all such bodies fall with the same acceleration. This, of course, is the result Galileo had already obtained by inference from experiments and from this result he proved mathematically that all bodies falling from the same height reach the ground in the same amount of time. Incidentally, the value of a is readily measured and is 32 feet per second each second.

Many more fascinating results can be obtained from the laws of motion and gravitation. To illustrate the power of mathematical reasoning we shall derive one more conclusion: Let us calculate the mass of the Earth. For this purpose we need the value of k, the constant of universal gravitation, which appears in formula (1). Since this quantity is always the same, regardless of which masses appear in formula (1), it can be obtained in a laboratory by using known masses m and M, a known distance r between them, and by measuring the force of attraction F between them. Hence k, the only unknown in the formula, can then be calculated. This experiment was performed by many physicists, the most famous of whom was Henry Cavendish (1731–1810). He came to the conclusion that k is the extremely small quantity 6.67×10^{-8}, or 6·67 divided by one hundred million, assuming that measurements are made in centimetres, grams, and seconds.

We may now use formula (5) wherein k is the quantity discussed above, M is the mass of the Earth, r is the radius of the Earth, and a is the acceleration of an object near the Earth. Since all these quantities except M are now known, we can calculate M. The result is $M = 6 \times 10^{27}$ grams, that is, 6 multiplied by 27 zeroes, or 6.6×10^{21} tons of mass.

An interesting by-product of this calculation is some information about the composition of the Earth. Since the radius of the Earth is known, its volume, assuming it is exactly spherical in shape, can be calculated from the formula for the volume of a sphere, $V = \frac{4}{3}\pi r^3$. Now the mass of a cubic foot of water can be measured and hence the mass of the Earth, were it composed entirely of water, could be calculated. The actual mass of the Earth given above is about 5½ times the mass it would have were it composed entirely of water. Hence geologists conclude that the interior of the Earth must be composed of heavy minerals.

Thus far Newton's contributions to the theory of gravitation may be summarized as follows. By studying the motion of the moon, he had inferred the correct form of the law of gravitation. He then showed that this law and the two laws of motion sufficed to establish valuable knowledge about the motions of objects on the Earth. He had therefore achieved one of the major goals in Galileo's programme because he had shown that the laws of motion and gravitation were fundamental. Like the axioms of Euclid, they served as the logical basis for other valuable laws. What a triumph indeed it would be to deduce in addition the laws of motion for the heavenly bodies.

This triumph was also reserved for Newton. A truly momentous series of deductions made by him showed that all three of Kepler's laws follow from the two basic laws of motion and the law of gravitation. We shall give the essence of one of these derivations, again with the purpose of illustrating the power of mathematics to obtain knowledge of the physical world by the deductive process. The derivation we shall present will be a somewhat simplified version of Newton's actual work, for we shall suppose the path of each planet to be circular rather than elliptical. Newton himself did treat the elliptical path but there is no need for us to follow this more difficult demonstration.

Kepler's third law states that the square of the time of revolution of any planet is proportional to the cube of its mean distance from the sun. As a formula this law is written $T^2 = KD^2$, wherein T is the time of revolution or length of the planet's year, D is the planet's mean (or average) distance from the sun, and K is a constant, that is, it is the same for *all* the planets. To derive Kepler's third law we shall need one more fact about motion, which is in itself easy to prove but which is aside from our main point. An object that moves in a circle is subject to some force which causes it to depart from the straight-line path Newton's first law of motion says it should otherwise follow. A measure of this force, commonly called the centripetal force, is given by the formula

$$(6) \qquad F = \frac{mv^2}{r},$$

wherein m is the mass of the body, v is its velocity, and r is the radius of the circular path. Such a force acts on each planet and is due to the gravitational pull of the sun. Formula (6), however, is a correct expression for centripetal force whether or not it arises from gravitation.

To proceed with the derivation of Kepler's law, we notice first that the velocity of a planet, assuming it travels at a constant speed along a circular path, is given by the circumference of the circle divided by the time of revolution. That is,

$$(7) \qquad v = \frac{2\pi r}{T}.$$

If we substitute this value of v in formula (6) we obtain an expression for the centripetal force F acting on a planet, namely,

$$(8) \qquad F = \frac{m}{r} \left(\frac{2\pi r}{T} \right)^2 = \frac{m}{r} \frac{4\pi^2 r^2}{T^2} = \frac{m 4\pi^2 r}{T^2}.$$

Now this centripetal force F is due to the gravitational force exerted by the sun whose mass we denote by M. That is,

$$(9) \qquad F = \frac{kmM}{r^2}.$$

By equating the two forces given in formulas (8) and (9) we obtain

$$(10) \qquad \frac{kmM}{r^2} = \frac{m 4\pi^2 r}{T^2}.$$

We may divide both sides of this equation by m, thus cancelling this factor on each side. If we multiply both sides by $T^2 r^2$ and divide by kM we obtain

$$(11) \qquad T^2 = \frac{4\pi^2}{kM} r^3.$$

We now observe that M, the mass of the sun, and k, the gravitational constant, do not change in this derivation no matter which planet, m, we consider. Hence, the quantity of $4\pi^2/kM$ is a constant and we shall denote it by K. Writing D for r, we may say that

$$(12) \qquad T^2 = KD^3,$$

and this result is Kepler's third law. Thus the famous planetary laws that Kepler obtained only after years of observation and trial and error can be proved in a matter of minutes by means of Newton's laws.

There is an important corollary to these laws that should be informative to the lay reader who seeks an explanation of the power of mathematics. The major value of the Newtonian laws lies, as we have just seen, in the fact that they apply to so many varied situations on heaven and Earth. The same quantitative relationships epitomize characteristics common to all. Hence knowledge of the formulas really represents knowledge about all the situations encompassed by the formulas. The person who looks at a mathematical formula and complains of its abstractness, dryness, and uselessness has failed to grasp its true value.

The work of Galileo and Newton was not the end but the beginning of a programme for science. Newton himself formulated the programme in the preface to his *Mathematical Principles of Natural Philosophy*, the classic which contains the work of his brilliant youth:

We offer this work as mathematical principles of philosophy [science]; for all the difficulty in philosophy seems to consist in this – from the phenomena of motions to investigate the forces of nature, and then from these forces to demonstrate the other phenomena; and to this end the general propositions in the first and second book are directed. In the third book we give an example of this in the explication of the system of the world; for by propositions mathematically demonstrated in the first book, we there derive from the celestial phenomena the forces of gravity with which bodies tend to the sun and the several planets. Then, from these forces, by other propositions which are also mathematical, we deduce the motions of the planets, the comets, the moon and the sea. I wish we could derive the rest of the phenomena of nature by the same kind of reasoning from mechanical principles; for I am induced by many reasons to suspect that they may all depend upon certain forces by which the particles of bodies, by some causes hitherto unknown, are either mutually impelled towards each other, and cohere in regular figures, or are repelled and recede from each other.

Like a rock rolling down a steep hill the movement to secure fundamental mathematical laws and to deduce their consequences gathered momentum and finally caused an avalanche. By procedures similar to those illustrated in this chapter, the mass of the sun and the mass of any planet with observable satellites were calculated. The idea of centrifugal force, the force that opposes the centripetal force discussed above, was applied to the motion of the Earth and produced the magnitude of the equatorial bulge of the Earth as well as the consequent variation of the weight of an object from point to point on the Earth's surface. From the knowledge of the observed departure from sphericity of the several planets it became possible to calculate their periods of rotation. The tides were shown to be caused by the gravitational attractions of the sun and moon. The paths of comets were computed and their reappearance predicted accurately. Also their sudden sweep past the Earth was explained as owing to the great eccentricity of their elliptical orbits. Incidentally, this mathematical work on the behaviour of the comets convinced people that the comets were legal members of a lawful, designed universe rather than visitations from God intended to strike terror into the hearts of men or to smash the Earth into bits. At the same time it gave indisputable evidence of the mathematical behaviour of nature and of the power of the quantitative approach.

The success of the search for laws extended far beyond the field of astronomy. The phenomenon of sound studied as a motion of molecules in air yielded now famous mathematical laws. Hooke measured the elasticity of solids. Boyle, Mariotte, Galileo, Torricelli, and Pascal measured the pressure and density of fluids and gases. Van Helmont used the balance to weigh substances, an important step in the direction of modern chemistry, and, with Hales, began quantitative studies in physiology such as the measurement of body temperature and blood pressure. Harvey proved by quantitative arguments that the blood pumped from the heart made a complete circuit of the body before returning to the heart. Quantitative studies extended to botany, too, where the rate of absorption and evaporation of water by plants was determined. Römer measured the velocity of light. The cold of winter and the heat of summer were found to be less or more excited motions of air molecules attracting each other according to the law of gravitation. Soon laws binding together separate branches of science were discovered. For example, chemistry, electricity, mechanics, and heat phenomena were all bound together by the law of conservation of energy.

All this was only the beginning of the vast and unparalleled scientific movement that has fashioned the modern world. The course of the movement continued to support Newton's conviction as to the possibility of deriving all the phenomena of nature from the laws of motion and gravitation. One or two examples chosen from the superlative accomplishments of the eighteenth century will indicate the extent to which his programme was carried.*

Though evidence was overwhelmingly in favour of the invariable, mathematical order of the heavens by the time of Newton's death in 1727, a number of irregularities in the motions of the heavenly bodies had been observed and were unaccounted for. For example, although the moon always presents the same face to the Earth, more or less of the region near the edges becomes periodically visible. In addition, increased observational accuracy had revealed that the length of the mean lunar month decreases by about $\frac{1}{30}$ of a second per century (such was the order of accuracy that observation and theory had come to handle). Finally, small changes in the eccentricities of the planetary orbits had also been observed.

These and other departures from perfect law and order added up to one large question: Is the solar system stable? That is, would these irregularities, small as they were, gradually increase and, by virtue of the complicated effects of the heavenly bodies on each other, tend to unbalance the solar system? Why should not a planet, under the cumulative effect of these irregularities, wander off into space some day, or why should not the Earth at some future date crash into the sun?

Newton was well aware of many of these irregularities, and in his own studies he had tackled the motion of the moon. This body follows an elliptical path, somewhat as a drunken man follows a straight line. It hurries and lingers, and reels from side

* See also Chapters xix and xx.

to side. Newton was convinced that some of this extraordinary behaviour was due to the fact that the sun as well as the Earth attracts the moon and causes departures from a truly elliptical path. However, since he had no proof that all the observed irregularities in the motions of the moon and the planets were due to gravitational pulls and since he could not show that the cumulative effect would not ultimately disrupt the solar system, he felt obliged to call on God's intervention to keep the universe functioning. But Newton's eighteenth-century successors decided to rely less on God's will and more on their own powers of deduction.

The path of each planet around the sun would be an ellipse only if the one planet and the sun were in the heavens. But the solar system contains eight planets, many with moons, all not only moving around the sun but attracting each other in accordance with Newton's universal law of gravitation. Their motions, therefore, certainly could not be truly elliptical. Their exact paths would be known if it were possible to solve the general problem of determining the motion of an arbitrary number of bodies, each attracting all the others under the action of gravitation. But this problem is beyond the capacity of any mathematician. Two of the greatest mathematicians of the eighteenth century did, however, make phenomenal steps along these lines. The Italian-born Joseph Louis Lagrange, in a brilliant exhibition of youthful genius, tackled the mathematical problem of the moon's motion under the attraction of the sun and Earth and solved it at the age of twenty-eight. He showed that the variation in the portion of the moon that is visible is caused by the equatorial bulges of both the Earth and the moon. In addition, the attraction of the sun and moon on the Earth was shown to perturb the Earth's axis of rotation by calculable amounts; and thus the wandering of the Earth's axis of rotation with the consequent precession of the equinoxes, an observational fact known at least since Greek times, was shown to be a mathematical necessity of the law of gravitation. Lagrange made another notable step in his mathematical analysis of the motions of the moons of Jupiter. The analysis showed that the observed irregularities there too were an effect of gravitation. All these results he incorporated in his *Mécanique analytique,* a work which extended, formalized, and crowned Newton's work on mechanics. Lagrange had once complained that Newton was a most fortunate man in that there is but one universe and Newton had already discovered its mathematical laws. However, Lagrange had the honour of making apparent to the world the perfection of the Newtonian theory.

The Frenchman Pierre Simon Laplace, who, like Lagrange, showed genius in his youth, devoted his life to the problem of applying Newton's law of gravitation to the solar system. One of Laplace's spectacular achievements was the proof that the irregularities in the eccentricities of the elliptical paths of the planets were periodic. That is, these irregularities would oscillate about fixed values and not become larger and larger and so disrupt the orderly motions of the heavens. In brief, the universe is stable. This result Laplace proved in his epochal work, the *Mécanique céleste*, which he published in five volumes over a period of twenty-six years.

The perfection of the mathematical order of the universe was, by the time of Laplace's death (exactly one hundred years after Newton's) now clearly evident. It reflected itself in the famous reply of Laplace to Napoleon who, on receiving a copy of the *Mécanique céleste*, chided Laplace for writing a work on the system of the universe that did not mention God. Laplace's reply was: 'I have no need of this hypothesis.' The world was stable and God was no longer required, as He had been by Newton, to correct its irregularities or to prevent errant behaviour.

One remarkable deduction from the general astronomical theory of Lagrange and Laplace is especially worth mentioning. This was the purely theoretical prediction of the existence and location of the planet Neptune. It had been conjectured that unexplained aberrations in the motion of the planet Uranus were due to the gravitational pull on Uranus of an unknown planet. Two astronomers, John Couch Adams in England and U. J. J. Leverrier in France, used the observed irregularities and the general astronomical theory to calculate the orbit of the supposed planet. Observers were then directed to search for the planet at the time and place determined mathematically by Adams and Leverrier. The planet was located. It was barely observable with the telescopes of those days and would hardly have been noticed if astronomers had not been looking for it at the predicted location. The problem Adams and Leverrier solved was an extremely difficult one because they had to work backwards, so to speak. Instead of calculating the effects of a planet whose mass and path were known, they had to deduce the mass and path of an unknown planet from its effects on the motion of Uranus. Their success was therefore regarded as a great triumph of theory and widely proclaimed as final proof of the universal application of Newton's law of gravitation.

By the middle of the eighteenth century the infinite wisdom of Galileo's and Newton's quantitative approach to nature was clearly established. Had they undertaken the perhaps unsolvable problem of analysing matter and forces qualitatively, they might have advanced science no farther than the medievalists did. The problem of the structure of matter is highly complex; indeed modern research in atomic theory is beginning to make us aware of the almost unbelievable degree of this complexity. Galileo and Newton avoided discussion of the structure of matter but showed how to measure its inertial and gravitational properties in terms of acceleration, which means in terms of distance and time. The force of gravity has also defied qualitative analysis. In fact, Newton admitted that the nature of this force was a mystery to him. Just how it could reach out 93 million miles and pull the Earth towards the sun seemed inexplicable to him and he framed no hypotheses concerning it. He hoped that others would study the nature of this force. People did try to explain it in terms of pressure exerted by some intervening medium and by other processes, all of which proved unsatisfactory. Later, all such attempts were abandoned and gravitation was accepted as a 'common unintelligibility'. But despite total ignorance about the physical nature of gravity, Newton did have a quantitative formulation of how it acted, and this was both significant and usable. The paradox of modern science is that though it is content with seeking so little it accomplishes so much.

There are other vital implications of Galileo's and Newton's work.* Copernican theory had brushed away some of the mysticism, superstition, and theology which veiled the heavens and

* See Chapters XVI, XVII, XVIII, and XXI.

enabled man to view them in a more rational light. Newton's law of gravitation cleared out the cobwebs from the corners, for it showed that the planets follow the same pattern of behaviour as do the familiar objects moving on the Earth. This fact provided additional and overwhelming evidence for the conclusion that the planets are composed of ordinary matter. The identification of the stuff of heaven with the crust of Earth wiped out libraries of doctrines on the nature of heavenly bodies. In particular, the distinction affirmed by the great Greek and medieval thinkers between the perfect, unchangeable, and incorruptible heavens and the decaying, imperfect Earth was now shown all the more clearly to be a figment of men's imaginations.

Over and above the identification of the Earth and the heavenly bodies, Galileo's and Newton's work established the existence of universal, mathematical laws. These laws described both the behaviour of a speck of dust and the most distant star. No corner of the universe was outside their range. Thus the evidence for the mathematical design of the universe was immensely strengthened. Moreover, the unvarying adherence of natural phenomena to the pattern of these laws spoke for the uniformity and invariability of nature and opposed the medieval belief in an active Providence to whose will the universe was continually subject.

The seventeenth century found a qualitative world subject to divine will and understood only in terms of the ways and purposes of the Creator. It bequeathed to mankind a mechanical universe operating unfailingly in accordance with invariable, universal mathematical laws. It will become clearer as we proceed that the change inaugurated in that period was no less than a cultural revolution.

There is a lesson to be learned from reviewing the major steps leading to this intellectual upheaval. The study of the heavens provided the first great scientific synthesis in the form of the astronomical theory of Eudoxus. This was followed by the quantitative, practically useful, and highly influential system of Hipparchus and Ptolemy. Further study of the heavens produced the revolutionary astronomy of Copernicus and Kepler. On the basis of a heliocentric theory the universal law of gravitation became a tenable hypothesis. The validity of the law was further attested to by the deduction from it of Kepler's laws. Finally, the astronomical work of Lagrange and Laplace removed all doubts about the reign of universal mathematical laws in nature. The lesson to be gathered from this history is that the curious stargazer can tell us more about our world than the practical 'man of affairs'. Our best knowledge of the behaviour of even those natural phenomena pervading our immediate environment has come from the contemplation of the heavens and not from the pursuit of practical problems. The sense of law that predisposes men to attribute all phenomena, even completely inexplicable ones, to regular rather than to abnormal behaviour of nature, this habit of substituting law for supernatural intervention, was developed by looking *away* from man's immediate problems and by studying the motion of the most distant stars.

The work of Copernicus, Kepler, Galileo, and Newton made possible the realization of many dreams. There was the dream and hope of ancient and medieval astrologers of anticipating nature's ways. There was also the plan that Bacon and Descartes had advanced of mastering nature for the improvement of human welfare. Man progressed towards both goals, the scientific and the technological. The universal laws certainly made possible prediction of the phenomena they comprised. And mastery is but a step away from prediction, for knowing the unfailing course of nature makes possible the employment of nature in engineering devices.

Still another programme for probing and understanding nature found fulfilment in the work of Galileo and Newton. The Pythagorean-Platonic philosophy that number relations are the key to the universe, that all things are known through number, is an essential element in the Galilean scheme of relating quantitative aspects of phenomena through formulas. This philosophy was kept alive throughout the Middle Ages though most often, as with the Pythagoreans themselves, it was part of some larger mystical theory of creation, with number as the form and cause of all created objects. Galileo and Newton divested the Pythagorean doctrine of all mystical associations and re-clothed it in a style that set the fashion for modern science.

XV

Grasping the Fleeting Instant:
The Calculus

When Newton saw an apple fall, he found
   In that slight startle from his contemplation –
'Tis said (for I'll not answer above ground
   For any sage's creed or calculation) –

A mode of proving that the earth turn'd round
   In a most natural whirl, called 'gravitation';
And this is the sole mortal who could grapple,
   Since Adam, with a fall, or with an apple.

Lord Byron

THE derivation of universal laws undoubtedly had to await an age disposed to think in such terms and leaders such as Descartes, Galileo, and Newton who could fashion the goals and methods of modern scientific activity. But it also had to await, and indeed would have been impossible without, the creation of an indispensable tool – the calculus. Of all the veins of thought explored by geniuses of the seventeenth century, this one proved to be the richest. Over and above its value in the derivation of many of the universal laws already discussed, the calculus provided the wealth to found many new scientific enterprises.

Contrary to the popular belief that genius breaks radically with its age, three of the greatest seventeenth-century minds, Pierre Fermat, Isaac Newton, and Gottfried Wilhelm Leibniz, each working independently of the other, became absorbed in the problems of the calculus. Fermat did his work in France, Newton in England, and Leibniz in Germany. This third member of the triumvirate of genius, who is new to our story, was born in Leipzig in 1646. At the age of fifteen he entered the University of Leipzig for the announced purpose of studying law and with the unannounced intention of studying everything. An essay on law written soon after he left Leipzig attracted the attention of the Elector of Mainz who thereupon decided to employ Leibniz as a diplomat. Unfortunately, his time for study during this period was limited because poverty forced him to continue to serve as errand boy extraordinary for German princes. In 1676 he was appointed councillor and librarian to the Elector of Hannover, and this job, though still requiring him to travel a great deal on diplomatic missions, allowed him some leisure. In his spare time, then, he managed to write articles, essays, and letters which fill more than twenty-five volumes with profound contributions to law, religion, politics, history, philosophy, philology, logic, economics, and, of course, science and mathematics. This man of universal gifts and interests has been called 'a whole academy in himself'.

Numerous capable mathematicians had already made progress in the direction of the calculus. The work of Fermat, Newton, and Leibniz was, therefore, the continuation and the culmination of a long series of efforts on the part of their predecessors. Apparently, no matter how great are the contributions of individual genius, the spirit and substance of its thoughts are confined to its own age. The contribution of genius is to sense and fructify the ideas which the particular age stirs up. It makes capital of society's cogitations and returns dividends for centuries thereafter.

Whatever conclusions may be drawn about the relation of genius to its age, there is no doubt that the concepts of the calculus were in the seventeenth-century air, so much so, in fact, that a quarrel arose between the friends of Newton and those of Leibniz over whether a breeze from England had not carried Newton's ideas to Leibniz. The feelings engendered by the quarrel were so bitter and the leading thinkers in this most rational of subjects were so partisan that the English and Continental mathematicians stopped the interchange of ideas and correspondence for about a hundred years after the deaths of Newton and Leibniz. Nor was the language used by either side to comment on the work of the other always sober and rational, or even polite. One exception to this state of affairs was a very generous remark by Leibniz, namely, that if one took mathematics from the beginning of the world to the time when Newton lived what the Englishman did was much the better half.

During the period in which Fermat, Newton, and Leibniz worked, the mathematicians of Europe were united in seeking to solve a whole group of problems involving a very special type of difficulty – the instantaneous rate of change of variables. Before examining the decisive contributions of the three men we must be clear about the nature of the problem they faced.

In dealing with variables, that is, quantities which change continually, it is necessary to distinguish between *change* and *rate of change*. As a bullet travels through the air, the distance and time it travels are continually increasing; at the instant it strikes a person, however, what is important is its speed, or rate of change of distance compared to time, and not the distance and time it has travelled. If that speed is one mile per hour, the bullet will drop harmlessly to the ground at the person's feet. If it is one thousand miles per hour, the person will drop to the ground safe from future harm. Obviously, the rates of change of varying quantities are at least as significant as the fact that they are changing.

Among rates of change of variables we must distinguish two kinds: *average* and *instantaneous*. If a person motors from New York to Philadelphia, a distance of ninety miles, in three hours, his average speed, that is, his average rate of change of distance compared to time, is 30 miles per hour. This number, however, obviously does not necessarily represent his speed at any particular instant of time during the journey, say at 3 o'clock. Suppose now that at this instant, exactly 3 o'clock, the traveller looks at the speedometer of his automobile and notices that it reads 35 miles per hour. This quantity is an *instantaneous* speed: that is, it is his rate of change of distance compared to time at 3 o'clock, but not necessarily the speed at any instant before or after. We might argue that there is no such thing as speed at an instant because no time elapses at an instant and hence there can be no motion. At present, we shall simply appeal to our physical experience to support the assertion that a person travelling in an automobile is moving at a definite speed at each instant. A collision with a tree at any one of these instants would surely convince the doubting reader.

The need to deal with instantaneous speed arises primarily when an object moves with varying speeds; otherwise the concept of average speed suffices. Now varying speeds are precisely what the seventeenth-century scientists encountered. Kepler's second law, for example, states that a planet moves, not at a constant speed as the Greeks and other pre-Renaissance scientists had believed, but at a continually varying speed. Similarly, according to Galileo, bodies rising or falling near the surface of the Earth travel at continually varying speeds. Pendulum and projectile motions, which were intensively studied at that time, also involve varying speeds. To treat such motions the scientists lacked a clear understanding of instantaneous speeds and, in addition, some method of calculating them.

It should be understood that we cannot obtain an instantaneous speed as we obtain an average speed, for at an instant zero distance is traversed and zero time elapses and to divide zero by zero is meaningless. A little thought about the matter will convince the reader that only an unusual solution of the problem of defining and calculating instantaneous speed could succeed. To this problem, Fermat, Newton, and Leibniz applied their genius.

Let us consider first a simplified description of their mathematical approach. We have already agreed that if an automobile leaves New York city at 2 p.m. and arrives in Philadelphia at 5 p.m., its average speed for the trip is the distance it travels, 90 miles, divided by the time it takes to travel this distance, 3 hours; that is, the average speed is 30 miles per hour. What can we say about the speed at 3 o'clock? It is clear that although the average speed is 30 miles per hour, the speed at 3 o'clock may have been 40 miles per hour or almost any other number. We can attempt to answer the question by considering the average speed for a brief period of time around 3 o'clock. Thus if the automobile travels ·6 miles in the minute following 3 o'clock the average speed for this minute is ·6 miles divided by 1 minute or 36 miles per hour. Is this the average speed at 3 o'clock?

Although one minute is a fairly short interval of time, it is still possible that the average speed for this minute may differ considerably from the speed at exactly 3 o'clock because the automobile could increase or decrease its speed during this minute. Let us then decrease the length of the time interval around 3 o'clock over which the average speed is computed. Now we can compute the average speed for 1 second, or $\frac{1}{10}$ of a second, or $\frac{1}{100}$ of a second, and so on. The shorter the interval for which the average speed is computed, the more closely the average speed in that interval should approximate the speed at 3 o'clock.

Suppose that the average speeds computed over smaller and smaller intervals of time turn out to be 36, 35½, 35¼, 35⅛, and so on, as we get closer to 3 o'clock. Because the average speeds for smaller and smaller time intervals around 3 o'clock should be closer and closer estimates of the speed at 3 o'clock, we define the instantaneous speed at 3 o'clock to be the *number approached by the average speeds as the time intervals approach* 0. In the case of the average speeds – 36, 35½, 35¼, 35⅛, and so on – these numbers are presumably approaching 35 and so we would take 35 to be the instantaneous speed at 3 o'clock. We should notice that the instantaneous speed is *not* defined as the quotient of distance divided by time. Rather we have introduced the idea of taking a number that is *approached* by average speeds.

We can now consider a more precise description of the method of obtaining instantaneous speed. Let us take the actual formula which relates the distance a dropped body falls to the time it falls and let us calculate the instantaneous speed of a ball exactly three seconds after it is dropped. According to Galileo the relation between distance fallen in feet and time elapsed in seconds is

(1) $$d = 16t^2.$$

The distance fallen by the end of the third second, indicated by d_3, is therefore obtained by substituting 3 for t in this formula. That is,

$$d_3 = 16.3^2 = 144.$$

Now instead of calculating average speeds for various intervals of time around the end of the third second, as we did for the automobile's speed around 3 o'clock, we can work more efficiently as follows.

Let h represent any interval of time. Then $3 + h$ represents a new interval of time larger than 3 seconds by the amount h. In order to find how far the ball falls in $3 + h$ seconds, we substitute this time value in formula (1). We know that the new distance will not be 144 but a different value of d. Let us call it $d_3 + k$ where k is the extra distance travelled during the additional h seconds. Then

$$d_3 + k = 16(3 + h)^2.$$

By multiplying $(3 + h)$ by itself we obtain

$$d_3 + k = 16(9 + 6h + h^2).$$

We now multiply each term in the parentheses by 16. The result is

(2) $$d_3 + k = 144 + 96h + 16h^2.$$

At the end of 3 seconds the distance fallen is

(3) $$d_3 = 144.$$

To get k, the change in distance during the h seconds, we subtract equation (3) from equation (2). This operation gives

(4) $$k = 96h + 16h^2.$$

Now, just as the average speed of the automobile was obtained by dividing 90 miles by 3 hours, so we divide k, the distance travelled, by h, the number of seconds it takes to travel that distance, to obtain the average speed during the h seconds. If, then, we divide both sides of formula (4) by h we obtain

(5) $$k/h = 96 + 16h.$$

We see from formula (5) that the average speed, k/h, in the interval of h seconds after the third second is a function of h, the function being $96 + 16h$. As h becomes smaller, k/h represents average speed over a smaller and smaller interval of time measured from the end of the third second. We agreed above to take the number approached by these average speeds as the instantaneous speed at the end of the third second. Hence we want the value approached by k/h as h approaches 0. As h approaches 0, $16h$ approaches 0; and, as we can see from the right-hand side of formula (5), k/h approaches 96 in value. Hence the instantaneous speed at the end of the third second is 96 ft/sec. This is the speed any body dropped in a vacuum attains after three seconds.

The reader should notice that to determine the number 96 as the instantaneous speed, we observe what happens on the right side of formula (5) as h *approaches* o. Our reasoning was that the smaller h becomes, the closer $96 + 16h$ approaches 96. The *thought process is not the same* as that of substituting o for h despite the fact that the same result could be obtained by the substitution *in the case of this simple function*.

Let us see why the thought process is not the same. When h is o, k is o because k is the distance travelled by the ball during the time h. Hence when h is o, $k/h = o/o$ and this is a meaningless expression. Thus it is incorrect to speak of obtaining the speed at the end of the third second by substituting o for h in the expression for k/h. However, to find the number *approached* by the average speeds as the intervals of time over which the average speeds are computed approach o is logically sound and this is just the idea introduced to get around the difficulty in the concept of instantaneous speed. There is no difficulty, of course, about computing the average speeds because they are all for nonzero intervals of time.

We now have a concept of instantaneous speed. It is the number approached by average speeds as the intervals of time over which the average speeds are computed approach zero. Equally important, *we have a method for calculating instantaneous speed* by working from the formula relating distance and time. Incidentally, we should notice that if we had calculated the speed at the end of t seconds instead of 3, our result would have been that the speed v equals $32t$. Thus we can obtain a formula for the speed at *any* instant t.

The process we have just examined is characteristic of mathematics. In order to treat the concept of instantaneous speed the mathematician has idealized space and time so that he can speak of something existing at an instant of time and at some point in space. He thereby obtains speed at an instant. The layman finds his imagination and intuition strained by the notions of instant, point, and speed at an instant and he might prefer to speak of speed during some very small interval of time. Yet mathematics produces through its idealization not merely a concept but a formula for speed at an instant that is precise and more readily applied than is the notion of average speed during some sufficiently small interval. The imagination may be strained but the intellect is aided. This is the paradox of mathematics, which we have already encountered in other guises, that by introducing seeming difficulties it simplifies and renders easy a truly complex problem.

The method of defining and calculating instantaneous speed is actually more widely applicable than has hitherto been apparent. Nothing in the *mathematics* of it requires that d represent distance and t represent time. These variables may have *any physical meaning whatsoever*, and we can calculate the rate of change of one variable with respect to the other at a value of the second one by the same mathematical procedure we used to calculate the rate of change of distance compared to time at an instant. For example, if d represents speed and t time, we can calculate the rate of change of speed compared to time at an instant; this instantaneous rate of change of speed is instantaneous acceleration. As another example, the pressure in the atmosphere varies with height above the surface of the Earth; for this function we can calculate the rate of change of pressure compared to height *at any given height*. Or, if the variable d represents price level of commodities and t represents time, then we can compute the rate of change of price compared to time *at any instant*. Thus our method enables us to define and calculate thousands of significant and useful rates of change of one variable with respect to another *at a value of the second variable*. Incidentally, all such rates are referred to as instantaneous rates, despite the fact that time may not be one of the variables involved, because the original calculus problems of speed and acceleration did involve time and were concerned with rates at an instant of time. The calculus may now be defined as the subject that treats the concept of instantaneous rate of change of one variable with respect to another and the various applications of this concept.

The instantaneous rate of change of one variable with respect to another is usually indicated by a special symbol. Thus if the two variables are y and x, one symbol commonly used is Dxy, which is read *the derivative of y with respect to x.* (Another common but misleading symbol is dy/dx.) Either symbol is an excellent example of the conciseness of mathematical language. In less than the space of a word the symbol describes the result of the entire operation of finding the instantaneous rate of change of some variable y compared to another related variable x. We know now how much is comprised in that reference. Evidently the use of such a symbol is quite a step beyond the use of the letter x to represent an unknown. Advanced mathematics differs from elementary mathematics partly in this very effective use of symbols for complex concepts.

In the application of the concept of instantaneous rate of change mentioned thus far we started with the formula relating two variables and then found the rate of change. Suppose we were given the rate of change of one variable with respect to another, would there be any value in the reverse process of finding the formula relating the two variables? Of course the value of reversing the process of finding the rate of change depends on knowing some important rates of change to start with. Fortunately, this information can be readily obtained in many natural and man-made phenomena. From this we proceed to the formula and to the solution of many problems. Let us examine one actual case.

Suppose we were interested in finding the formula which relates the two variables, namely, the distance a body falls and the time the body takes to fall this distance. It is a logical consequence of Newton's laws, as shown in the preceding chapter, that the acceleration of a falling body is constant. That is, the rate of change of speed compared to time is the same at each instant of time. Simple experiments such as were made by Galileo show that the value of this constant is 32 ft/sec.[2]. In symbols, if a stands for acceleration,

(6) $a = 32.$

All bodies in the air above the Earth, the airplane flying over the Rockies, the bullet shot from a gun, and the ball thrown up into the air, possess this downward acceleration.

Now a is the instantaneous rate of change of speed compared to time; hence we can think of it as coming from a formula relating speed v and time t. If we could find this formula it would give the expression for the speed in terms of the time. We can obtain it by reversing the process of finding the rate of change. The reader may accept the fact that the formula relating speed and time is

(7) $v = 32t,$

or he can check it by finding the rate of change of v with respect to t and see that formula (6) results from his check. But formula (7) is not the answer to our problem, for this gives us the speed at each instant the body falls in terms of the time it has been falling, whereas we are seeking the relation between distance and time. However, the speed is the rate of change of distance compared to time. Therefore, in order to find the distance the body falls in t seconds we must find a new formula for which formula (7) represents the instantaneous rate of change. Again by reversing the process of finding a rate of change we get the formula relating d, the distance the body falls, to t, the number of seconds. The result is

(8) $d = 16t^2.$

The reader can confirm this result by showing that the rate of change of d compared to t is formula (7). Thus by twice reversing the process of finding instantaneous rates we can find the formula relating the distance and time a dropped body travels.

One more illustration of a class of problems in which the rate of change is the most readily obtained information may suffice to indicate the importance of the process of finding the formula from the rate of change. Newton's second law of motion, a law used as a basis for the most fundamental investigations in physics, is a statement about a rate of change. It says that the force acting on a body equals the mass of the body multiplied by the acceleration of the body's motion. When the force is known,

Plate I. Praxiteles: *Aphrodite of
Cnidos*, Vatican

Plate II. Myron: *Discobolus*,
Lancellotti Collection, Rome

Plate III.
Augustus from Prima Porta,
Vatican

Plate IV. *The Parthenon at Athens*

Plate V. *The Orbits of the Planets as Determined by the Five Regular Solids.* From Kepler's *Harmony of the Cosmos* (1596)

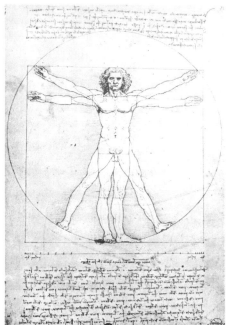

Plate VI. Leonardo da Vinci: *The Proportions of the Human Figure*, Academy, Venice

Plate VII. Early Christian Mosaic: *Abraham With Angels*, San Vitale, Ravenna

Plate VIII. Simone Martini: *The Annunciation*, Uffizi, Florence

Plate IX. Duccio: *Madonna in Majesty*, Opera del Duomo, Siena

Plate x. Duccio: *The Last Supper*, Opera del Duomo, Siena

Plate xi. Giotto: *The Death of St Francis*, Santa Croce, Florence

Plate XII. Giotto: *Salome's Dance*, Santa Croce, Florence

Plate XIII. Ambrogio Lorenzetti: *Annunciation*, Academy, Siena

Plate XIV. Masaccio: *The Tribute Money*, Church of S.M. del Carmine, Florence

Plate XV. Uccello: *Pawning of the Host*, a scene from the *Desecration of the Host*, Ducal Palace, Urbino

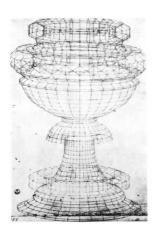

Plate XVI. Uccello: *Perspective Study of a Chalice*, Uffizi, Florence

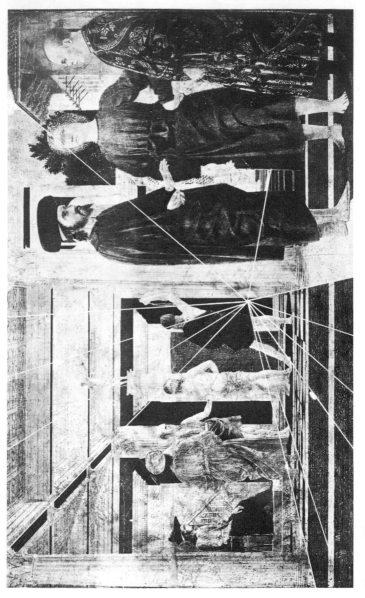

Plate XVII. Piero della Francesca: *The Flagellation*, Ducal Palace, Urbino

Plate xviii. Piero della Francesca: *Resurrection*, Palazzo Communale, Borgo San Sepolcro

Plate xix. Leonardo da Vinci: *Study for the Adoration of the Magi*, Uffizi, Florence

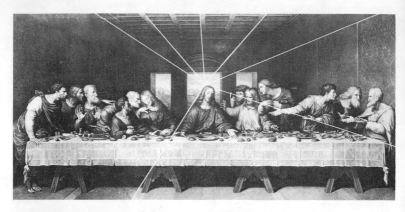

Plate xx. Leonardo da Vinci: *Last Supper*, Santa Maria delle Grazie, Milan

Plate xxi. Botticelli: *The Calumny of Apelles*, Uffizi, Florence

Plate XXII. Mantegna: *St James Led to Execution*, Eremitani Chapel, Padua

Plate XXIII. Raphael: *School of Athens*, Vatican

Plate XXIV. Tintoretto: *Transfer of the Body of St Mark*, Palazzo Reale, Venice

Plate xxvi. Hogarth: *False Perspective*

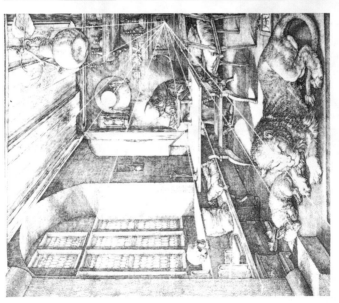

Plate xxv. Dürer: *St Jerome in His Study*

Plate XXVII. Picasso: *The Three Musicians*, Courtesy of Philadelphia
Museum of Art, A. E. Gallatin Collection

the law becomes a statement about acceleration or the rate of change of speed compared to time. Then by proceeding somewhat as we did above in going from formula (6) to formula (8) we can find the formula relating the distance and time in the situation where the force applies. Very often a formula obtained by reversing a rate of change could not have been obtained in any other way.

Expressions involving instantaneous rates of change are usually written in the form of equations, as (6) and (7) are written, and are called *differential equations*. A differential equation expresses some fact about the instantaneous rate of change of one variable with respect to another. The process of finding the formula that relates these variables by working from the differential equation is called solving that equation. It was by solving a famous differential equation that Newton was able to deduce Kepler's laws so readily. Because differential equations have proved to be the most effective means of formulating and developing whole branches of science, nature and God are often credited with 'speaking' in terms of such equations.

Were we concerned with the practical uses of the calculus it would be profitable to see how reversal of the process of finding an instantaneous rate of change could be applied to finding the lengths of curves, areas bounded by curves, volumes bounded by surfaces, and numerous other quantities not otherwise obtainable. Perhaps we should see, at least, how the calculus is involved in such applications.

As a simple illustration let us consider the area in fig. 50. We may think of this area as being swept out by a moving vertical line segment AB which starts at P (with zero length) and moves to the right. For any position of AB the area swept out is the shaded area of the figure. Now as AB moves to the right, the area swept out increases at a rate that is equal to the length of AB. Since AB changes in length from one position to another, the area swept out varies from point to point and the concept of instantaneous rate of change enters the picture. It would take us too far into the purely technical aspects of the calculus to complete the story of how the area of this figure and other areas can actually be found. The recognition of the relationship between the general concept of rate of change, on the one hand, and the determination of lengths, areas, and volumes, on the other, is the greatest single discovery made by Newton and Leibniz in the calculus.

While the efficient production of tin cans is sufficient motivation in our civilization for the study of a mathematical idea, the calculus, like other branches of mathematics, warrants attention because it has played larger roles in the creation of modern

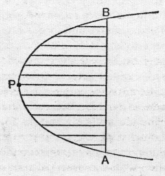

Figure 50. Area generated by a moving straight-line segment of variable length

civilization and culture. Of course, the use of calculus techniques in the derivation of scientific laws has already been described. Moreover, Newton's success in obtaining universal laws governing motion stimulated the scientists to seek such laws in other branches of physics. As a consequence, basic laws each embracing a large class of natural phenomena were found in such fields as electricity, light, heat, and sound. But we have not yet touched upon the most significant development that followed the creation of the calculus.

Scientists, like all men, are not readily satisfied. Once they obtain some success, they immediately desire a greater one. The eighteenth-century scientists, with courage high because they possessed the powerful weapon of the calculus, with appetites whetted by initial successes, and with tastes for scientific progress cultivated by their experience, dared to speculate about whether even the universal laws of the several branches of physics could all be deduced from *one* single law which perhaps underlies the design of the entire universe. At the very least, they hoped to unify several branches of science under one general mathematical law from which the separate laws of the several branches could then be deduced. Daring and ability won the day. The mathematicians and scientists did discover an entirely new principle which has not only guided the course of vast scientific developments but has been accepted as a basic doctrine on the design of the universe. The connection between the calculus and cosmic plan needs some elucidation.

Suppose a ball is thrown straight up into the air and we wish to find the maximum height it reaches above the ground. By means of the calculus this question is readily answered. Suppose, for example, that the height h of the ball above the ground is given by the formula

$$(9) \qquad\qquad h = 128t - 16t^2,$$

wherein t is the number of seconds from the instant the ball is thrown up. Since the ball rises when it first starts out, this means that h increases with t. The speed of the ball, however, decreases because gravitation opposes the upward velocity. The ball will continue to rise until its speed is zero. This must occur at the highest point of its flight or else the ball would continue to rise. This argument suggests that if we find the instant at which the speed is zero we will know at least the instant at which the ball is at the maximum height. By applying the process for finding the instantaneous rate of change of h with respect to t to formula (9) we should find that the speed is given by the formula

$$(10) \qquad\qquad v = 128 - 32t.$$

We agreed that the speed v equals zero at the instant when the ball is highest. Hence we let $v = 0$ in formula (10) and notice that the time t at which the ball is highest satisfies the equation

$$0 = 128 - 32t.$$

Evidently $t = 4$ satisfies this equation and so the ball is highest 4 seconds after it leaves the ground. How high is the ball at that

time? Formula (9) gives the height above the ground at any instant of time. We substitute 4 for t in this formula and find
that

$$h = 128.4 - 16.4^2 = 256.$$

Thus the maximum height attained by the ball is 256 feet above
the ground. The point of this illustration is that the calculus
enables us to find the maximum value of a variable, h in the
example above, through the concept of instantaneous rate of
change. The same procedure when applied to a variable that has
a minimum value would enable us to find that minimum.

By this time in the eighteenth century, scientists had observed
that in numerous phenomena nature behaves so that some quantity is either a maximum or a minimum. For example, a light ray
that goes from a point A to a mirror and then to a point B (see
fig. 16) could conceivably take many paths. But, as the Greeks
discovered, the ray takes the shortest path. Since light travels at a

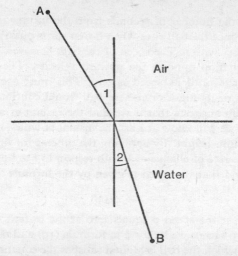

Figure 51. Refracted light takes the path requiring least time

constant speed in a uniform atmosphere, the shortest path is also
the path that requires the least time. In this phenomenon, there-

fore, nature behaves so that both the distance and time involved are a minimum.

When light travels from one medium into another, as from air to water, not only does its velocity change from a quantity c_1, say, to a quantity c_2, but the direction of the light ray changes (fig. 51). Again the light ray could take many paths in going from A in the first medium to B in the second. Both Willebrord Snell, a professor of mathematics at the University of Leiden, and Descartes showed, however, that the path the light ray does take is the one for which c_1 divided by c_2 equals *sine* 1 divided by *sine* 2. Fermat then showed that this path is also the one requiring the least time.

Light also follows the path requiring the least time when passing through a medium of variable characteristics such as the

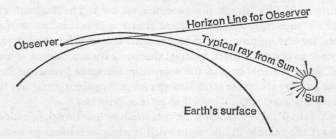

Figure 52. Light passing through a variable atmosphere takes the path requiring least time

atmosphere above the Earth. This behaviour of light can be attested to almost daily. The atmosphere near the surface of the Earth is denser than it is far from the surface of the Earth. But the speed of a light ray is slower in a dense atmosphere than in a rare one. Hence light coming from the sun to us stays in the rarer atmosphere as long as possible, presumably to take advantage of the higher speed that is possible there. The resulting curved path of the light rays permits us to see the sun after sunset, that is, when the sun is actually below the geometrical horizon (fig. 52).

On the basis of such evidence Fermat affirmed his Principle of

Least Time, which says that a ray of light travelling from one point to another will always take the path requiring the least time. Since the true path is the one for which the time is a minimum and since the calculus can be applied to determine the value of one variable that minimizes or maximizes a related variable, Fermat's principle tells us, in effect, how the calculus may be used to determine the paths of light rays. Fermat's principle, however, applies only to the behaviour of light rays. What about other phenomena?

Other instances wherein nature obeys a minimum principle were sought and soon found. A balloon made of uniform rubber takes a spherical shape when blown up. So does a soap bubble. It is a mathematical theorem that of all surfaces containing a given volume the sphere has the least surface area. (A Greek of the classical period would have given his life's blood to be able to prove this fact about his precious sphere.) The balloon and bubble, therefore, assume a shape that requires the least surface area for the volume of air blown into them. Why should they choose to obey this mathematical theorem? By assuming a spherical shape the rubber and the soap film are spread over the least area and are therefore stretched the least. Apparently nature, like human beings, extends herself as little as possible.

Could all these examples be included in one broad principle? About the middle of the eighteenth century a famous physicist, Pierre L. M. de Maupertuis, announced the Principle of Least Action. This Principle, which Maupertuis discovered while working with the theory of light, asserts that nature behaves so as to make as small as possible a certain complex mathematical quantity known technically as action and amounting to the product of mass, velocity, and space traversed. By applying the calculus to the formula for action, Newton's first two laws of motion, as well as other laws of mechanics and light, can be deduced. Hence bodies moving in accordance with Newton's laws, the planets for example, can be said to be obeying a minimum principle. Moreover, Maupertuis had succeeded in bringing the laws of mechanics and light under one minimum principle.

Maupertuis sought and advocated his principle for theological reasons. He believed that the laws of behaviour of matter must reveal the perfection worthy of God's creation. The Principle of Least Action satisfied this criterion because it showed that nature was economical. He therefore not only proclaimed it as a universal law of nature but also as the first scientific proof of the existence of God, for it was 'so wise a principle as to be worthy only of the Supreme Being'.

The great eighteenth-century Swiss mathematician, Leonhard Euler, like Maupertuis, believed that the existence of a minimum principle such as Least Action was no accident, and so he defended all of Maupertuis' claims for it. The principle was evidence of God's conscious design. Apparently, the God who was formerly merely the geometer of the Greek and Renaissance scientists was now being educated. He was shortly to become not merely a geometer but a rounded mathematician, proficient in all its branches.

As a matter of fact, Fermat, Maupertuis, and Euler were wrong in supposing that nature always behaves so as to make some function *least*. There are situations, for example, when a light ray takes a path requiring the most time compared to the time required for other possible paths. Hence the correct formulation of the principle these men sought is that nature behaves so that some function is either a maximum or a minimum. Maupertuis should not have said that nature is economical. He could have said, instead, that nature often runs to extremes.

Yet, although Maupertuis and his colleagues may have been mistaken about a detail or two, their nineteenth- and twentieth-century successors have been confident that these men were on the right track. Stripped of theological associations, a maximum and minimum principle now dominates physical science. One of the outstanding physicists of the last century, Sir William Hamilton, showed that nearly all gravitational, optical, dynamical, and electrical laws can be obtained by maximizing or minimizing a function created by him and known technically as the time integral of kinetic potential. Hamilton's function is valued partly because so many physical laws are encompassed by it and also because these laws must be deduced by applying a maximizing or minimizing process. Moreover, the outstanding mathematical physicist of this century, Albert Einstein, achieved his greatest success within his greatest creation, the theory of relativity, by showing that the natural path of bodies in space-time is one that maximizes a function called the interval. The importance of this statement lies in the fact that it accounts for the observed paths of the planets. The goal of embracing all phenomena in one principle, namely that the actual behaviour of nature minimizes or maximizes some very general mathematical quantity, is being actively pursued today. Einstein himself is still engaged in this task of compressing all electrical and mechanical knowledge into one mathematical sentence from which the laws of nature would be deduced by a minimizing or maximizing process.

We see, then, that the emphasis of scientists on a maximum-minimum principle has not diminished. The only change is that whereas such principles were formerly attributed to the providence of God, they are now accepted and welcomed because they are aesthetically appealing and scientifically helpful. Even so, such famous twentieth-century scientists as Eddington and Jeans have continued to look upon God as the First Cause, the ultimate *raison d'être.*

Though the great mathematicians and scientists lost no time in applying the calculus to the architecture of the universe, they were baulked for generations in their attempts to erect an adequate logical basis for the subject. Just as the gap between the conception of a horseless carriage and the modern automobile was bridged by a hundred major inventions and several hundred minor ones, so the gap between the calculus of Newton and Leibniz and what is now regarded as a satisfactory account of the subject was bridged by the work of hundreds of mathematicians, great and small. It took about a hundred and fifty years of work to produce a logical presentation of the calculus.

The major difficulty arises in the very step that gives the instantaneous speed. We may recall that from the formula $d = 16t^2$ we obtained the expression

$$k/h = 96 + 16h$$

for the average speed during the time interval of h seconds. The instantaneous speed was then taken to be the number approached by this expression as *h* approaches zero, or the *limit* as it is now called in the calculus. It may seem obvious to the reader that the number approached is 96; perhaps this fact is evident in this simple example but the concept of a limit is, nevertheless, a subtle and elusive one. Let us examine some of the difficulties involved.

The numbers of the sequence 0, ¼, ⅜, ⁷⁄₁₆, ¹⁵⁄₃₂ ... are increasing and approach 1 but evidently they do not get close to 1 because no term of this sequence is even as much as ½. If the values of k/h as h approaches zero made up this sequence, what would the limit or the number approached by k/h be? Evidently something more must be said about how the approach is made. It might be said that the sequence of values must come very close to the limit. But the word *close* is vague. The planet Mars comes close to the Earth when it is 50 million miles away. On the other hand, a bullet comes close to a person if it gets within a few inches of him.

The difficulty with which the founders of the calculus struggled was precisely this matter of giving some satisfactory definition of what they meant by instantaneous speed or the number approached by the quantity k/h. The attempts of some of the early seventeenth-century workers on the calculus to understand and justify their fragmentary contributions to the subject are ludicrous by modern standards. Despite the long tradition of rigorous proof in mathematics some mathematicians were ready to abandon the standard just because they knew they had their fingers on a valuable idea which they wished to advance but could not justify. Rigour, said Bonaventura Cavalieri, a pupil of Galileo and a professor at Bologna, is the concern of philosophy and not of geometry. Pascal argued that the heart intervenes to assure us of the correctness of some of the mathematical steps. The proper 'finesse' rather than logic is what is needed to do the correct thing, just as the appreciation of religious grace is above reason.

Though Newton and Leibniz made the most significant advances in the technique of the calculus, they did not contribute much to the rigorous establishment of the subject. No one can read the details of their writings on the calculus without being amazed by the variety of ways in which they stabbed at, around,

and about the correct version of the limit concept without actually striking it. Several times they changed their approaches and
contradicted their earlier statements. Neither man succeeded in
doing more with the limit concept proper than confusing himself, his contemporaries, and even his successors. At one place in
his *Principles*, Newton does state the correct version of the
notion of instantaneous rate of change but apparently he did not
recognize this fact, for in later writings he gave poorer explanations of the logic of his procedure. Leibniz did attempt a
justification of his work on rates by philosophical arguments
about the nature of the quantities h and k which appear in the
ratio k/h when h is allowed to approach zero; yet he believed
that, metaphysical considerations aside, the calculus was only
approximately correct but useful because the errors involved
were too small to matter practically. In his mathematical exposition of the calculus, Leibniz gives only rules and no proofs.
To describe the values of k and h which make up the number
approached by k/h as h approaches zero, he speaks of the h as
being the difference in two values of the time t that are *infinitely*
near each other; similarly k is the difference in two such values of
the distance d. In some of his writings he refers to the limiting
values of k and h as quantities that are infinitely small, or vanishing quantities, or quantities that are incipient as opposed to
the usual existing quantities. Newton used the phrase 'prime and
ultimate ratio' for the limit of k/h. But all such phrases do no
more than gloss over the difficulty involved.

Because of the lack of rigour in the early works on the calculus, conflicts and debates on the soundness of the whole subject
were prolonged. The mathematician, Michel Rolle, a contemporary of Newton, taught that the calculus was a collection
of ingenious fallacies. Shortly after Newton's death a good mathematician, Colin MacLaurin, decided that he would rigorize the
calculus. His book, published in 1742, was undoubtedly profound but also unreadable. Many other eighteenth-century ex-
positions of the calculus were written for the precise purpose of
supplying the logic. Their accomplishments may be epitomized
by Voltaire's description of the status of the calculus as 'the art of
numbering and measuring exactly a Thing whose Existence

cannot be conceived'. Two of the greatest mathematicians of all times, Joseph Louis Lagrange and Leonhard Euler, both of whom did their best work about a hundred years after that of Newton and Leibniz, still believed that the calculus was unsound but gave correct results only because errors were offsetting each other. Near the end of the eighteenth century, D'Alembert advised students to keep on with their study of the subject; *faith* would eventually come to them. It was a very fortunate circumstance that mathematics and science were closely linked in the Newtonian era and that physical reasoning could guide the mathematicians and keep them on the right track. Because the results they obtained were useful and sound in application, they maintained confidence in their methods and the courage to proceed farther. In fact, the calculus procedures worked so well and to such great advantage that at times mathematicians willingly closed their minds to the problem of rigour.

We know now that intuition and physical arguments rather than logic guided Newton and Leibniz along the proper paths. Incompleteness in the thinking of the creators of major ideas is almost to be expected. Pioneers in intellectual adventures make their great strides along paths briefly illuminated by flashes of brilliance. Were they to delay for smaller, time-consuming observations, their progress might be limited to the dainty, mincing steps of near-sighted academicians. Nevertheless, the history of the calculus is most revealing because it shows how progress is made in mathematics. The popular conception of a mathematician who reasons perfectly and directly to a conclusion is nowhere more sharply at variance with history than in the case of the creators of the calculus. Of course, many mathematical proofs have had to be corrected because some error was made unconsciously. It is a professional secret which should not be allowed to go farther than the reader that even Euclid made mistakes which were not discovered until the latter part of the nineteenth century. In the case of the calculus, however, we find an extensive body of mathematics applied to the most profound problems of science and producing the weightiest laws of the eighteenth century, while all the time the mathematicians, scientists, and other intellectuals were aware of the unsatisfactory foundations of the subject and were even dubious about its soundness. It should also be comforting to our egos to remember that almost all the best mathematicians of two centuries concentrated hard on the problem of rigorizing the calculus and failed miserably.

Fortunately for mathematics and the world this comedy of errors ended happily. The brilliant French mathematician, Augustin-Louis Cauchy, succeeded in formulating the limit concept correctly and in providing theorems about limits that were needed to justify the techniques. Cauchy published a definitive work, *Cours d'Analyse*, in 1821. We should be wrong to infer that mathematicians thereupon cast off the nonsense that had been written for a hundred and fifty years before that date and adopted Cauchy's ideas. The calculus textbook most widely used in the United States during the last fifty years and the one that is still the most popular might well have been written in 1700.

Contrary to common belief, the calculus is not the height of the so-called 'higher mathematics'. It is, in fact, only the beginning. Soon after it was created it became the cornerstone of *analysis,* a branch of mathematics, far vaster than algebra and geometry, that has served, guided, and led science so remarkably. Subjects such as ordinary and partial differential equations, infinite series, the calculus of variations, differential geometry, the calculus of functions of a complex variable, and potential theory are only some of the domains of analysis. With such tools the scientists continued their search for the mathematical laws of nature and strengthened their mastery of vast portions of it. Some of these achievements await our inspection.

While these branches of mathematics were being created a new culture was being fashioned on the basis of the contributions of the sixteenth- and seventeenth-century mathematicians. Abandoning the dried-up stalks of medieval knowledge which had previously supplied them with sustenance, science, philosophy, religion, literature, art, and aesthetics sought nurture from the fruitful mathematical contributions to a new interpretation of the cosmos. The directions pursued by these revivified branches of our culture will be our concern in the next few chapters.

XVI

The Newtonian Influence: Science and Philosophy

Expiate free o'er all this scene of Man;
A mighty maze! but not without a plan;

Alexander Pope

A SEVENTEENTH-CENTURY poll to select the most influential 'man' of that age would surely have been won by the devil. According to the science of demonology developed and preached by the theologians, the devil and his assistant evil spirits caused wars, famines, plagues, and storms. They amused themselves by frightening children and keeping churned cream from turning into butter. Also aiding the devil in his work were witches, 'anointed' human beings who derived their powers from him. Witches could infect people, transform themselves into wolves and devour their neighbours' cattle, and even have carnal relations with the devil himself. In idle moments they rode broomsticks up and down chimneys or through the air.

The evil perpetrated by the devil and his collaborators, in spite of God's omnipotence, was so monstrous that the political and spiritual representatives of God could find no mission more weighty or more sacred than the elimination of these enemies of mankind. Among these self-appointed protagonists of society were such confirmed believers in witchcraft as King James I of England, Luther, Calvin, some of the popes, John Wesley, and, in our own New England, Cotton Mather. On the basis of the most flimsy evidence old men, young men, women, and children were accused of being witches. In order to be certain that no suspects were overlooked, anonymous accusations were solicited even during church services. Boxes were passed around regularly into which worshippers could insert names. The accused were imprisoned, tortured, and urged to confess. Whether or not the accused confessed, torture went on until death, for failure to confess was interpreted as obstinacy, while confession obviously called for punishment. To ease the infinitesimal consciences of the judges some of those who did not confess were awarded certificates of innocence – posthumously.

With almost unbelievably firm adherence to doctrines that are now regarded as fantastic, secular judges and churchmen coldly condemned witches and sorcerers to death. The hold of the witch menace on seventeenth-century Europe may be estimated from one of the 'reform' measures. Pope Gregory XV decreed prison rather than death for those witches who, by their magic powers, had produced divorces, sickness, or impotence, or had harmed animals or crops.

Although the pursuit of witches was responsible for the death of many thousands of innocent people in the seventeenth century, it was by no means the only black aspect of life in that violent age. People lived in continual terror of what they were told awaited them after death. Priests and ministers affirmed that nearly everyone went to hell after death, and described in greatest detail the hideous, unbearable tortures that awaited the eternally damned. Boiling brimstone and intense flames burned victims who, nevertheless, were not consumed but continued to suffer these unabating tortures. God was presented not as the saviour but as the scourge of mankind, the power who had fashioned hell and the tortures therein and who consigned people to it, confirming His affection to only a small section of His flock. Christians were urged to spend their time meditating upon eternal damnation in order to prepare themselves for life after death. The credulous, unthinking people for whom religion was the only outlet, next to slavery, accepted this account of their fate as literally true. No wonder men felt impelled to 'justify the ways of God'!

Religious freedom was a rarity during the seventeenth century; but even worse, wars were fought regularly to stamp out heterodox opinion within the state and in neighbouring states. Total uniformity of belief was insisted upon to the point that any independent thought was eliminated. The Spanish, Roman, and Mexican Inquisitions, the Saint Bartholomew's Day massacre in

France, the Piedmont massacre in Italy, and the Thirty Years War in Germany were just a few of the 'inspired' efforts to educate mankind. Heresy, which included any act displeasing to the particular church dominant in any one country or even idle words against the pope in Catholic states, was immediately and ruthlessly stamped out. Even here in America, Quakers were hanged just for daring to come to Puritan Boston. Not only was there almost no freedom of religion, but religion held men in fear: fear of punishment, fear of damnation, fear of the devil, fear of God, and fear of torture after death.

In such a reactionary atmosphere it might be expected that freedom of the press would be as little known as freedom of religion. From 1543 on it was a penal offence in Catholic states to print, sell, own, convey, or import any literature not expressly sanctioned by the Inquisition. An Index of Prohibited Books listed those forbidden to the faithful. No sharper dagger for the assassination of letters was ever devised. Even where there was a degree of religious freedom, as in Prussia under Frederick the Great, freedom of the press was considered dangerous to the ruling class. While Frederick did agree that 'every man must go to heaven his own way', he stoutly maintained that man should have nothing to say about the government ruling his life on Earth. Hence censorship of books and articles was rigidly imposed. Governments ostensibly urged their citizens to search for the truth but punished them for finding it. As a result of the restrictions on the diffusion of knowledge the ignorance of the masses was as profound as it was widespread, and the traditionally learned class still 'solved' theological problems and dabbled in Aristotle.

Democracy itself was confined to an Aristotelian concept in speculative philosophy rather than proclaimed as a goal to be achieved in this world. The servile, chattel-like common man had not yet learned that he should challenge the divine right of his royal masters. In addition, the masses enjoyed no civil rights. People were thrown into prison without specific charges being lodged against them and waited years for trial. The most trivial offences, such as stealing a sheep or a small sum of money, were punished by death, and prison for indebtedness was usual. The favourite sport of 'ladies and gentlemen' of quality in England and elsewhere was to watch the torture and execution of criminals by the cruelest methods. To be drawn and quartered was not merely a figure of speech in those days.

Fortunately, these manifestations of intellectual, social, and moral depravity were the death throes of a passing culture. By the seventeenth century the medieval civilization had become completely disrupted. Its place in the Western world was to be taken by a more enlightened civilization which was just being fashioned. And the manner in which mathematics and science contributed to the moulding of this new civilization is certainly as worthy of examination as the manner in which they brought about such modern 'miracles' as radio and television.

The religious and social upheavals of the Renaissance and the rapid accumulation of knowledge from geographical explorations and mathematical and scientific research produced, at first, only intellectual confusion. Throughout this period, however, a small group of scientists and mathematicians, beginning with Copernicus, and including Kepler, Galileo, Descartes, Fermat, Huygens, Newton, and Leibniz, had been working steadily. While the ultimate effect of their work was to replace medieval decadence by a new cultural order, the goal of these men, as they themselves envisioned it, was a relatively limited one. In accordance with Galileo's new conception of the task of science and in accordance with the explicit statement by Newton in his *Mathematical Principles of Natural Philosophy,* it called for discovering the mathematical relations that hold for the physical universe.

Towards this end, the laws of motion and gravitation were the major contributions of Newton. These laws in themselves were found to embrace an amazing variety of phenomena. The laws of Kepler, based theretofore on observation, were recognized as immediate deductions from Newton's mathematical laws. When Newton and others following him found that light could be successfully studied as a motion of corpuscles and sound as a motion of air molecules, the Newtonian laws proved effective in these studies, too. Many other fields of science began to yield to mathematical formulation. Quantitative laws were discovered in the fields of electricity and heat, for the forces acting in liquids and gases, and for many chemical phenomena. Though the victories were largely in the fields of astronomy and physics and, to a lesser extent, in chemistry, their significance was heightened by the promise of things to come.

The improvement of the telescope and microscope, through mathematical and physical studies of light, quite literally opened up a new world to biologists. The success of the quantitative approach, along with analysis in terms of force and motion, suggested to the physiologists and psychologists that they look for explanations of their problems in these mechanical terms instead of in terms of astrological portents, soul, mind, spirits, humours, and other vague notions. Quantitative studies of the flow of water in pipes, they believed, would cover the case of blood flowing in arteries and veins. Indeed, Harvey's proof that the blood circulates around the body before returning to the heart reinforced this mechanistic view because it likened the body to a pumping plant with the heart as the pump. The work on light would explain much of the bodily function of sight while the study of sound would clarify the problems involving the sense of hearing. Two great works, *Man a Machine* by the celebrated French physician, Julian O. de la Mettrie, and *The System of Nature* by the French radical, Baron Paul Heinrich d'Holbach, went so far as to 'explain' consciousness, the bodily processes, and all human thoughts and actions in terms of matter and motion. Not long after Newton studied the heavens, La Mettrie claimed to have discovered the calculus of the human mind and the French economist, François Quesnay, announced equations for economic and social life. It seemed to be only a question of time before all phenomena, natural, social, and mental, would be reduced to mathematical laws.

The secret of the successes already attained and those confidently awaited was clear to the eighteenth-century thinkers. Men such as the Comte de Buffon, the leading French naturalist, and the Marquis de Condorcet, the famous metaphysician, discerned that the introduction of quantitative methods into science invested it with a new power to rationalize and master nature. Kant declared, in fact, that the progress of a science could be determined by the extent to which mathematics had entered into its methods and contents. Mathematics thus became the celebrated key to knowledge, the 'Queen of the Sciences'.

Appreciation of the already amazing power of the allies, mathematics and science, imbued thinking men with enthusiasm for a sweeping reorganization of all knowledge along the following lines. First of all, they exalted human reason as the most effective instrument for the attainment of truths. Second, because they regarded mathematical reasoning as the embodiment of the purest, deepest, and most efficacious form of all thought, the perfect justification of the claims made for the mental faculties of human beings, they urged the use of mathematical methods and mathematics proper for the derivation of knowledge. Third, investigators in each field of inquiry were to search for the relevant natural, mathematical laws. In particular, the concepts and conclusions of philosophy, religion, politics, economics, ethics, and aesthetics were to be recast, each in accordance with the natural laws of its field.

The chief characteristic of this new approach to knowledge was unbounded confidence in reason and in the validity of the extension of mathematical methods throughout the physical and formal sciences and beyond them to all fields of knowledge. This brave programme, as we shall see, was not entirely successful. Not all problems yielded to mathematical methods, despite the expectations and efforts of many great men. Yet the rationalistic temper of the period permanently altered the course of thought in almost all fields. And, as the reason-intoxicated leaders of the eighteenth-century Enlightenment anticipated, mathematics served as the fulcrum of the lever that overturned the existing world order and as the chief instrument to forge a new one.

It was almost to be expected that one of the first major efforts of the eighteenth-century thinkers should be to formulate a mathematical approach to all problems. Descartes, we saw, had sought to reconstruct all knowledge on an unquestionable basis and had singled out the deductive method of mathematics as the only reliable one. Though he envisioned a 'universal mathematics', he offered no symbolism or technique, however, with which to attack non-mathematical problems comparable to his introduction of algebra for the study of curves.

In pursuit of the same broad goal as Descartes', the mathematician and philosopher Leibniz launched a more ambitious programme. He sought to devise a universal, technical language and a calculus that would be adequate to embrace and prosecute effectively all inquiries. Thereby he hoped that *all* questions facing mankind would be readily answered. Mathematics was not merely the inspiration for Leibniz's plan but was also the starting point for its execution. This subject already had an ideal language and modes of operation suited to its purposes. Why not, reasoned Leibniz, broaden the scope of the mathematical language and mathematical machinery to include all studies? He therefore proposed as a first step towards his universal deductive science the decomposition of all ideas employed in thought into fundamental, distinct, and non-overlapping ones just as a composite number such as 24 is decomposed into the prime factors 2 and 3. At first, he used prime numbers as symbols for the fundamental ideas but later he decided to construct a special language with symbols similar to Chinese ideograms. Complex ideas were to be represented by combinations of the basic symbols just as a quantity such as $a(b + c)$ represents a complex algebraic quantity. He then intended to codify the laws of reasoning so that a person could apply them to the symbols and combinations of symbols in order to deduce conclusions as mechanically and as efficiently as mathematics does in algebra.

At first blush Leibniz's plan seems preposterous. The expectation that all questions in all fields can be settled strikes a modern person as rather far-fetched. Yet this much can be said in Leibniz's behalf. The history of mathematics shows that the introduction of better and better symbolism and operations has made a commonplace of processes that would have been impossible with the unimproved techniques. To take the simplest example, the use of the Hindu-Arabic symbols for our numbers and of positional notation makes it possible for elementary school children today to perform operations beyond the capacities of learned mathematicians of Greek, Roman, and medieval times.

Nevertheless Leibniz was too ambitious. Not only did he himself never complete his efforts in these directions, but his belief that all ideas could be decomposed into relatively few fundamental ones has not been substantiated.

Yet his programme did lead to some action and results in the nineteenth century. Logic itself has adopted his methods to the extent of employing symbols for the fundamental ideas and operations that occur in reasoning and to the extent of carrying out its investigations into the nature and forms of valid reasoning in this purely symbolic language. Thus Leibniz is the founder of the science known today as symbolic logic, a science actively pursued in our own century by such learned men as Bertrand Russell and Alfred North Whitehead.

If the attempt to solve all problems by a universal calculus was abortive, the same cannot be said for the other revisions of knowledge which the Newtonian age undertook. The most pervasive change was made, naturally enough, in the sciences themselves. When Descartes, Galileo, and Newton decided that the goal of science was to find the mathematical laws of nature, the two fields, mathematics and science, merged forces. The greater were the successes that attended the combined efforts, the closer the alliance became. Mathematical branches were created to further science and science supplied the principal mathematical problems. In fact, the best mathematical accomplishments and the best scientific ones were realized by the same people. It is indeed impossible to judge whether Newton, Leibniz, the Bernoulli family, D'Alembert, Legendre, Lagrange, and Laplace were more accomplished as mathematicians or as scientists. Gradually, however, one of the partners began to dominate the alliance and the eighteenth century witnessed the beginning of a new phase of the relationship, for in this century mathematics began to absorb science. Though science held steadily to its objective of studying and fathoming nature, it became more and more mathematical in content, language, and method.

Of the branches of science that became more and more mathematical at the same time that they attained what appeared to eighteenth-century minds as perfection in the representation and explanation of nature, the most outstanding and the most developed branch was the science of mechanics. Galileo and Descartes had proposed a programme and philosophy, namely, that nature consisted of matter in motion and that science had but to discover the mathematical laws of these motions. One hundred years later this programme had been converted into a solid and most impressive reality. Through the work of the initiators and of dozens of other leading lights, the study of the motions of bodies on Earth and especially of the heavenly bodies had attained a completeness and apparent finality that convinced the men of the Enlightenment of the truth and value of this philosophy of science. It was in the eighteenth century that the two monumental works on mechanics, Lagrange's *Mécanique analytique* and Laplace's *Mécanique céleste* 'proved' that nature is governed by precise and eternal mathematical laws which comprehended every phenomenon of motion observed by scientists.

At the same time, these scientific classics reduced mechanics to pure equations. The science of mechanics proved to be a paradise wherein mathematicians could roam freely and happily. In this domain no major unsolved problems troubled their minds and nature's phenomena were but fruit to be had for the picking. Whereas the seventeenth century could pride itself on brilliant mathematical creations, the eighteenth century could boast of the successful mechanical philosophy of nature and could be properly described as the age of mathematical mechanics.

Concomitant with the change in the content of science were changes in language and *modus operandi*. The language became more and more the language of mathematics with its precise, unambiguous, convenient, and universal symbolism. Science also began to make much more extensive use of abstract or ideal concepts. Actually, all of us are continually abstracting ideas from experience though, like the Molière character who failed to realize that he had been talking prose all his life, we are often unaware of it. The force of gravity was one of the notable abstractions of the seventeenth century. Space-pervading ether, another concept employed widely since the seventeenth century, and mass, as a scientific concept, were other important abstractions. Among famous abstractions introduced since the 1600s, we might mention also the concepts of power and energy.

Science has become more mathematical in its methods by its wider use of deduction. By this we mean that it has adopted axioms, as mathematics did during the Greek period, and that it uses these axioms in conjunction with the axioms and theorems of mathematics to deduce its own theorems. We may ask: what axioms, in addition to the purely mathematical ones, are at the basis of reasoning in physics, for example? Newton's laws of motion and gravitation are such axioms, and we have seen them used as such in preceding chapters. Another example of what may be termed a physical axiom is the statement asserting the conservation of energy. This axiom is suggested by the observation that when energy is expended in one form it reappears in another. If muscular energy is used to saw wood, energy in the form of heat appears in the saw and wood. The energy latent in coal is used to create energy in the form of electricity. Upon the basis of these observations and many precise measurements physicists are willing to accept as axiomatic the fact that in physical and chemical processes energy is never lost but merely changes form.

The conversion of one whole branch of science into an essentially mathematical discipline, as well as the growing use by science of the language, conclusions, and processes of mathematics such as abstraction and deduction, has been characterized as the mathematization of science. It seemed clear in the eighteenth century that it was only a matter of time before all of science would be mathematized and indeed that the progress of science would be more and more rapid as the absorption of science by mathematics continued.

In their investigation of nature the scientists of the Renaissance sought and found mathematical truths. Of course, mathematics had been recognized as a source of such truths since Greek times. Only after the Renaissance, however, did mathematical laws begin to make such sweeping affirmations about the universe that they jeopardized the titles of the traditional philosophic and religious rulers of the realm of truth. Indeed mathematics was revealing a new order and plan in the universe more majestic than any ever offered before. And with mathematics in the ascendancy and pointing to a zenith of accomplishment beyond man's imagination, both philosophy and religion had to discard long established systems of thought and rebuild in the light of the new mathematical and scientific knowledge.

The philosophers began reconstruction by reopening the question: how does man come to know truths? Theologians were equally concerned with this question, for the new mathematics and science had destroyed so much of what had passed for knowledge that among intellectuals, at least, the orthodox religious belief in God was fast disappearing. Since proof of God's existence was not likely to result from a mathematical theorem or a scientific experiment, some saw the necessity for founding that belief in a new theory of knowledge. Perhaps the concept of a deity could be shown to be inborn in man and thereby placed above all doubts.

How does man come to know truths? How does man obtain the knowledge he is willing to swear by? How does he account for the conviction that accompanies such knowledge? The philosophers pondered these problems and brought forth answers which disappointed the theologians to the same degree that they reflected the new outlook of the age.

Responding directly to the knowledge being acquired by mathematics and science, the philosopher Thomas Hobbes affirmed first in his *Leviathan* (1651) that external to us there is only matter in motion. External bodies press against our sense organs and by purely mechanical processes produce sensations in our brains. All of our knowledge is derived from these sensations.

A sensation may linger in the brain because, like all matter, it possesses inertia. The sensation is then called an image. When a train of images arrives, it recalls others already received as, for example, the image of an apple recalls that of a tree. Thought is the organization of chains of images. Specifically, names are attached to bodies and properties of bodies as they appear in images, and thought consists in connecting these names by assertions and in seeking the relations that necessarily hold among these assertions. Knowledge consists of regularities discovered by the brain as it organizes and relates the assertions. Now mathematical activity produces just such regularities, for through mathematics the brain singles out and abstracts necessary relations that are not immediately apparent in physical objects as such. Hence the mathematical activity of the brain produces genuine knowledge of the physical world and mathematical knowledge *is* truth. In fact, reality is accessible to us only in the form of mathematics.

So strongly did Hobbes defend the exclusive right of mathematics to truth that even the mathematicians objected. In a letter to a leading physicist of the age, Christian Huygens, the mathematician John Wallis wrote of Hobbes:

> Our Leviathan is furiously attacking and destroying our Universities (and not only ours but all) and especially ministers and the clergy and all religion, as though the Christian world had no sound knowledge, none that was not ridiculous either in philosophy or religion, and as though men could not understand religion if they did not understand philosophy, nor philosophy unless they knew mathematics.

The emphasis placed by Hobbes on the purely physical character of sensation and of the action of the brain in reasoning shocked many philosophers to whom the mind was more than a mass of matter acting mechanically and who sought support for religious concepts such as God and the soul. In his *Essay Concerning Human Understanding*, published in 1690, John Locke began somewhat as Hobbes did, but unlike Descartes, by asserting that there are no innate ideas in men; they are born with minds as empty as blank tablets. Experience, through the media of the sense organs, writes on those tablets and produces simple ideas. Some simple ideas are exact resemblances of qualities actually inhering in bodies. These qualities, which he called primary, are exemplified by solidity, extension, figure (shape), motion or rest, and number. Such properties exist whether or not anyone perceives them. Other ideas that arise from sensations are the effects of the real properties of objects on the mind but these ideas do not correspond to actual properties. Among such secondary qualities are colour, taste, smell, and sound.

Although the mind cannot invent or frame any simple ideas, it does have the power to reflect on, compare, and unite simple ideas and thus form complex ideas. Here Locke departed from

Hobbes. In addition, the mind does not know reality itself but only ideas of reality and works with these. Knowledge concerns the connection of ideas such as their agreement or inconsistency. Truth consists in knowledge that conforms to the reality of things.

Demonstration connects ideas and thereby establishes truths. Of the certainties reached by demonstration the mathematical ones are perfect. Locke preferred mathematical knowledge because, first, he felt that the ideas with which it deals were the clearest and most reliable. Furthermore, mathematics relates ideas by exhibiting necessary connections among them and the mind understands such connections best.

Locke not only preferred the mathematical knowledge of the physical world produced by science but he even rejected the direct physical knowledge. He argued that many facts about the structure of matter are not clear, such as the physical forces by which objects attract or repel each other. Moreover, since we can never know the real substance of the external world but only ideas produced by sensations, physical knowledge can hardly be satisfactory. He was convinced, nevertheless, that the physical world possessing the properties described by mathematics does exist, as do God and we ourselves.

Locke's philosophy is an almost perfect reflection of the contents of Newtonian science. Consequently his influence on popular thought was enormous. His philosophy pervaded the eighteenth century much as Descartes' did the seventeenth.

In their theories of knowledge both Hobbes and Locke put primary emphasis on the existence of a world of matter external to human beings. While all knowledge stemmed from this source, the truths about this world finally obtained by the mind, or brain, were the laws of mathematics. Bishop George Berkeley, famous as a philosopher as well as a churchman, recognized in this emphasis on matter and mathematics the threat to religion proper and to concepts such as God and the soul.* With ingenious and trenchant arguments he proceeded to attack both Hobbes and Locke and to offer his own theory of knowledge.

* See also the next chapter.

In his chief philosophical work, *A Treatise Concerning the Principles of Human Knowledge, wherein the chief causes of error and difficulty in the sciences, with the grounds of scepticism, atheism and irreligion, are inquired into*, Berkeley made a frontal assault. Both Hobbes and Locke had maintained that all we know are ideas, but these ideas are produced by the action of external, material things upon our minds. Berkeley granted the sensations or sense impressions and the ideas derived from them but challenged the belief that they are caused by material objects external to the perceiving mind. Since we perceive only the sensations and the ideas, there is no reason to believe that anything is external to ourselves. In response to Locke's argument that our ideas of the primary qualities of material objects are exact copies, Berkeley retorted that an idea can be like nothing but an idea.

Berkeley strengthened his position with an argument unintentionally supplied by Locke. The latter had distinguished ideas of primary qualities from those of secondary qualities. The former corresponded to real properties whereas the latter existed only in the mind. Berkeley asked: can anyone conceive of the extension and motion of a body without including other sensible qualities, such as colour? Extension, figure, and motion *per se* are inconceivable. If, therefore, the secondary qualities exist only in the mind, so do the primary ones.

In brief, Berkeley argued that since we know only sensations and ideas formed by these sensations but do not know external objects themselves, there is no need to assume an external world at all. That world does not exist any more than do the stars we see when we are hit on the head. *An external world of matter is a meaningless and incomprehensible abstraction.* If there were external bodies, we should never be able to know it; and if there were not, then we should have the same reasons as now to think that there were such bodies. Mind and sensations are the only realities. Thus Berkeley disposed of matter.

The reader may protest this conclusion and perhaps attempt to rebut it as Samuel Johnson did by kicking a very solid-appearing stone but the logic of Berkeley's position is not thereby invalidated. At best it may be rejected for the reason described by the Earl of Chesterfield in a letter to his son:

Doctor Berkeley, Bishop of Cloyne, a very worthy, ingenious, and learned man, has written a book to prove that there is no such thing as matter, and that nothing exists but an idea; that you and I only fancy ourselves eating, drinking, and sleeping. . . . His arguments are, strictly speaking, unanswerable; but yet I am so far from convinced by them that I am determined to go on to eat and drink, to walk and ride, in order to keep that *matter*, which I so mistakenly imagine my body at present to consist of, in the best plight possible. Common sense (which, in truth, is very uncommon) is the best sense I know of.

It must be confessed that even Berkeley himself was not above an occasional sortie into the very physical world whose existence he denied. His last work, entitled *Siris: A Chain of Philosophical Reflections concerning the virtues of Tar-Water*, recommended the drinking of water in which tar had been soaked as a cure for smallpox, consumption, gout, pleurisy, asthma, indigestion, and many other diseases. Such occasional mis-steps must not be held against Berkeley. The reader who consults his delightful *Dialogues of Hylas and Philonous* will find an extremely able and entertaining defence of his philosophy. At any rate, by depriving materialism of its matter Berkeley believed he had disposed of the physical world and with it Newtonian science.

But Berkeley had yet to reckon with mathematics. How was it that the mind was able to obtain laws which not only described but predicted the course of the external world? What could he do to counter the strongly established eighteenth-century belief in the truths about an external world proffered by mathematics?

He proceeded to demolish mathematics and was shrewd enough to attack it at its weakest point. The fundamental concept of the calculus is that of the instantaneous rate of change of a function; but, as we stated earlier, this concept was not clearly understood and therefore not well presented by either Newton or Leibniz. Hence Berkeley was able in his day to attack it with justification and conviction. In *The Analyst* of 1734, addressed to an infidel mathematician, he did not mince any words. Instantaneous rates of change he condemned as 'neither finite quantities, nor quantities infinitely small, nor yet nothing.' These rates of change were but 'the ghosts of departed quantities. Certainly . . . he who can digest a second or third fluxion [Newton's technical name for instantaneous rate of change] ... need not, methinks, be squeamish about any point in Divinity.' That the calculus proved useful nevertheless, Berkeley accounted for on the grounds that somewhere errors were compensating for each other. Though Berkeley had made a criticism of the calculus which was warranted at that time, he had not actually disposed of all the truths mathematics had produced about the physical world. Nevertheless, having given his opponents something to think about, he rested his case against mathematics at this point.

It would seem that Berkeley's philosophy was about as radical as thought can be on the subject of man's relation to the physical world. But the sceptic Scot, David Hume, thought Berkeley had not gone far enough. Berkeley did accept a thinking mind in which the sensations and ideas existed. Hume denied mind too. In his *Treatise of Human Nature* (1739–40), he maintained that we know neither mind nor matter. Both are fictions. We perceive neither. We perceive impressions (sensations) and ideas such as images, memories, and thoughts, all three of which are but faint effects of impressions. There are, it is true, both simple and complex impressions and ideas, but the latter are merely combinations of simple ones. Hence it can be asserted that the mind is *identical* with our collection of impressions and ideas. It is but a convenient term for this collection.

As for matter, Hume agreed with Berkeley. Who guarantees to us that there is a permanently existing world of solid objects? All we *know* are our own impressions of such a world. By association of ideas through resemblance and contiguity in order or position, the memory orders the mental world of ideas as gravitation presumes to order the physical world. Space and time are only a manner and order in which ideas occur to us. Similarly, causality is but a customary connection of ideas. Neither space nor time nor causality are objective realities. We are deluded by the force and firmness of our ideas into believing in such realities.

The existence of an external world with fixed properties is really an unwarranted inference. There is no evidence that anything exists beyond impressions and ideas which belong to nothing and represent nothing. Hence there can be no scientific laws concerning a permanent, objective physical world; such laws signify merely convenient summaries of impressions. Moreover, since the idea of causality is based not on scientific proof but merely on a habit of mind resulting from the frequent observation of the usual order of 'events', we have no way of knowing that the sequences we have observed will recur.

Man himself is but an isolated collection of perceptions, that is, impressions and ideas. He exists only as such. Any attempt on his part to perceive himself reaches only a perception. All the other men and the supposed external world are just perceptions to any one man and there is no assurance that they exist.

Only one obstacle stood in the way of Hume's thoroughgoing scepticism, namely, the existence of the generally acknowledged truths of pure mathematics itself. Since he could not demolish these, he proceeded to deflate their value. The theorems of pure mathematics, he asserted, were no more than redundant statements, needless repetitions of the same facts in different ways. That 2×2 equals 4 is no new fact. Actually, 2×2 is but another way of saying or writing 4. Hence this and other statements in arithmetic are mere tautologies. As for the theorems of geometry, they are but repetitions in more elaborate form of the axioms, which in turn have as much meaning as 2×2 equals 4.

Hume's solution, then, of the general problem of how man obtains truths is that he cannot obtain them. Neither the theorems of mathematics, the existence of God, the existence of an external world, causation, nature, nor miracles constituted truths. Thus Hume destroyed by reasoning what reasoning had established while, at the same time, he revealed the limitations of reason.

Hume's work not only vitiated the efforts and results of science and mathematics but challenged the value of reason itself. Some philosophers, such as Rousseau, drew the obvious inference. They urged the abandonment of reason in favour of an imaginative and intuitive approach to life. To them reason was a form of self-deception, an unfortunate delusion. The thinking man was after all nothing but a sick animal.

But such a conclusion, such a denial of man's highest faculty, was revolting to most eighteenth-century thinkers. Mathematics and other manifestations of human reason had accomplished too much to be so easily cast aside as aberrations. The supreme philosopher Immanuel Kant actually expressed his revulsion for Hume's unwarranted extension of Locke's theory of knowledge. Reason must be re-enthroned. It appeared indubitable to Kant that man possesses ideas and truths beyond mere amalgamations of sense experience.

Kant thereupon undertook an entirely new approach to the problem of how man obtains truths. His first step was to distinguish between two kinds of statements or judgements that give us knowledge. The first kind, which he called analytical and which is exemplified by the statement *All bodies are extended,* does not really contribute to knowledge. The statement that bodies are extended is merely an explicit statement of a property that bodies have by the very fact that they are bodies and says nothing new. Hence we learn nothing by being informed that they are extended, though the statement may perhaps serve for emphasis. On the other hand, the statement that all bodies have colour does add something new to our knowledge because it adds to our information about bodies a fact not inherent in their nature as bodies. This type of judgement Kant called synthetic. Kant also distinguished between knowledge obtained directly from experience and knowledge somehow obtained by the mind independently of experience. The latter type he called *a priori.*

According to Kant, truth cannot come from experience alone, for experience is a *mélange* of sensations, devoid of concepts and organization. Mere observations therefore will not furnish truths. Truths, if they exist, must be *a priori* judgements and, moreover, in order to be genuine knowledge, they must be synthetic judgements. To combat Hume and Rousseau, Kant showed first that man does possess truths, that is, he does have *a priori* synthetic judgements.

Patent evidence was at hand in the body of mathematical knowledge. Almost all of the axioms and theorems of mathematics were to Kant *a priori* synthetic judgements. The statement that the straight line is the shortest distance between two points is certainly synthetic, since it combines two ideas, straightness and shortest distance, neither of which is implied by the other. Also, it is *a priori* since experience with straight lines or even measurements could not ensure the invariable and universal truth Kant believed this statement to be. Hence to Kant there was no question that man does have *a priori* synthetic judgements, that is, genuine truths.

Kant probed still deeper. Why, he asked, was he willing to accept as a truth the statement that the straight line is the shortest distance between two points? How is it possible for the mind to know such truths? This question could be answered if we could answer the question of how mathematics is possible. The answer Kant gave is that our minds possess, *independently of experience*, the forms of space and time. Kant called these forms intuitions. Space is therefore an intuition through which the mind necessarily 'views' the physical world in order to organize and understand sensations. Since the very intuition of space has its origin in the mind, certain axioms about space are at once acceptable to the mind. Geometry then goes on to explore the logical implications of these axioms.

Why, then, do the theorems of geometry, which are mental constructs, apply to the physical world outside mind? Kant's answer is that the form of space which the mind inherently possesses is the only way in which it can comprehend spatial relations. We perceive, organize, and understand experience in accordance with this spatial form; that is, experience fits into this form as dough into a mould. For this reason there must be agreement between Euclidean geometry and our experience with physical figures.

More generally, Kant argued that the world of science is a world of sense impressions arranged by the mind in accordance with innate principles or categories. These sense impressions do originate in a real world but unfortunately this world is unknowable. The mind itself provides the organization and understanding of experience. Actuality can be known only in terms of the subjective categories supplied by the perceiving mind.

It is evidence from the above sketch of Kant's theory of knowledge that he made the existence of mathematical truths a central pillar of his philosophy. In particular, he relied on the truths of Euclidean geometry. His inability to conceive of any other geometry convinced him there could be no other. Thereby the truths of Euclid and the existence of *a priori* synthetic propositions were guaranteed.

Alas, the nineteenth-century creation, non-Euclidean geometry, demolished Kant's arguments. Nor was the problem of how man obtains truths definitively answered by subsequent contributions to philosophic thought. Indeed, as we shall see later, the subject was thrown into even greater confusion by the non-Euclidean geometers. Nevertheless, though the great eighteenth-century philosophers did not succeed in answering the question of how man comes to know truths, they did at least open up the dams of thought and permit new ideas to flow freely through the mind of mankind.

Although the eighteenth-century philosophers disagreed vigorously on the question of how man comes to know truths, there was very little disagreement among them on what was true. As the laws of motion and gravitation extended their sway over more and more phenomena, and as planets, comets, and stars continued to pursue paths so precisely described by mathematics, the assumption of Descartes and Galileo that the universe is interpretable in terms of matter, force, and motion became a conviction in the mind of almost every thinking European.

Because matter in motion was the key to a mathematical description of falling bodies and planetary motion the scientists themselves attempted to fit such a materialistic explanation to phenomena whose nature they did not understand at all. Heat, light, electricity, and magnetism were regarded as imponderable kinds of matter, imponderable meaning merely that the densities of these kinds of matter were too small to be measured. The matter in heat, for example, was called caloric. A body being heated soaked up this matter just as a sponge soaks up water. Electricity was, similarly, matter in the state of a fluid and this fluid flowing through wires was the electric current.

Of the three concepts, matter, force, and motion, force acted on matter and motion was a property of matter; hence matter was fundamental. The philosophers thereupon proclaimed matter behaving in accordance with fixed mathematical laws as the sole reality. This is the doctrine of *materialism*. As expressed by Hobbes in its crudest form it asserted:

> The universe, that is, the whole mass of all things that are, is corporeal, that is to say, body, and hath the dimensions of magnitude, namely, length, breadth, and depth; also, every part of body is likewise body, and hath the like dimensions, and consequently every part of the universe is body, and that which is not body is no part of the universe; and because the universe is all, that which is no part of it is nothing, and consequently nowhere.

Body, he continued, is something that occupies space, is divisible and movable, and behaves mathematically.

Materialism, then, may be said to assert that reality is merely a complex machine, a mechanism of objects moving in space and time. Since man himself is part of physical nature, all of man, too, must be explainable in terms of matter, motion, and mathematics. In the language of Hobbes all that exists is matter; all that occurs is motion; consciousness is simply the impact of material particles on brain substance. Other exponents of the new philosophy such as La Mettrie, whose *Man a Machine* stated the thesis bluntly, and Baron d'Holbach, whose book *The System of Nature* was called the Bible of materialism, went even further. Thought, as well as consciousness, was considered to be molecular motion. The mind cannot be distinguished from the brain and perishes with it. The concept of a non-physical 'substance' such as soul must be completely rejected. Man's moral state is only a special aspect of his physical state, a particular mode of action caused by his organization and physical environment. Prejudices alone prevent us from examining the influences that determine our moral behaviour. In brief, matter is the cause and explanation of all phenomena and is the startling alternative to God.

Before we examine the devastating consequences of materialism we should be a little clearer about the precise source of its strength. The scientific activity of observation and experimentation, as well as the scientific concepts of matter, force, and motion, combined with pure mathematics to produce the evidence for the doctrine. It would seem, however, that a doctrine which asserted the fundamental reality of matter should rely more heavily on scientific rather than on mathematical foundations. Yet, as Newton clearly realized, the strength of the materialistic movement lay not in solid matter but in the immaterial abstractions of ethereal mathematics. The entire system of natural science, founded by Galileo and Descartes and erected by Newton, rested upon the universal force of gravitation. Though Newton had made the theory of this universal force indispensable, he admitted that he did not know its nature. In fact, he stressed the problem of investigating its physical nature and its *modus operandi*, rejecting as unripe and vague his own conjectures on the subject. With prophetic insight he adhered strictly to the mathematical formulation of the action of gravity and to the mathematical consequences of the formulation. Mathematics was the sign in which Newton conquered.

Of course Newton and his successors expected that the physical explanation of the action of gravity would be forthcoming some day. At the very least, such famous scientists as Huygens, Leibniz, and Johann Bernoulli realized that the explanation was missing, and that because of this deficiency in physical theory they too were compelled to treat the behaviour of gravity entirely mathematically. In the meantime lesser scientists referred to it as 'action at a distance' as though the phrase somehow supplied a physical explanation. Gradually the endless repetition of the words 'action at a distance', which merely slurred over the problem, dulled critical sensibilities into accepting the phrase as a substitute for an explanation. Physical meaning was bound and sacrificed on the altar of mathematical fertility. The nature and *modus operandi* of the force of gravitation have never been explained.

For this reason the materialists who talked so glibly and assuredly for centuries after Newton about the solid, tangible, observable phenomena of nature were actually proclaiming in the very same breath the importance of a notion more mystic and obscure than transubstantiation. In boasting of the progress of the materialistic outlook in science they were unconsciously boosting the importance of mathematical laws, for the materialistic philosophy which appears to derive its strength mainly from the scientific treatment of matter actually derives it from mathematics, the most abstract of scientific abstractions. Pythagoreanism, with its emphasis on number relations as the ultimate reality, was being vindicated in the guise of materialism.

Despite the inadequate material basis for materialism, the belief that the universe can be completely explained in terms of the mechanical concepts of force, matter, and motion and their mathematical relations acquired such a hold on the minds of men that it became a fashionable commonplace. It still is a conviction possessed by many who follow consciously or unconsciously the point of view of Newton's immediate successors. This conviction is often voiced today despite the fact that it is now realized that nature is far more complex than the mechanically minded eighteenth-century scientists believed it to be. It is this conviction that is the basis for the nineteenth century belief in scientific perfectibility and in the ultimate solution of all problems, such as a cure for cancer and the creation of life by chemical means.

The corollary of eighteenth century materialism was determinism. Mathematical formulas provided veridical descriptions of so many phenomena and proved so useful in applications as to make inescapable the conclusion that the world was carefully planned and that it operated in accordance with these formulas. The world's course appeared completely determined by harmonious, mathematical laws that prescribe for each event a necessary consequent event. The leading exponents of this view were the brilliant eighteenth century mathematicians Lagrange and Laplace. To Laplace the future was as clearly readable as the past.

We may regard the present state of the universe as the effect of its past and the cause of its future. An intellect which at any given moment knew all the forces that animate nature and the mutual positions of the beings that compose it, if this intellect were vast enough to submit its data to analysis, could condense into a single formula the movement of the greatest bodies of the universe and that of the lightest atom: for such an intellect nothing could be uncertain; and the future just like the past would be present before its eyes.

The Age of Reason is gone. Philosophically speaking we have

progressed since the eighteenth century. Determinism, however, is still the most popularly accepted point of view. The prevailing opinion is that the world is fashioned in accordance with mathematical laws and that its future is determined by them. Some appreciation of the conviction people still feel for this doctrine may be gained from our own behaviour, which is very much a reflection of eighteenth-century thought. Consider, for example, the modern reaction to an eclipse. Unlike primitive people, we do not rush out into the open, fall on trembling knees, and pray to the gods to avert the calamity the impending event presages. Instead we go out with stop watches in our hands to check to the fraction of a second the scientists' *prediction* of the eclipse. And at the end of such occurrences we are all the more convinced of the regularity and the lawfulness of nature's behaviour.

The deterministic point of view was held so firmly that the materialists applied it at once to the actions of man as part of nature. Determinism applied to man mercilessly declares: There is no free will. The human will is determined by external physical and physiological causes. On this subject Hobbes was blunt and direct: Free will is a meaningless conjunction of words, insignificant nonsense. And Voltaire stated in his *Ignorant Philosopher:*

It would be very singular that all nature, all the planets, should obey eternal laws, and that there should be a little animal, five feet high, who, in contempt of these laws, could act as he pleased, solely according to his caprice.

This conclusion was so disturbing that even materialists sought to modify its severity. Some said that though the body's actions are determined its thoughts are not. This resolution was not too comforting for it vitiates thoughts that lead to actions. According to this view man is still an automaton. Others reinterpreted the meaning of freedom so as to retain some semblance of it. Voltaire hedged: 'To be free means to be able to do what we like, not to be able to will what we like.' Apparently, we must like what is willed for us in order to be free. This unhappy position was also held by Leibniz.

By an act of will we shall interrupt our discussion of this problem. The most significant arguments for and against free will that have since been offered by philosophers can be understood only after we have surveyed some of the more recent mathematical developments. For the moment we can obtain some indication of the success of mathematical inroads into the territory of philosophy from the remark of the famous nineteenth-century physicist Lord Kelvin that 'Mathematics is the only good metaphysics.'

XVII

The Newtonian Influence: Religion

> And as of old from Sinai's top
>   God said that God is One,
> By Science strict so speaks He now
>   To tell us, There is None!
> Earth goes by chemic forces; Heaven's
>   A Mécanique Céleste!
> And heart and mind of human kind
>   A watch-work as the rest!
>
> *Arthur Clough*

GIORDANO BRUNO had declared that 'man is no more than an ant in the presence of the infinite'. This challenge to the Christian doctrine that man is at the apex of creation and the chief object of God's ministrations and solicitude could be answered in only one way in the sixteenth century – burning at the stake. The world of science came to Bruno's support a century too late.

As law after law was unearthed during the intervening century, the more nature appeared glorified and man humbled. The mathematical and mechanical realm of extension and motion, with man as an accidental offshoot and an irrelevant spectator, stood forth as the real world. The fact that the mind of man had penetrated to the core of phenomena and had devised the mathematical laws that described and rationalized nature passed unnoticed. Instead, emphasis was placed on the existence of laws, while man was deprecated because his limited mind was only gradually able to read them. In effect, nature was being read slowly but man was being read out of nature. It was certainly apparent that the universe took no account of human goals, desires, or needs. The good intent of God towards man in His plan and organization of the universe appeared to be a baseless notion having no more support than a myth.

As with man, so with God. The Newtonian era created celestial mechanics but destroyed heaven, the seat of God and the eventual dwelling place of privileged human souls. The work of Copernicus, Kepler, and Galileo on the heliocentric theory showed not only that the heavens were guided by simpler mathematical laws than the Ptolemaic theory indicated but also compelled the abandonment of the naïve conceptions of the cosmos that had become imbedded in Aristotelian and Thomist philosophy and that had been taken over by Christianity. Later Newton showed that heavenly bodies followed the same laws as do bodies on Earth. It seemed certain, then, that the heavenly bodies were made of the same stuff as was the Earth. By this discovery more mystery, as well as the fears and superstitions associated with the planets, were abolished.

God lost not only His home but His importance. To Descartes it was already clear that God, the omnipotent, could not abolish extension or the laws of motion. Newton, along with Descartes, credited God with the act of creation but restricted His daily functions. God prevented the stars from falling into each other and corrected irregularities that arise in the motions of the planets and comets. Huygens and Leibniz curtailed God's role even more. They too credited Him with the initial act of creation, at which time He established the mathematical order of the universe. Thereafter, however, His active relations with it ceased. In fact, it was an insult to God to believe that His creations would need repair.

Actually, Huygens and Leibniz were ignoring astronomical observations of irregularities unexplained in their time. These seeming aberrations from mathematical law, which Newton thought might become disruptive and require God's intervention, were shown later by Lagrange and Laplace to be periodic and hence part of the order of nature. The universe was stable; there was no room for whim or chance. By this masterly mathematical achievement even the corrective measures formerly required of God were made unnecessary and God was deprived of one more duty. In fact, the intervention of Providence in the affairs of nature became impossible while prayer for intercession on behalf of man certainly appeared futile.

It was not long before God Himself was completely dispensed with, for Hume attacked causality and, consequently, the need for a creator, or Prime Mover, of the universe. The world became an eternal, infinite, self-moving machine which existed before and would exist after insignificant man, serving no apparent purpose except perhaps to delight the mathematician who was slowly but surely uncovering the controlling principles. Events took place not because God ordered them individually and after due consideration but rather because they were predetermined by fixed, already existing mathematical laws. Thus God, who in medieval thought was not merely a cosmic carpenter but the end of all thought, activity, and purpose in the universe, was reduced to, at best, only a means to an end; the end itself became the regular, exact working of all the processes of the universe.

Not only the content but the spirit which suffused the great mathematical and scientific works of the seventeenth and eighteenth centuries threatened religious thinking. Since reason was exalted, faith was discredited as a meaningless guarantee of truth and was labelled credulity. In addition, under the scrutiny of rationalism much of the mystery and emotional appeal of the orthodox religions was dispelled. Emotion itself was frowned upon and considered suspect. Materialism submerged spiritualism, destroyed the soul and its after-life, and rendered pointless the Christian emphasis on preparation for an after-life. Determinism challenged free will, excused man's sins, and thereby removed the need for salvation. On all battle fronts religion and Newtonianism met as combatants.

This course of thought was contrary to the wishes and intentions of the great seventeenth-century scientists, for they were God-fearing men. Their very scientific work was an expression of religious feeling in that they studied nature to perceive God's law and order. To paraphrase James Thomson, they sought from motion's simple laws to trace the secret hand of Providence, wide-working through this universal frame. Each of the great intellects possessed a combination of mathematical or scientific genius and religious orthodoxy which today are regarded as incompatible and possible only in a period of transition. When these men did, however, become cognizant of the threat their work posed for religious beliefs, they attempted to reconcile their intellectual and spiritual affirmations. Robert Boyle, famous father of modern chemistry, devoted most of his time outside of the laboratory to religion. Even his experimental work he regarded as service to God. In his will he left funds to combat atheists, sceptics, and other infidels. Isaac Barrow, Newton's teacher, resigned his professorship to turn to divinical studies. Newton, too, devoted himself to theology, and considered the strengthening of the foundations of religion more important than his mathematical and scientific achievements, for the latter were restricted to uncovering God's design of the natural world only. Towards this end he contributed some studies which attempted to prove that the prophecies of Daniel and the poetry of the Apocalypse made sense and to harmonize the dates of the Old Testament with those of history. He often justified the hard and, at times, dreary scientific work only because it supported religion by providing evidence of God's order in the universe. It was as pious a pursuit as study of the Scriptures.

Most eloquent is Newton's statement of the classic argument for the existence of God:

The main business of natural philosophy is to argue from phenomena without feigning hypotheses, and to deduce causes from effects, till we come to the very first cause, which certainly is not mechanical. . . . What is there in places almost empty of matter, and whence is it that the sun and planets gravitate towards one another, without dense matter between them? Whence is it that nature doth nothing in vain; and whence arises all that order and beauty we see in the world? To what end are comets, and whence is it that planets move all one and same way in orbs concentric, while comets move all manner of ways in orbs very eccentric, and what hinders the fixed stars from falling upon one another? How came the bodies of animals to be contrived with so much art, and for what ends were their several parts? Was the eye contrived without skill in optics, or the ear without knowledge of sounds? How do the motions of the body follow from the will, and whence is the instinct in animals? . . . And these things being rightly dispatched, does it not appear from phenomena that there is a being incorporeal, living, intelligent, omnipresent, who, in infinite space, as it were in his sensory, sees the things themselves intimately, and throughly perceives them; and comprehends them wholly by their immediate presence to himself?

In the second edition of his *Principles*, Newton answers his own questions:

This most beautiful system of sun, planets, and comets could only proceed from the counsel and dominion of an intelligent and powerful Being. This Being governs all things, not as the soul of the world, but as Lord over all.

Joseph Addison's 'Hymn' framed Newton's argument in poetical terms:

> The spacious firmament on high,
> With all the blue ethereal sky,
> And spangled heavens, a shining frame,
> Their great Original proclaim.
> Th' unwearied sun from day to day
> Does his Creator's power display;
> And publishes to every land
> The work of an Almighty hand . . .
> What though in solemn silence all
> Move round the dark terrestrial ball;
> What though no real voice nor sound
> Amidst their radiant orbs be found?
> In Reason's ear they all rejoice,
> And utter forth a glorious voice;
> Forever singing as they shine,
> 'The Hand that made us is divine.'

Newton was convinced too that God was a skilled mathematician and physicist. He says in one of his letters,

To make this [solar] system, therefore, with all its motions, required a cause which understood, and compared together the quantities of matter in the several bodies of the sun and planets, and the gravitating powers resulting from thence; the several distances of the primary planets from the sun, and of the secondary ones [i.e. moons] from Saturn, Jupiter, and the earth; and the velocities with which these planets could revolve about those quantities of matter in the central bodies; and to compare and adjust all these things together in so great a variety of bodies, argues that cause to be not blind or fortuitous, but very skilled in mechanics and geometry.

Leibniz wrote many articles and books to combat the spreading apostasy. His *Testimony of Nature Against Atheists* tries to prove that the assumption of the existence of God explains some aspects of natural phenomena better than the scientific description in terms of matter, force, and motion, while his *Essais de Théodicée* rephrases the familiar argument that God is the intelligence who created this carefully designed world.

The defence of religion by Boyle, Newton, Leibniz, and others was not without some effect. Those people who were favourably disposed to religion experienced a tremendous exhilaration. God, the creator, had built a vaster heaven and Earth than man had previously dreamed of, a universe that operated unfailingly in accordance with wonderfully precise mathematical laws. Moreover, these laws revealed new aspects of God's nature in both senses of the phrase. Such manifestations of God's majesty could only renew faith and give additional reason for exultation in that faith.

Nevertheless, the efforts of these men were doomed to failure. Though mathematicians and scientists affirmed and defended the existence of God and the soul, these concepts were presented as intellectual abstractions rather than as deeply felt convictions. To accept such entities the mind had to know them as clearly and as distinctly as it knew mathematical conclusions. Since God was not known with such distinctness it followed that He did not exist. At least, history chose this implication rather than the one Boyle, Newton, and Leibniz had intended to support in their theological writings.

Their works failed to stem the tide that engulfed huge portions of the existing religious edifices. The fond hope that the mechanistic philosophy of nature as advanced by Descartes and Galileo and developed by Boyle, Newton, and Leibniz would furnish a lasting proof of the existence of a Divine Creator and thereby buttress Christianity was dashed by their successors. The mathematical and scientific work of the age was made the foundation of an intellectual crusade against the orthodox religions and helped to support all shades of opposition to these faiths. The name of Newton, in particular, became a symbol for the spirit of revolt against religion.

Desertions from the ranks of the religious became widespread. For example, whereas all the great French intellects of the seventeenth century were warmly attached to Catholicism, all those of the next century were opposed to it. The position of these intellectuals passed successively from a defence of orthodoxy to a rationalization of orthodoxy, from belief to Christian Deism, and then to 'scientific Deism', to scepticism, and finally to atheism.

To survey the influence of Newtonianism on religion we shall follow these leading eighteenth century currents. Faith had been the main support of religion but the new science and mathematics made the age partial to reason. Therefore, religion somehow had to be associated with reason. Towards this end some men maintained that the aim of theology should be to found the Christian religion on reason rather than revelation. Such a basis would guarantee its truth and, since reason and nature were wont to be identified in that age, would also provide a natural religion.

The movement to re-establish Christianity on rational principles is sometimes referred to as rationalistic supernaturalism, and one of its most famous exponents was John Locke. In his *Reasonableness of Christianity* and *Discourse on Miracles* he argued that religion is essentially a science; that is, from a set of reasonable axioms further propositions can be deduced that are not only reasonable but useful. Among reasonable axioms he proposed three: the existence of an omnipotent God, an axiom well supported by knowledge of our existence and the wisdom manifested in nature; virtuous living in obedience to the will of God; and the existence of a future life in which God will reward the virtuous and punish the wicked. It follows from these axioms that man will live so as to merit and attain reward in heaven.

Since all of Christianity could not be rationalized, some hedging was to be expected. In addition to the truths that were in accordance with reason or deducible from reasonable axioms, Locke admitted truths that are above reason and that are supplied by revelation. Resurrection of the dead is such a truth. We must be sure, however, that a revelation really comes from God;

also, no revelation must be contrary to our clear intuitive knowledge. *Reason must judge.* Reason is, in fact, revelation whereby
God communicates to us as much truth as lies within the reach
of our natural faculties. In any case reason is the last judge and
best guide. Unfortunately, vice and the craftiness of priests prevent reason from being heard in matters of religion.

Locke hedged further. Religion by its very nature must involve
man's relations with a superior power and hence must contain
some supernatural elements such as miracles. Evidently, if the
supernatural itself could not be rationalized at least the ad-
mission of such elements might be.

It is perhaps evident from Locke's defence of orthodoxy that
two difficulties were central, namely, the justification of both
revelation and miracles. Some who were not content with Locke's
arguments defended revelation on the ground that it was not
inconsistent with reason. Others adopted the negative defence
that nature and, therefore, reason contain unexplained phenomena and hence are as baffling as revelation. Neither explains evil,
for example. Still others argued that God seeks to test our capacity for understanding by acts of revelation and this accounts
for their lack of clarity.

Miracles, which in former ages were the best proof of the
existence of God, now had to be rationalized, for they were inconsistent with the order of nature. Some thinkers chose to
accept the miracles that were 'within' reason or at least not contrary to reason. For example, the dead could be brought back to
life but women could not reasonably be turned into pillars of salt.
To many, miracles were actually natural events only seemingly
unreasonable, just as the phenomenon of snow would appear to
be unreasonable to a native of the tropics.

As might be expected, the attempts to defend orthodoxy by
reasoning did not satisfy everyone. Most of the enlightened
wanted a completely rational religion, Christian or not, and since
Christianity could not be made completely reasonable to them,
these men proceeded to define and erect a new religion –
Deism.

It has sometimes been remarked that for the Deists, Reason
was God, Newton's *Principles*, the Bible, and Voltaire, the

Prophet. The Deists believed there is a natural religion just as there are natural mathematical laws of the heavens and Earth. It was not necessary, however, to resort to revelation or the Bible in order to seek the doctrines of this religion. They could be found by studying the sea, the sky, the flowers, the earth, and men. The study of creation is the best study of the Creator. From these natural sources rather than from the Scriptures some fundamental principles would immediately be apprehended and others then obtained by rational demonstration. Human reason, successful in the physical sciences, would be successful with this problem too.

By arguments too detailed to warrant repetition here, the Deists arrived at several positive principles. God held his place as the designer of the universe. He was the source of the universal laws discovered by Newton. There was a future life in which each man would be dealt with according to his merits. The worship of God and repentance were encouraged because they fostered a better life on Earth. Sin was disobedience to the dictates of reason. As these doctrines indicate, the Deists believed that the essence of religion was morality.

Such doctrines did not constitute too much of a deviation from Christianity. But the Deists also maintained that only those Christian doctrines were valid which could be defended by reason. Any that carried a taint of superstition, irrationality, or myth had to be rejected. Since the virgin birth, the divinity of Christ, and the concept of original sin were rationally inexplicable, these were among the first to be thrown out. Miracles, special providences, and supernatural revelation were also discarded as false. Rejection of such beliefs naturally brought Deism into direct conflict with Christianity, and despite the fact that the Deists accepted a God, albeit one suited to the role of master of the Newtonian universe, they were called atheists by the orthodox Christians.

Voltaire, the leading spirit and genius of the Enlightenment and a devoted follower of Newtonian mathematics and physics, was the chief advocate of the Deist movement. Championed by his lively and prolific writings, Deism became, among the educated people, the strongest of all religious movements of the eighteenth century. Here in America Thomas Jefferson and Benjamin Franklin were converted. So great was the influence of this rational religion that no one of our first seven presidents professed Christianity, though of course many made references to the Christian God in political addresses. Deism died down as a formal movement after the eighteenth century, but it is still the essence of the prevailing religious attitude among educated people of the twentieth century.

Many thinkers who wished to found a natural religion and were therefore essentially Deists even dispensed with God. They argued that natural theology was really a branch of science. The existence of God as an agent essential to the workings of the universe was rejected as extra-experimental. Furthermore, since the universe may always have been as it is there was no need to assume a creator. The argument for God as a First Cause was regarded as inferring the action of an 'inconceivable Being performing an inconceivable operation upon inconceivable materials'. Any explainable phenomenon, on the other hand, did not require the existence of God.

With or without God, Deism attempted to be completely rational. Actually, it catered somewhat to man's desire for faith and mystery, apropos of which it has been remarked that the Deists were rationalists with a nostalgia for religion. For this reason it did not satisfy some of the great thinkers who were completely sceptical. These men, among them the philosophers Hobbes, Hume, Montaigne, and Diderot, the mathematician D'Alembert who was Diderot's chief assistant in the writing of the *Encyclopédie*, and the historian Edward Gibbon, preferred to see in religion nothing more than a historical phenomenon that arises naturally with any people though it need not be indispensable. Hobbes explained the existence of formal religions as merely accepted superstition. 'Fear of power invisible feigned by the mind or imagined from tales, publicly allowed, is religion; not allowed, superstition.' To 'the infidel Hume', for example, religion was merely a mode of human behaviour. No supernatural elements in any faith were to be given the slightest credence. His contempt for the vast bodies of theology which the leading faiths had gradually created was unmistakable.

If we take in hand any volume of divinity, or school metaphysics, for instance, let us ask, Does it contain any abstract reasoning concerning quantity or number? No. Does it contain any experimental reasoning concerning matter of fact and existence? No. Commit it then to the flames: for it can contain nothing but sophistry and illusion.

Scepticism is most often an intermediate and transitory phase. Human beings find it beyond their strength to balance on a fence indefinitely. In eighteenth-century France scepticism was but a prelude to atheism. Early in the century the denial of religion was so rare and daring a view that the question of whether a disbeliever could die in peace was much discussed and the atheist who died with a jest on his lips was a sensation because he had not yielded to remorse; later on, however, atheism gained many followers. The French materialists, who started, as did the rational supernaturalists and the Deists, from the Newtonian structure of the universe, reasoned directly to a complete denial of religion.

It was Laplace, France's leading mathematician, who, in dispensing with a creator, drew the ultimate conclusion from the Newtonian cosmology. We have already related that when asked by Napoleon why he made no use of God in his book on the heavenly bodies, *Mécanique céleste*, Laplace replied that he had no need for such a 'hypothesis'. He was able to describe the motions of the heavenly bodies by recourse to mathematics and the Newtonian laws alone. Laplace thus excelled Newton in parsimony with hypotheses, though it was Newton who enjoined scientists not to include unnecessary ones.

It is seemingly paradoxical that Newton could 'prove' God's existence on the basis of his mathematical discoveries whereas Laplace with even more marvellous confirmation of the mathematical design of the universe at hand saw no need for God. But this paradox is readily explained by the remark of Pascal that nature proves the existence of God only to those who already believe in Him.

Laplace's stand on religion was shared by many other leading French thinkers. The idea of God, according to Baron d'Holbach, does not correspond to anything real. Originating in fear and calamity, it is created by the imagination to conciliate fictitious powers. On this extrapolation from ignorance are founded bodies of dogma and vast organizations. Religion diverts men's minds from the evils rulers inflict upon them and only serves to continue their misery by promising happiness in another world if they agree to be unhappy in this one. Ignorance, d'Holbach said, begets Gods; enlightenment destroys them. God is but nature; soul is just body.

. A large part of d'Holbach's widely read *System of Nature,* which was called the Bible of atheism, argues against the existence of God. Thoroughly in agreement with this view, the physician Julian O. de la Mettrie declared further that religion is useful only to priests and politicians. Since man understood nature there was no need for the primitive, superstitious account furnished by the established religions. Though La Mettrie was willing to admit the existence of God, he regarded this existence as pure hypothesis and of no practical use. In fact it was dangerous and evil. Far from guaranteeing morality it permitted religious leaders to stir up wars in the name of God. Thus the culmination of materialism in France was a revolt against what was regarded as the spiritual tyranny of all religions.

In the advocacy of atheism, religious thinking reached a pinnacle too high for many people. Some who made the climb intellectually were dizzied by the view and uncomfortable in the cold, rarefied atmosphere. Still others who attempted the climb could not find their way up and preferred to have a kindly light lead them where it would. The appeal for guidance was poignantly stated by Tennyson:

> Strong Son of God, immortal Love,
>   Whom we, that have not seen thy face,
>   By faith, and faith alone, embrace,
> Believing where we cannot prove . . .
> We have but faith; we cannot know,
>   For knowledge is of things we see;
>   And yet we trust it comes from thee,
> A beam in darkness; let it grow.

While some of the perplexed merely voiced their despair, others acted. The Wesley brothers, Cardinal Newman, and the leaders of the Oxford Movement saw the return to religious orthodoxy as the only salvation for civilization. The motivation for their actions as well as for those of the other religious movements of the eighteenth and nineteenth centuries can be best understood as a reaction against mathematical and scientific influences.

Many of us may regard the eighteenth-century trend to atheism as an evil. But one concomitant of the trend, the rise of tolerance and free thought, has been a superlative good. No one can read the history of the medieval and early modern periods without being struck by the power the religions wielded. In the name of God men were kept poor, dirty, and uneducated; men were trampled upon, tortured, burned, and killed; independence of thought and action was discouraged, repressed, or stifled.

The history of persecution for religious differences, which was by no means confined to the acts of Christians of the Renaissance, is indeed a horrible and shameful segment of human history. Men who had only their own faith to support their religious beliefs dared to murder dissenters by the most ingenious and devilish tortures: the boot, the rack, public whippings, slow fires, brandings, and nails driven into the body. Long and hard must these bigots have pondered to have been able to invent means of torture so 'ingenious' that they warrant the museum displays now given to them. Relying only on private judgement men dared to affirm their exclusive access to truth and to compel public acceptance by fire and sword. Montaigne described the situation with nice irony: 'It is setting a high value on one's opinions to roast men on account of them.'

Toleration is not the direct contribution of mathematics. Rather the movement owes its birth to the rationalistic spirit of the seventeenth and eighteenth centuries. The triumphs of human reason in the form of universal mathematical laws, however, constitute the fibre of rationalism. Moreover, mathematics, which stands or falls on the most rigorous reasoning man is capable of, is diametrically opposed to authority, blind faith, miracles, and the unreasoning acceptance of 'truths'. Finally, science, which teaches observation of nature, the continual verification of its conclusions, and a receptiveness to any theory that fits the facts even though it have the seeming wildness of the heliocentric and relativistic theories, rests largely upon math-ematics. This body of learning has, therefore, contributed vitally, if in some respects indirectly, to the propagation of a beneficent spirit.

The war between free thought and religion was declared in the days of Copernicus. The dust of battle is not yet fully settled but we have at least reached the point of recognizing the importance of freedom of worship, free speech, free press, and free inquiry. Fortunately, 'Liberty and not theology is the enthusiasm' of our age.

One more freedom was gained through the mathematical ac-complishments of the Newtonian era – freedom from super-stition. Most people in Western civilization are now convinced that the course of nature cannot be affected by mysterious devils, spirits, or ghosts, by incantations, or by mis-steps of human beings. Conviction of the sway of natural law has practically abolished the belief that certain trivial acts of man can insure good fortune or prevent calamities.

It is not usually recognized and seldom acknowledged that religions evolve. There is, however, no question but that the rise of rationalism has had salutary effects on religion itself. Religion no longer pre-empts the domain of science. Consequently, the work of mathematicians and scientists is relatively unhampered and the findings of science are recognized as the best source of our knowledge of nature. Theology and science are now regarded by religionists as mutually compatible and as reinforcing each other, while the conquests of science are accepted as a basis for rational theological speculations. Nowadays theologians repeat the very arguments advanced by Newton and Leibniz to prove the existence of God, and the services rendered by science in this demonstration are freely acknowledged. The mathematical laws of nature are held forth as evidence of a harmoniously designed universe, with God as the creator and lawgiver. As more and more laws are discovered science is hailed as revealing God to an ever-increasing degree.

Before the eighteenth century, moral laws had generally found their sanction in religion. The weakening and denial of religion

left these laws suspended in a vacuum. Moreover, the materialistic emphasis on worldly pleasures opposed the very substance
of Christian ethics, and determinism vitiated the doctrines of sin
and salvation, for it argued that the will is bound fast in the
determined behaviour of matter. Since, according to this view,
man is not a free agent, he is not responsible for his actions. The
denial of sin in turn re-opened the question of why evil exists on
this Earth, a problem, incidentally, as acute for the rationalists as
for the theologians. Christendom had accounted for evil by the
story of man's sin and fall, but this 'explanation' collapsed with
the collapse of sin.

Under rational scrutiny many ethical doctrines certainly
seemed baseless. Once the nature of God was examined with
scrupulous care a question arose – why should He favour virtue
rather than vice? The cultured Third Earl of Shaftesbury ridiculed the theory that virtue is the product of a bargain with
supernatural powers who would reward good and evil. Even
more radical was La Mettrie's conviction that pleasure is not a
sin but an art. Sensual pleasures in particular were approved.

Could the moral code survive religion? Some tentative answers
were proposed by eighteenth-century thinkers. Reason itself was
urged as the guide to conduct. Locke, for one, believed that the
principles of morality were capable of mathematical demonstration. Follow reason, the God within us, in order to determine
the proper conduct. We have reason enough to guide us provided
we take the trouble to be reasonable. Of those who urged the
application of reason some added that man has a moral sense
which operates in harmony with his reason. This natural sense of
right and wrong is independent of religion. It is not necessary to
fear God or seek rewards in heaven. In fact such a motive is
unchristian. The moral sense enables man to avoid evil and
choose the good just as his aesthetic sense predisposes him to
beauty.

Others, following the eighteenth-century identification of
reason with nature, said we should study man in his natural state
and imitate him. Hence the ways of primitive people, known to
Europe through the great explorations, were held up as ideals.
Because Magellan had written that the Brazilians had no civi-

lized vices and lived to be 140 years old, the Brazilian way of life was extolled. Since the Chinese way of life was more primitive than the European, it followed that the Chinese were more moral and that their society was an exemplary one. And when Bougainville, the explorer, published a glowing account of the life of the Tahitians, some Europeans became convinced that imitation of these people would restore the Garden of Eden. Even the Jesuit missionaries praised the virtues of unspoiled natural man, the noble savage.

Many philosophers decided that the position of ethics in relation to religion should be the reverse of the historical one. Locke said that the Scriptures confirm the moral laws which reason discovers. Kant, among others, believed that our morals are the basis for religion rather than the other way around. The Bible is valuable only in so far as it coincides with and supplements the moral code, and religion is useful only in so far as it sugar-coats the moral pill man must swallow in order to live as a decent member of society. Christianity becomes, in his view, no more than an 'admirable auxiliary to the police force'. Matthew Arnold expresses a similar view in evaluating religion as 'morality tinged with emotion'.

The code of ethics was so devastated by the weakening of religion that it required complete rebuilding. Mathematics made amends by providing a plan. A new Euclid was born who wrote the moral laws for all of society. This story, however, must be reserved for a later chapter.

XVIII

The Newtonian Influence: Literature
and Aesthetics

All Nature is but Art, unknown to thee;
All Chance, Direction, which thou canst not see;
All Discord, Harmony not understood . . .

Alexander Pope

DURING his travels in Laputa, Gulliver encountered several professors engaged in projects to improve the language of the country. One project was to shorten discourse by cutting polysyllables down to one syllable and by leaving out verbs and participles because in reality all things imaginable are but nouns. Another sought to dispense with all words whatsoever simply by having people carry objects about with them to exhibit instead. Though this latter plan was advocated as a great boon to brevity and even health, the women of Laputa objected because it did not allow them to use their tongues.

In this passage, as well as in numerous others, Jonathan Swift used his strongest weapon – satire – to ridicule the thoroughgoing influence on literature exercised by the mathematics of his day. Just as the successful businessman in twentieth-century America has become the authority in our time, so mathematicians, successful in revealing and phrasing the order in nature, became the arbiters of the language, style, spirit, and content of seventeenth- and eighteenth-century literature. The biggest literary figures of the age decided that their writings were inferior in all respects to the mathematical and scientific works and that prose and poetry could be improved by following these examples.

The writers began reconstruction by standardizing the language. Arbitrary symbols intended to remain fixed for all time were adopted for ideas just as the mathematicians use x as an arbitrary, fixed symbol for an unknown quantity. The standardization of the English language could be seen in the constant references to girls as nymphs, lovers as swains, lawns as dewy, fountains and streams as mossy, and water as limpid, while particular words were associated *ad nauseam*.

In further imitation of mathematics, ordinary discourse began to use abstract concepts. A gun became a levelled tube; birds were a plumy band; fish were a scaly breed or a finny race; the ocean became a watery plain; and the sky, a vault of azure. The poets in particular indulged in abstract terms such as virtue, folly, joy, prosperity, melancholy, horror, and poverty, which they personified and wrote in capital letters. Both standardization and the preference for abstractions stripped the language of concrete, colourful, picturesque, and succulent words.

The movement towards standardization culminated in one of the landmarks of the English language, Samuel Johnson's *Dictionary*. Johnson undertook to regulate a language which had been 'produced by necessity and enlarged by accident'. From a more or less inclusive explanation of the meanings of words Johnson converted the dictionary into an authoritative standard of good usage and the arbiter of verbal fashions. By careful distinctions clearly set forth, often with the aid of quotations, he established exact meanings and proper use of words. It was his intent that these meanings and usages were to be fixed for all time just as the word *triangle* has meant precisely the same thing for thousands of years.

This change in the concept of a dictionary appears radical in the history of dictionaries but it was almost to be expected in the eighteenth century. Johnson set about to do for the English language what had already been started in all spheres of activity, namely, to determine and establish the most reasonable, most efficient, and most permanent standards. Philologists since his day have learned that despite rules and definitions language is necessarily a fluid and evolving phenomenon. Words change in meaning from year to year and from place to place, as the modern dictionary clearly demonstrates by its inclusion of archaic meanings.

Standardization of language was accompanied by a critical

examination of the efficacy of ordinary language. Jeremy Bentham, distinguished for his ethical and political philosophy, concerned himself with this problem too. Nouns, he said, are better than verbs. An idea embodied in a noun is 'stationed on a rock'; one embodied in a verb 'slips through your fingers like an eel'. The ideal language would resemble algebra; ideas would be represented by symbols as numbers are represented by letters. Thereby ambiguous or inadequate words and misleading metaphors would be eliminated. Ideas would be associated by the smallest possible number of syntactical relationships just as all numbers are associated by just the few operations – addition, multiplication, equality, and so forth. Two statements could then be comparable in the same way that two equations are – for example, when one equation is obtained from another by multiplication by a constant. The movement to use symbols for nouns and connectives, in which Bentham participated, was related to the Leibnizian plan for the symbolization of language. Whereas Leibniz sought to facilitate reasoning, Bentham and others were concerned, however, with attaining precision.

The reform of language itself was but a minor mathematical influence on literature. Style was radically altered. It was well recognized in the Newtonian age that statements in a mathematical discussion or demonstration are concise, unambiguous, clear, and exact. Many writers believed that the success enjoyed by mathematics could be credited almost entirely to this naked and pristine style, and therefore resolved to imitate it.

In the seventeenth century the Fellows of the Royal Society decided that the reformation of English prose was within the province of that august body. A committee, including Sprat, Waller, Dryden, and Evelyn, was appointed to study the language. With furtive glances at the Académie Française the committee suggested founding an English academy for the 'improvement of speaking and writing'. It urged the members of the Society to avoid eloquence and extravagance of expression in the description of their experiments. They were to reject all 'amplifications, digressions, and swellings of style' and to seek a 'return to primitive purity, and shortness, when men delivered so many things in almost an equal number of words'. They were to use a 'close, naked, natural way of speaking; positive expressions, clear senses, a native easiness; *bringing all things as near the mathematical plainness as they can;** and preferring the language of artisans, countrymen, and merchants before that of wits and scholars'.

One of the great intellectuals of the age and its most famous popularizer of science, Le Bovier de Fontenelle (1657–1757), wrote in an essay 'On the Utility of Mathematics and Physics',

> The geometrical spirit is not so tied to geometry that it cannot be detached from it and transported to other fields of knowledge. A work on ethics, politics, or criticism, perhaps even a work of eloquence will be finer, other things being equal, if it is done by the hand of a geometrician. The order, the neatness, the precision, the exactness prevailing in good books for some time may well have arisen in that geometrical spirit now more widespread than ever.

Men we have met as outstanding mathematicians in preceding chapters were set up as literary models in the eighteenth century. Descartes' style was extolled for its clarity, neatness, readability, and perspicuity, and Cartesianism became a style as well as a philosophy. The elegance and rationality of Pascal's manner, especially in his *Lettres Provinciales*, were hailed as superb attributes of literary style. Writers in almost all fields began to ape as closely as their subject matter permitted the works of Descartes, Pascal, Huygens, Galileo, and Newton.

Under such influences numerous changes in prose style took place. Metaphors were banished in favour of accurate language describing objective realities. Locke said, in this connection, that metaphor and symbolism are agreeable but not rational. The pedantic, florid, scholarly style with complex Latinized constructions was abandoned in favour of a simple, more direct prose. Banished, also, were impetous flights of imagination, vigorous, emotionally charged expressions, poetic exuberance, enthusiasm, and sonorous and highly suggestive phrases. The writer's job, said Pope, is

> more to guide than spur the Muse's steed;
> Restrain his fury, than to provoke his speed.

* Italics mine.

The concern of writers was to communicate facts in a style that would accord with the high standards of logical thought. Clarity, proportion, the architectural instinct for form, rhythm, symmetrical structure, and cadences, and rigid adherence to set patterns were qualities of the new prose style. Prose became sober, terse, precise, and epigrammatic. A demand for easy intelligibility and clarity required that each phrase or group of words be readily grasped by the mind. Hence brief sentences became fashionable. Inversion was frowned upon; within the sentence the order of words was dictated by the thought. Also, sentences were organized to link with each other so as to show clearly and at once whence the thought came and where it went. The aim and law of prose style became the 'easy intelligible intercourse of minds'.

The emphasis on the rational elements in style at the expense of the emotional fostered the qualities appropriate to fine rhetoric, reasoning, and narrative, and discouraged the expression of the strong emotion and passion that inspire great poetry. The Age of Reason expressed itself, therefore, most characteristically in prose, the novel, diary, letter, journal, and essay lording it over the lyric and drama. In fact, the novel pretty much replaced poetry as the outlet for imaginative writing while the lyric poetry of the age became prosaic, 'poetized prose'.

Among prose forms satire became a favourite. The worship of reason made the unreasonable conspicuous and therein writers found a new theme. Since nature and reason were identified in the eighteenth century, the ways in which man had departed from the state of nature, for instance, in his grasping for power, wealth, and position, were readily singled out and attacked. The supreme satirist of the age, Jonathan Swift, is still widely read and what he wrote is still pointed. Each account of what Gulliver discovered in strange lands is a satire on some phase of the eighteenth-century European civilization. The puny Lilliputians appear at first to be amusing, uninformed, and helpless people, and we condescend to laugh at them until we realize that the joke is on ourselves. Gulliver's attempts to explain the customs and ways of Europeans to the Houyhnhnms, the elite members of a society of horses, succeeds only in ridiculing the Europeans.

As we have just seen, the Age of Reason favoured prose above poetry. In addition, the Newtonian spirit forced a sharp separation between prose and poetry, between what a person thought as a man of sense and judgement and what he felt as a poet, between knowledge of nature, on the one hand, and the colours of rhetoric, the devices of fancy, and the deceit of fables on the other. Prose dealt with facts, poetry with pleasure and fancy. A man might feel in poetic terms but he must think in prose. Thus emphasis on the reasoning faculty discredited the concepts, images, and values of poetry. Furthermore since truth in the Newtonian age consisted in knowledge of the clear and distinct mathematical properties of objects and since these are not the truths of poetry, the latter were rejected as fictional. In fact, in order to obtain truths men had to exorcize the phantasms of the imagination. Poetry could at best only decorate and render agreeable the abstract truths of mathematics and science.

Leading figures deprecated and some actually declared war on poetry. Locke said that poetry offers merely pleasant pictures and agreeable visions but these do not conform to truth and reason. Poetry is not really needed by people who have seen the light of reason; hence no labour or thought should be expended to examine the truth in a poem. In fact, the pleasure of poetry would be spoiled by the application of reason to its contents. He said, in addition, that if a child has poetic leanings the parents should labour to stifle them. Newton gave his opinion of poetry by citing his teacher Barrow, who said that poetry is a kind of ingenious nonsense. Hume was more brutal. According to him, poetry is the work of professional liars who seek to entertain by fictions. Bentham distinguished poetry from prose by the criterion that in prose all the lines except the last extend to the margin whereas in poetry some of them fall short. Poetry, he continued, proves nothing; it is full of sentimentalism and vague generalities. The silly jingling might satisfy the ears of a savage but would make no impression on a mature mind.

Even the poets themselves seem to have been browbeaten into accepting an inferior status. Dryden wrote in his *Apology for Heroic Poetry and Poetic License* that we should be pleased with the images of poetry but not cozened by the fiction. The best that

Addison could say in defence of poetry was that if the material world were endowed with only those qualities *which it actually possesses* it would make a joyless and uncomfortable poetic figure. Fortunately, a kindly Providence has given matter the power of producing in us a whole series of delightful *imaginary* qualities so that man may have his mind cheered and delighted with agreeable sensations. The dictator of eighteenth-century literature, Samuel Johnson, damned with faint praise. Poetry, he said, is the art of uniting pleasure with truth by calling imagination to the help of reason.

Of course poetry suffered. The opinion prevailed that the art required only a limited outlook, a little imagination, and a few rules to attain perfection. Poets accepted the belief, too, that their creations were not truths but just agreeable fictions. They merely catered to the delight of people; they provided embellishments that appealed to the fancy but that were not significant as reality even to the poets.

The art sank until it was considered to be only a minor amusement; then it sought to justify its existence by becoming more philosophic or more useful. Consequently, some poets decided that the functions of poetry should be didacticism, ratiocination, and argumentation in rhyme. Though it should not stir the feelings, poetry might refine the passions, moderate the fears, and propose examples of great virtues.

Not content with reducing poetry to a minor activity, the critics of the period strove to achieve mathematical objectivity by suppressing all personal or individualistic efforts in this medium. They ruled, first, that a poet should be something of a mathematician. Dryden declared: 'A man should be learned in several sciences, and should have a reasonable, philosophical and in some measure, a mathematical head to be a complete and excellent poet. . . .' The young American nation also fell under the new influences; in the words of Emerson:

We do not listen with the best regard to the verses of a man who is only a poet, nor to his problems if he is only an algebraist; but if a man is at once acquainted with the geometric foundation of things and with their festal splendour, his poetry is exact and his arithmetic musical.

Presumably mathematicians would appreciate that art, like science, possessed natural laws which could be derived by studying nature. Dryden said, in fact, that those things which delight all ages are an imitation of nature. Pope also expressed his belief in natural laws for poetry. In his *Essay on Criticism* he says,

> First follow Nature and your judgement frame
> By her just standard, which is still the same.
> Unerring Nature, still divinely bright,
> One clear, unchanged and universal light,
> Life, force, and beauty, must to all impart,
> At once the source, and end, and test of Art.

Curiously enough, to 'follow nature' did not mean precisely what it meant in the physical sciences, that is, to obey nature's mathematical laws. Rather, through the historically justifiable association of the Greeks with nature, to follow her meant to imitate the form of the Greek classics. Hence, said Pope:

> Those rules of old discovered, not devised,
> Are nature still, but nature methodized;
> Nature, like liberty, is but restrained
> By the same laws which first herself ordained . . .
> When first young Maro [Virgil] in his boundless mind
> A work t'outlast immortal Rome designed,
> Perhaps he seemed above the critic's law,
> And but from Nature's fountains scorned to draw;
> But when t'examine every part he came,
> Nature and Homer were, he found, the same.

Nevertheless when Pope translated Homer's *Iliad* he rendered Pope and not Homer. As Sir Leslie Stephen points out in *English Literature and Society in the Eighteenth Century*, 'When we read in a speech of Agamemnon exorting the Greeks to abandon the siege

> Love, duty, safety summon us away;
> 'Tis Nature's voice, and Nature we obey,

we hardly require to be told that we are not listening to Homer's Agamemnon but to an Agamemnon in a full-bottomed wig.' We need hardly be told, too, that we are listening to the voice of the eighteenth century attuned to its basic assumptions; the validity of rationalism and the prevalence of natural law. Thus the rules or laws of poetry grew out of an identification of nature, the ancients, and reason, so that to follow one was to follow all. The rules of art were 'nature methodized'.

Pope, Addison, and Johnson dictated poetic style in accordance with the philosophy described above. Strict rules were derived from a study of the ancients, while Dryden's translations of the Latin classics prescribed the laws of metrical translation into English. Verse, it was thought, could be written by rule; lyric, epic, sonnet, epistle, didactic verse, ode, and epigram could be built up by observance of the laws that established their forms; and order, lucidity, and balance were the goals to be sought in the process. Attention to grammatical rules and sentence structure was even recommended. The principles of form in poetry were likened to mathematical axioms because the axioms determined the form as well as the content of the theorems. The heroic couplet won favour because of its balance and symmetry, and, extreme as this view may seem to us, because the form was analogous to a series of equal proportions. The heroic couplet was regarded as the essence of cadenced regularity. Beauty, to the literary critics of the age, consisted in adherence to these strict rules of versification.

The poets adopted a code that was laid down as a series of mathematical propositions, and they followed the rules of the critics meticulously. Great poetry was reduced to correct writing, that is, obedience to the code. Poetry did become temperate, well regulated, and intellectual. The poets adopted Pope's formal and strictly regulated versification and emphasized such neo-classical ideals as lucidity, moderation, elegance, proportion, and universality. Decorum, which meant harmony of theme, matter, and form, was also observed.

Since spirit as well as form was prescribed, the poets, though often ironic, suppressed feeling. Enthusiasm was abhorred; emotion, abandon, rapture, and mystical contemplation were outlawed. Imagination was limited severely by reason, coolness, and discretion, in accordance with Dryden's injunction that imagination 'is a faculty so wild and lawless that like a high-ranging spaniel, it must have clogs tied to it, lest it outrun the judgement'. Thus the great tragedies became the tragic victims of the new literary atmosphere of common sense. The union of heart and head, the synthesis of thought and feeling, was destroyed.

The conception of poetry as something awful, spiritual, and divine was almost forgotten during the eighteenth century. Those few writers who persisted in writing the poetry of passion had to smuggle their works into the literary world either by disguising them or by pretending to ridicule their very efforts while offering them. Only a few men, notably Collins, Smart, Cowper, and Blake, some supposedly having traces of insanity in their make-ups, dared to violate the rules and to write according to their own dictates.

If the spirit of eighteenth-century poetry was impoverished, the substance, at least, was enriched. The major seventeenth-century poets of Newton's youth wrote devotional poetry or love lyrics. Almost all of them ignored mathematics and science. The few who happened to touch upon these subjects seemed unaware of the tremendous import of current developments. Still others even ridiculed mathematics. In 1663 Samuel Butler wrote in *Hudibras*,

> In Mathematicks he was greater
> Than Tycho Brahe, or Erra Pater:
> For he, by Geometrick scale,
> Could take the size of Pots of Ale;
> Resolve by Signs and Tangents streight,
> If Bread or Butter wanted weight;
> And wisely tell what hour o' th' day
> The Clock doth strike, by Algebra.

After Newton's work, ridicule changed to unbounded admiration. Poetry became filled with appreciation and praise of the new mathematics and science. The writers found reason, mathematical order and design, and the vast mechanism of nature themes so moving that these replaced the concern for the birth, love, and death of insignificant man. No one was so unrestrained in his enthusiasm for the new wonders of the world as Dryden.

> From harmony, from heavenly harmony
> This universal frame began;

> From harmony to harmony
> Through all the compass of the notes it ran,
> The diapason closing full in Man . . .

> As from the power of sacred lays
> The spheres began to move,
> And sung the great Creator's praise
> To all the blest above;

Famous also are the lines Alexander Pope intended as an epitaph for Newton's tomb in Westminster Abbey:

> Nature and Nature's laws lay hid in night;
> God said, 'Let Newton be,' and all was light.

Unfortunately it is impossible to survey here the contents of the great poetry of the Newtonian age. The occasional quotations that occur elsewhere in the book and the chapter mottoes may perhaps give some indication of the new subject matter.

However much the critics of the eighteenth century defended cold, mechanical, and impersonal literature, they could not abolish the hearts of sensitive people. During the nineteenth century the realization came that the code for poetry was hopelessly inadequate and that the images of poetry had worn thin. The rules of descriptive geometry produce a draughtsman's sketch, not a work of architecture. As Robert Burns put it, in reference to the imitation of the ancient classics, the poet could not 'hope to reach Parnassus by dint o' Greek'.

Suppression of the spirit had been so drastic that the early nineteenth-century poets felt all beauty had been banished. Keats execrated Descartes and Newton for cutting the throat of poetry and Blake damned them. At a dinner party in 1817 Wordsworth, Lamb, and Keats, among others, drank a toast that ran: 'Newton's health, and confusion to mathematics.' Though Blake, Coleridge, Wordsworth, Byron, Keats, and Shelley understood what mathematics and science had accomplished and admired the accomplishments, they protested nevertheless against what had happened to the essence of poetry. Shelley said in reference to the restriction of the imagination that, 'Man, having enslaved the elements, remains himself a slave.' Coleridge rejected the mechanical universe as a dead world. William Blake called reason the devil whose high priests were Newton and Locke. 'Art is the Tree of Life. ... Science is the Tree of Death.' He felt that the mechanical account of nature is hopelessly inadequate to render nature.

> Tiger, tiger, burning bright
> In the forests of the night,
> What immortal hand or eye
> Could frame thy fearful symmetry?

Wordsworth contended that reason alone produces immoral monsters and he attacked scientists who dive rather than soar and who pry apart nature and soul and thereby miss the grandeur and mystery.

> Man now presides
> In power, where once he trembled in his weakness,
> Science advances with gigantic strides;
> But are we aught enriched in love and meekness?

Reaction and revolt, conscious and unconscious, set in against the material, colourless physical machine that was called nature in the eighteenth century. Emotions repressed for a century broke the restraining bonds and rebelled against the domination of thought and feeling by mathematics and science. The perfect order of the universe proclaimed in the eighteenth century was declared to be an illusion, for mysteries and contradictions unresolved by reason were still there. Poets asserted the importance of the senses, feelings, and man's own consciousness. Nature, they said, is to be lived with rather than apprehended through the inadequate mathematical account given by scientists. Let us enjoy nature directly, said Wordsworth, rather than indulge in rationalistic orgies.

> Great God! I'd rather be
> A pagan suckled in a creed outworn;
> So might I, standing on this pleasant lea,
> Have glimpses that would make me less forlorn;

Poetry was delivered from the fetters of the mechanical tradition. Emotions were revived and expressed; myth and symbol were revitalized. Imagination was set above reason or, by some,

proclaimed as the highest form of reason since it supplied intuitive truths. The poet was enjoined to be more than a rational
commentator. He was directed to exercise his own genius and to
express the deity who resided in his breast. By the bond between
nature and the soul of man the dead world might be brought to
life and enjoyed directly.

> Paradise, and groves
> Elysian, Fortunate Fields – why should they be
> A history only of departed things,
> Or a mere fiction of what never was?
> For the discerning intellect of Man
> When wedded to this goodly universe
> In love and holy passion, shall find these
> A simple produce of the common day.

The universe is not cold but active and capable of being moulded
by the power within man. Poetry records the ennobling action,
and the poet's spirit transforms the inanimate world into life.
Therefore am I still, said Wordsworth,

> A lover of the meadows and the woods,
> And mountains; and of all that we behold
> From this green earth; of all the mighty world
> Of eye, and ear – both what they half create,
> And what perceive; well pleased to recognize
> In Nature and the language of the sense,
> The anchor of my purest thoughts, the nurse,
> The guide, the guardian of my heart, and soul
> Of all my moral being . . .
> Knowing that Nature never did betray
> The heart that loved her;

Thus nature remained the major theme of the poets but it was
nature suffused with feeling, alive and vibrant, rich in colour,
appealing to the senses, and mysterious, rather than nature
confined in the chains of abstract laws. The poets of the nineteenth century chose the concrete experiences of nature, 'sensations sweet, felt in the blood and felt along the heart.' They
enjoyed the sounds, light, smells, and sights of life itself. Sunrise
and sunset bedazzled the mathematical analysis of light; the

living fire of the sun overpowered its gravitational attraction of other masses; and the wild west wind – the 'breath of autumn's being', the uncontrollable spirit moving everywhere, and the wakener of the blue Mediterranean – swept away the regular, mechanical motion of the air molecules.

Though the romantic poets rebelled, they did not free themselves completely from the chains that bound down their spirit. In fact, the progress in mathematical and scientific thought made during the nineteenth century reinforced the conceptions of the universe so ardently advanced by the eighteenth-century rationalists and, of course, the poets were keenly aware of this fact. When their passionate outbursts had subsided somewhat they again faced the problem of the meaning of the universe. Throughout the nineteenth century they meditated upon and were torn between the account of nature furnished by mathematics and science and the account furnished by the senses. Matthew Arnold spoke for his contemporaries,

> . . . for the world, which seems
> To lie before us like a land of dreams,
> So various, so beautiful, so new,
> Hath really neither joy, nor love, nor light,
> No certitude, nor peace, nor help for pain;
> And we are here as on a darkling plain
> Swept with the confused alarms of struggle and flight,
> Where ignorant armies clash by night.

The conflict between heart and mind is still the major theme of poetry. The more that reason has accomplished, the more troubled have the poets become.

Literature was not the only art to be strongly influenced by the flourishing and almost domineering mathematical spirit of the Newtonian age. Eighteenth-century painting, architecture, landscape gardening, and even furniture design became subject to rigid conventions and explicitly set standards. The precepts of the painter Sir Joshua Reynolds illustrate the artistic temper of the times. He stressed fidelity to the object painted, subservience of colour to idea, and the sacrifice of details to the general and everlasting elements. Moreover, the painter was asked to address himself to the mind and not the eye. In architecture and the minor arts, order, balance, symmetry, and strict adherence to well-known simple geometrical forms ruled the day. Art academies formed on the pattern of the successful scientific academies promulgated the criteria of art and exerted great influence in setting and in securing adherence to the fashion. Unfortunately, our brief survey of the Newtonian influences does not permit extensive excursions into the history of these arts.

Following the changes in the character of literature, painting, and the other arts came the change in the philosophy of aesthetics, which rationalized and justified the new attitudes. The new thesis of aesthetics was that art like science was derived from the study and imitation of nature and hence, like nature, was susceptible to mathematical formulation. According to Sir Joshua Reynolds,

It is the very same taste which relishes a demonstration in geometry that is pleased with the resemblance of a picture to its original and touched with the harmony of music. All these have unalterable and fixed foundations in nature.

Moreover, said Sir Joshua, the essence of beauty is the expression of universal laws.

Just as observation had produced Kepler's laws, so the study of nature would reveal the laws of art. Some, however, believed that reason alone, independently of observation, could deduce by the *a priori* geometrical method the mathematical laws of aesthetics, for beauty like truth is apprehended by rational faculty.

And so men studied nature or applied their rational faculties to reduce art to a system of rules, and beauty to a series of characterizing formulas. Precepts for attaining beauty were laid down, and analyses were made of the nature of the sublime. It was expected that the search for beauty in nature would produce not only an abstract ideal of beauty but its chief characteristics. With this knowledge, beautiful works could then be created almost at will, though only by observance of the rules of art so discovered. Unfortunately great art is still not being made on a mass-production basis; perhaps this is because no modern industrial tycoon has capitalized on the findings of the eighteenth century.

The last three chapters have given some indication, it is hoped, of the revolution in culture caused by Newtonian mathematics.* At the time of Newton's death the changes were already immeasurable and the impress was just beginning to be exerted. The implications and amplifications of Newtonian mathematics are still shaping our thoughts, as well as our mode of life. Indeed the eighteenth-century Age of Reason marked merely the inauguration of an essentially modern culture as opposed to an earlier ecclesiastical and feudal one.

In general the accomplishment of Newton and his contemporaries was to initiate a vast intellectual inquiry into the nature of the world, man, society, and almost every institution and custom of man. The age passed on to its successors the ideal of general, all-embracing laws. It also launched our civilization on a quest for omniscience, stimulated the desire to organize thought into systems built on the mathematical pattern, and distilled a faith in the power of mathematics and science. The greatest historical significance of the seventeenth- and eighteenth-century mathematical creations is that they animated the rationalistic spirit that has suffused almost all branches of our culture.

On the basis of the striking successes achieved by Newtonian mathematics and science in the fields of astronomy and mechanics, the eighteenth-century intellectuals asserted the conviction that all of man's problems would soon be solved. Had these men known of the additional marvels science and mathematics were soon to reveal, they would have been even more unreserved, were that possible, in their expectations. It is now evident that these thinkers were indulging in unwarranted optimism. Their conviction, however, was prophetic at least to the extent of a half-truth, for mathematics and science did go on to re-make the world if not to solve all its problems. Even in those domains where very little progress has been made towards the solution of basic problems, the ideals of the Age of Reason still provide the goals and the driving force.

---

* See also the chapter called 'The Science of Human Nature'.

XIX

The Sine of G Major

Music is the pleasure the human soul experiences from
counting without being aware that it is counting.

Gottfried Leibniz

It may have been, to improvise on history, that Pythagoras
spent many an hour, seated in the shade of his native olive trees,
plucking the strings of a lyre. In some such way he discovered
that the pitch of a sound from a plucked string depends on its
length and that harmonious sounds are given off by strings, the
ratio of whose lengths are simple whole numbers. From the time
of Pythagoras, the study of music was regarded as mathematical
in nature and grouped with mathematics. This association was
formalized in the curriculum of the medieval system of education wherein arithmetic, geometry, spherics (astronomy), and
music comprised the famous quadrivium. The four subjects were
linked further by being described as pure, stationary, moving,
and applied number, respectively.

During the many years from the age of Pythagoras to the
nineteenth century, mathematicians and musicians alike, Greek,
Roman, Arabian, and European, sought to understand the nature
of musical sounds and to extend the relationship between mathematics and music. Systems of scales and theories of harmony
and counterpoint were dissected and reconstructed. The climax
of this long series of investigations, from a mathematical standpoint, came with the work of the mathematician Joseph Fourier,
who showed that all sounds, vocal and instrumental, simple and
complex, are completely describable in mathematical terms. Because of Fourier's work not even the elusive beauty of a musical
phrase escapes submission to mathematical formulation. Whereas
Pythagoras was content to pluck the strings of a lyre, Fourier
sounded the whole orchestra.

Though Joseph Fourier, born at Auxerre, France, in 1768, did exceedingly well as a student of mathematics, he set his heart on becoming an artillery officer. When he was denied a commission because he was the son of a tailor, he reluctantly turned to study for the priesthood. He abandoned this career when his mathematical ability won for him a professorship in the very military school he had attended, social status being unnecessary for so lowly a position.

It was in 1807, after years of political and scientific service to Napoleon, that he presented to the French Academy a theorem of unprecedented importance for the progress of the physical sciences. This theorem advanced the mathematical mastery of the motions of the air waves as much as Newton's work had forwarded the study of the motions of the heavenly bodies. Evidently the nineteenth century was rushing to fulfil the grandiose expectations of the eighteenth.

We shall now see how Fourier's work made possible a thorough mathematical analysis of musical sounds. Suppose a violinist stands on the stage of a large theatre and draws his bow over the strings of his instrument. Some of the notes he plays sound for only a fraction of a second; others are long drawn out; some are loud; some are soft; some are high-pitched; others, low. People seated a hundred feet away hear all these sounds exactly as they are played. What happens physically when the violinist plays and how does his music reach the audience?

By way of explanation let us consider first the simple sound given off by a tuning fork. If a prong of a tuning fork is struck, the fork will oscillate very rapidly. As the prong moves to the right for the first time it crowds together the air molecules alongside (fig. 53). This crowding is called a *condensation*. Because air pressure tends to equalize itself the crowded air particles move farther over to the right where there is not so much crowding. There the process is repeated and the condensation moves still farther to the right.

In the meantime, however, the prong has moved back to the left beyond its original position. This leaves a comparatively vacant region in the prong's former position. The air molecules situated to the right of this region rush into this less crowded space, thereby creating another rarefied region in their former position. The molecules farther over to the right now move to the left into this rarefied region, and so on. If we call the creation of a rarefied region a *rarefaction*, we can say that what is now happening is that a rarefaction is moving to the *right*, away from the prong. Each movement of the prong to the right and to the left sends a condensation and rarefaction to the right.

We have considered the motions that take place to the right of the prong. Actually, condensations and rarefactions move off in

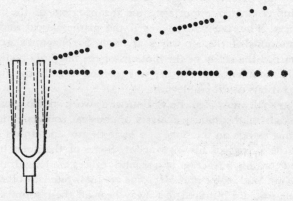

Figure 53. Motion of molecules of air due to a vibrating tuning fork

all directions. When these condensations and rarefactions reach our eardrums, the vibrations they induce there cause the sensation of a sound.

It is important to notice that the air molecules do *not* move from the tuning fork to the ear. Each molecule moves back and forth in a limited region around the position it occupied before it was disturbed. What is transmitted is the succession of condensations and rarefactions, and these constitute the sound wave.

Strictly speaking, all the air molecules in a particular region do not move in exactly the same way; it is the net effect of their collective movements, however, that interests us. This can be described in terms of the motion of a typical molecule. Suppose this molecule is originally at *O* (fig. 54). A condensation causes it to be displaced to the right to *A*. The ensuing rarefaction then causes it to move back past its original position to *B*; the next condensation causes it to move back to *O*. It has now made one complete vibration. Without stopping at *O*, however, the mol-

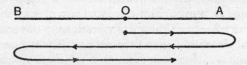

Figure 54. Motion of a typical molecule of air

ecule goes through the whole set of motions again and again under the successive impulses arising from the tuning fork. Thus the displacement of the molecule from its original position varies continually with the time during which it moves.

The movement of the typical air molecule is vividly shown by a very delicate instrument called a *phonodeik*. When a sound is

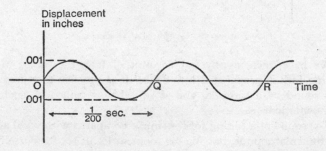

Figure 55. Graph of displacement versus time of a typical molecule of air

produced near this instrument it records the air vibrations in the form of a graph which pictures the displacement of the typical air molecule. The molecule moves back and forth along a straight line. The graph exhibits the displacement from the original rest position as a *vertical* distance, however, while the horizontal axis on the graph shows the time elapsed from the commencement of the motion. The portion of the curve from *O* to *Q* (fig. 55) represents the motion of the typical molecule during one complete vibration of the tuning fork; that from Q to R, the motion during another complete vibration; and so on. If the tuning fork is struck so as to cause the typical air molecule to move a maximum of ·001 inch first to one side of its original position and then to the other, the phonodeik records a graph with an amplitude, that is, a maximum displacement, of ·001 inches. If the tuning fork makes 200 complete vibrations in one second, so will the typical air molecule; and the phonodeik will record 200 complete portions, such as that from O to Q, in one second.

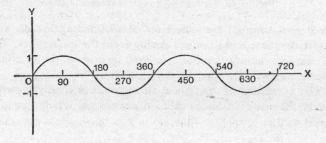

Figure 56. Graph of $y = sine\ x$

We have, then, a physical account of how the sound of a tuning fork is propagated into space. Is it possible to represent this sound by a formula, and if so, what is gained by such a representation?

The sound of a tuning fork is simple compared with vocal and instrumental sounds, but for the moment let us set for ourselves the task of representing this simple sound mathematically. What we seek, then, is a formula that relates the displacement and time of travel of the typical molecule, just as a formula relates the distance an object falls to the time it takes to fall.

The mathematician has the formula ready-made. He has in his store of relations among variables the formula $y = sine\ x$, with whose properties we can best become acquainted by means of a graph. As fig. 56 shows, the y-values of this function increase from 0 to 1 as x increases from 0 to 90; as x increases further the y-values decrease to 0, become negative until they reach the value

−1, and then increase to 0 as *x* reaches the value 360. In the interval from *x* = 360 to *x* = 720, the *y*-values repeat their behaviour from *x* = 0 to *x* = 360. In each succeeding 360 units of *x*-values, the *y*-values again repeat their behaviour of the first 360-unit interval. In other words, the function is regular, or *periodic*; or, we may say, the *cycle* of *y*-values repeats itself after every 360-unit interval of *x*-values.

The reader may have noticed the word *sine* here and he may have recalled its use earlier in connection with the mathematics of Alexandrian Greek times. The *y*-values of the function *y* = *sine* *x*, as *x* varies from 0 to 90, are precisely the values of the trigonometric ratio *sine x*, as *x* varies from 0° to 90°. During the centuries from Hipparchus to the time of the Swiss mathematician Euler, the trigonometric ratios, which were originally defined for angles in right triangles, were divorced from angles and came to be regarded solely as relationships between variables. Thus *y* = *sine x* became a relationship between two variables *y* and *x*. During these same centuries this relationship was broadened so that to each *x*-value, no matter how large, a *y*-value was

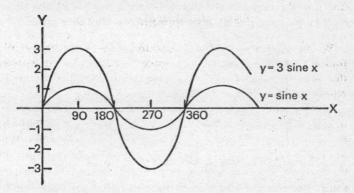

Figure 57. Graphs of *y* = *sine x* and *y* = 3 *sine x*

assigned as shown by figure 56. The formula *y* = *sine x* is then an old foe with a new face that has returned to torment us. Because of its origin in the ratios introduced for triangle measurement, *y* = *sine x* is called a trigonometric function.

This function does not quite represent the sound of a tuning fork but a very simple modification of it does. A little effort will produce the proper modification. Consider $y = 3\ sine\ x$. This formula differs from $y = sine\ x$ in that for the same x-value the y-value of the former is three times the y-value of the latter. Figure 57 shows the behaviour of $y = 3\ sine\ x$ and compares it with $y = sine\ x$. We can describe the curve of $y = 3\ sine\ x$ by saying that it is like the ordinary sine curve in shape; however, its amplitude, that is, its maximum y-value, is 3 units, where the amplitude of $y = sine\ x$ is 1. Similarly, the graph of $y = a\ sine\ x$, where a is an arbitrary positive number, has the general shape of the sine curve but has an amplitude of a units.

Another simple variation of the sine function is illustrated by $y = sine\ 2x$. We might suppose that this function is the same as $y = 2\ sine\ x$ and therefore that this function is another example of the type just analysed. We shall see in a moment, however, that this is not the case. The effect of the 2 in the formula $y = sine\ 2x$ is most readily appreciated in a graph. Fig 58 shows that in the interval from 0 to 180, $sine\ 2x$ takes on the full cycle of y-values that $sine\ x$ assumes in the interval from 0 to 360. By the time x reaches 360, $y = sine\ 2x$ goes through two complete cycles of y-

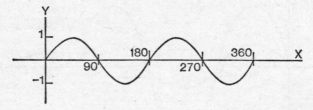

Figure 58. Graph of $y = sine\ 2x$

values whereas $y = sine\ x$ goes through only one. Therefore the frequency of the former function in 360 x-units is said to be 2. The amplitude of $y = sine\ 2x$ is 1 since the greatest numerical value of the sine of any quantity is 1.

We can generalize the result above to the case of the function $y = sine\ bx$ where b is an arbitrary positive number. The frequency of $y = sine\ 2x$ is 2. Similarly, the frequency of $y = sine\ bx$ in the x-

interval of 360 units is *b* – which means that the *y*-values repeat the full cycle of changes *b* times as *x* varies from 0 to 360. As in the case of *y* = *sine* 2*x* the amplitude of *y* = *sine* *bx* is 1.

A variation of the sine function that differs both in amplitude and in frequency from the behaviour of *y* = *sine* *x* is exemplified by *y* = 3 *sine* 2*x*. The *y*-values in this function are three times the

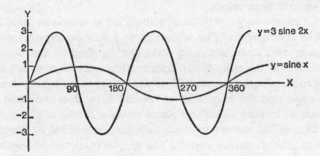

Figure 59. Graphs of *y* = *sine* *x* and *y* = 3 *sine* 2*x*

values obtained from *y* = *sine* 2*x* for the same values of *x*. Hence the amplitude of 3 *sine* 2*x* is 3, and its frequency in 360 units of *x*-values is 2 (fig. 59).

The results we have obtained thus far can be summarized by the statement that the function *y* = *a* *sine* *bx*, where *a* and *b* are any positive numbers, has the amplitude *a* and the frequency *b* in 360 units of *x*-values.

We are now prepared to represent the sound of the tuning fork mathematically. A comparison of the graphs we have just been discussing with the actual graph of the tuning-fork sound suggests what theoretical reasoning can confirm. The function relating the displacement and time of the typical vibrating air molecule is of the form *y* = *a* *sine* *bx*. We have merely to determine the proper *a* and *b* to suit the behaviour of the tuning fork.

If the amplitude of the motion of a typical air molecule, when acted upon by the tuning fork, is ·001, then this number should be the value of *a* in the formula *y* = *a* *sine* *bx*; and if the tuning fork, and therefore the typical air molecule, makes 200 vibrations per second, then the graph of this molecule's motion has a frequency of 200 per second. But the frequency of *y* = *a* *sine* *bx* is *b* in 360 units, or $b/360$ in one unit.* Hence $b/360$ should equal 200. Then $b = 360 \cdot 200$ or 72,000. Therefore the formula that describes the sound of the tuning fork is

$$y = \cdot 001 \ sine \ 72,000t,$$

wherein we have written t for x to remind us that this variable represents time values.

Of course very few musical sounds are as simple as those given off by tuning forks. The sounds from a flute do approximate the simple ones from a tuning fork, but the flute is the exception rather than the rule. What does mathematics have to say about the more complex sounds? How does it account for the sweetness of some and the harshness of others? Why does the same note given off by both violin and piano sound different to the ear?

Part of the answer to these questions is learned by observing the graphs of various sounds. The graphs of all musical sounds – the ordinary sounds of the human voice are included in this term – display regularity. That is, each graph of displacement against time repeats itself exactly many times a second. This periodicity is exemplified by the graphs of sounds of the violin and clarinet, as well as by the graph of the sound of a as in the word 'father' (fig. 60).

Sounds that possess this graphical regularity are, on the whole, pleasing to the ear and are to be distinguished from, say, the noise of a tin can being bounced along the street – this noise has a highly irregular graph. All sounds that possess graphic regularity or periodicity are called, technically, musical sounds no matter how such sounds are produced.

In 'graphic' terms we have, then, the distinguishing feature between pleasing and displeasing sounds, between musical sounds in the broad sense and noise. Unfortunately, such a bewildering variety of musical sounds possesses this feature of regularity that further analysis and characterization is necessary – and yet this had seemed impossible until the nineteenth cen-

* The frequency of actual sounds refers to the number of vibrations in *one* unit of time, usually a second. The frequency in 360 units is called the circular frequency.

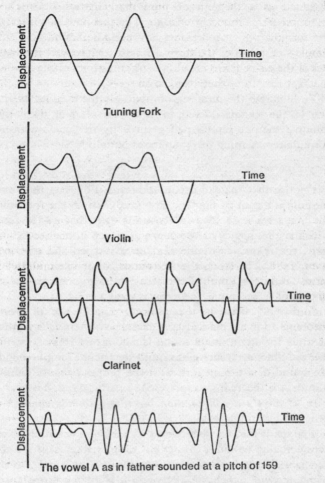

Tuning Fork

Violin

Clarinet

The vowel A as in father sounded at a pitch of 159

Figure 60. The periodicity of instrumental and vocal sounds (Courtesy of Dayton C. Miller)

tury. Then Fourier entered the scene and dispelled the confusion.

Stated as a theorem of pure mathematics Fourier's contribution seems innocent enough. The theorem says merely that the formula which represents any periodic sound is a sum of simple sine terms of the form *a sine bx*. Moreover, the frequencies of these sine terms are all integral multiples of the lowest one, that is, twice, three times, and so on.

To illustrate the meaning of Fourier's theorem let us analyse one of the sounds offered to us by an obliging violinist, for example the one represented graphically in fig. 60 above. The formula representing this graph is essentially*

$$y = \cdot06 \; sine \; 18,000t + \cdot02 \; sine \; 360,000t + \cdot01 \; sine \; 540,000t.$$

We notice first that, in accordance with Fourier's theorem, the formula is a sum of simple sine terms. Second, the frequency of the first term is 180,000 in 360 units of *t,* that is, 360 seconds, which is a frequency of 180,000/360 or 500 in one second. Similarly, the frequency of the next term is 1000 and of the third term, 1500. Therefore, the frequencies of the second and third terms are integral multiples of the lowest frequency. The graph of each of these simple sine terms is shown in fig. 61.

And now, what is the physical significance of Fourier's theorem? In mathematical language the theorem tells us that the formula for any musical sound is a sum of terms of the form *a sine bx*. Since each such term could represent a simple sound, say the sound of a tuning fork with the proper frequency and amplitude, the theorem says that *each musical sound, however complex, is merely a combination of simple sounds* such as those given off by tuning forks.

The mathematical deduction that any complex musical sound can actually be built up from simple sounds is physically verifiable. Experiments show that a vibrating string, as in the piano and violin, behaves as though it is giving off many simple sounds simultaneously. Each of the simple sounds can actually be detected by special instruments.

* For the sake of simplicity we have neglected the relatively unimportant matter of phase.

Even more remarkable evidence of the composite nature of musical sounds is furnished by the fact that any musical sound can be duplicated with the proper combination of the simple sounds of tuning forks. For example, a tone whose quality is practically indistinguishable from the quality of the violin tone discussed above can be produced by sounding simultaneously,

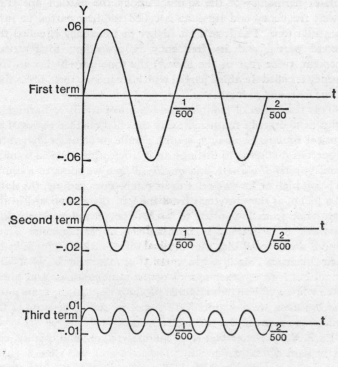

Figure 61. Graphs of the component *sine* terms in the sound of a violin

with suitable relative loudness, three tuning forks with frequencies of 500, 1000, and 1500 vibrations per second. These three tuning forks simultaneously impose their own vibrations on a typical air molecule so that their effect on the air molecules is recorded by the phonodeik as a single graph. If each of the forks is started at the proper instant the phonodeik will record the same graph it records for the complex violin sound. It is theoretically possible, therefore, to play Beethoven's Ninth Symphony (including the Choral Ode) entirely with tuning forks. This is one of the startling implications of Fourier's theorem.

Any complex sound, then, can be built up by a suitable combination of simple sounds. The simple tones are called the partials or harmonics of the sound. Among the partials one is of lowest frequency and this one is called the first partial or fundamental tone. The tone next higher in frequency is called the second partial and its frequency is, according to Fourier's theorem, twice that of the lowest; the tone next higher in frequency is called the third partial with a frequency three times that of the first; and so forth.

This resolution of complex sounds into partials or harmonics helps us to describe mathematically the chief characteristics of all musical sounds. Each such sound, simple or complex, has three properties that serve to distinguish it from other musical sounds, namely, *pitch*, *loudness*, and *quality*. When we speak of a sound as being high or low we refer to its pitch. For example, the notes of a piano, as they progress from the left of the keyboard to the right, rise from low pitch to high. The second property, the loudness of a sound, is immediately understandable. Some sounds are so weak that they are inaudible; others frighten us by their intensity. Finally, the quality of a sound is what distinguishes it from other sounds of the same loudness and pitch. Even when a violinist and flautist produce tones of the same pitch and loudness, we recognize a difference in quality because of the differences between the two instruments.

Each of these characteristics, loudness, pitch, and quality, can be 'explained' mathematically. The louder of two sounds has a graph with a greater *amplitude*. Since the amplitude of the graph is the maximum displacement of the air molecules conveying the sound, it follows that the loudness of a sound depends on the maximum displacement of the vibrating air molecules; the greater the displacement, the louder the sound. This conclusion is readily acceptable, for we know from experience that the loud twang of a guitar requires a greater displacement of the string than does a gentle strumming.

Sounds with the same pitch produce graphs having the same frequency, while the graphs of high-pitched sounds have larger frequencies than those of low-pitched sounds. Thus the graph of the sound known as middle C on the piano has a frequency of 261·6 a second, and the sound pitched one octave higher has a frequency of 523·2 a second.

The pitch of a complex sound, or the frequency of its graph, is always that of the fundamental tone. Consider the formula for the violin sound as an example. There the partials have the frequencies 500, 1000, and 1500 respectively. This means that the graph of the second partial will go through two complete cycles while the graph of the fundamental tone goes through its first one. Similarly, the graph of the third partial will go through three cycles while the fundamental tone goes through its first one. The composite graph, however, repeats its behaviour when and only when the graph of the fundamental does, that is, after $\frac{1}{500}$ of a second. This means that the air molecules will begin to repeat their behaviour after $\frac{1}{500}$ of a second. Since it is this

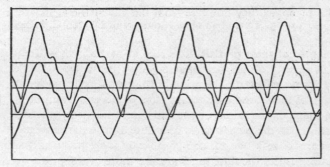

Figure 62. Different notes on the flute (Courtesy of Dayton C. Miller)

frequency that determines the pitch of a sound we see why the pitch of the complex sound is determined by the fundamental tone.

The quality of a musical sound affects the *shape* or form of the graph. If a sound of the same pitch and loudness is produced successively by a tuning fork, violin, and clarinet, the graphs for the different instruments have the same period and amplitude but differ in form (cf. fig. 60), while the graphs of different notes on the *same* instrument always have the same general shape (fig. 62). This means that each instrument has its characteristic quality.

The shape of the graph depends, in turn, partly on which partials are present in the sound and partly on the relative

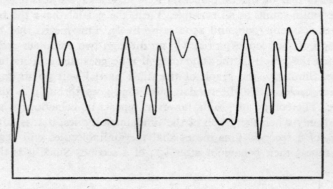

Figure 63. Tone of an oboe (Courtesy of Dayton C. Miller)

strengths of these partials. The second partial, whose frequency is twice that of the fundamental, may be so weak that it has almost no effect on the sound. Mathematically speaking, the graph of the second partial may have such a small amplitude that the shape of the graph of the full sound is hardly affected. For example, in the higher notes of a flute all the partials except the first are so weak that the composite sounds are practically simple. In this respect the flute tones are like those of a soprano voice of similar pitch. The flute is therefore often used to accompany a soprano in operatic arias and the combination produces a most pleasing effect. In the sound of a baritone voice the partials are generally strongest in the order six, seven, five, three, eight, and so forth. Such a sound is illustrated in fig. 60, the letter *a* there having been uttered by a baritone at a pitch of 159 cycles per second. In some tones of the oboe (fig. 63) the fourth, fifth, and sixth partials are stronger than the first three. In the sound of the clarinet shown in fig. 60 the eighth, ninth, and tenth partials predominate after which rank the seventh, first, and third partials.

It should now be apparent that not only the general nature of musical sounds but their structure and chief properties can all be characterized mathematically. In one stroke of Fourier's pen an endless variety of sounds – the human voice, the tones of a violin, and the wailing of a cat – is reduced to elementary combinations of simple sounds and each of these is, in turn, no more complex mathematically than a simple trigonometric function. Those dull, abstract formulas, which have endlessly bored high-school and college students, are really all about us. We give voice to them whenever we open our mouths and we hear them whenever we prick up our ears.

The nature of individual musical sounds is now clear to us, thanks to Fourier. But what does mathematics have to say about harmonic combinations of sounds, about the essence of beautiful musical compositions, about the 'soul' of music? The answer is a voluminous one, so that all we can do here is to read the first page.

The most pleasing chords or combinations of tones, as the Pythagoreans discovered, are made up of sounds the ratio of whose frequencies are the ratios of simple whole numbers. The *major third*, for example, is a pair of tones, or interval as it is called, whose frequencies are in the ratio of 4 to 5; the *fourth* is a pair of tones whose frequencies are in the ratio of 3 to 4; and the *fifth* consists of frequencies in the ratio of 2 to 3. No explanation of the ear's ready acceptance of these harmonies has gone much beyond the recognition of the numerical relation between the pitches involved.

Because the ear accepts only certain combinations of notes as harmonious, the construction of a satisfactory musical scale is a rather complicated problem. In order to play harmonious chords the scale must provide tones with the proper frequency ratios. In addition to this requirement, the introduction of polyphonic music or counterpoint and the desirability of utilizing different keys to achieve different emotional effects impose other requirements on the scale. Various musicians and mathematicians have attempted to satisfy all these demands.

Since it is not possible to have an unlimited or even a large range of frequencies on instruments such as the piano, wherein the frequency of each note is fixed, the difficulties were resolved by the construction of the equal-tempered scale. The advocacy of this scale by J. S. Bach and his son, Karl Philipp Emanuel, led to its permanent adoption in Western civilization.

The equal-tempered scale contains twelve notes; thus, from C, say, to C′, which is one octave higher, there are twelve intervals. The frequencies of the eleven intermediate notes are fixed so that each bears a constant ratio to the one preceding it. Since there are twelve intervals from C to C′ and the ratio of the frequencies of these two notes is 2, the ratio of the frequencies of consecutive notes is $1 \cdot 0594$, for $(1 \cdot 0594)^{12} = 2$. Thus each interval in the equal-tempered scale, called a semitone, is the same. Consequently, any note may be used as the key of a composition. The intervals that may be formed with the notes of this scale, however, are not exactly those which have been found to be the most pleasing. In order to produce the fifth, in which the ratio of the frequencies of the two notes is 3 to 2, the best that can be done on the equal-tempered scale is to select two notes whose frequency ratio is $1 \cdot 498$. The interval of the fourth, which should call for a frequency ratio of 4 to 3, can be approximated by the ratio $1 \cdot 335$. These differences, seemingly insignificant, can be detected, nevertheless, by a good ear. Of course, the violinist, by adjusting the length and tension of his strings, and the singer need not limit themselves to the frequencies of the equal-tempered scale. Nevertheless, because the piano is a basic instrument it has dictated the scale for Western music of the last two hundred years.

The role of mathematics in music extends to composition itself. Masters such as Bach and Schoenberg have constructed and advocated vast mathematical theories for the composition of music. In such theories cold reason rather than an ineffable, spiritual feeling supplies the creative pattern.

But subjects such as chords, scales, and theories of composition lie beyond our present goal. Our survey of the cultural bearings of mathematics does not permit too long a glance in these directions. The few remarks just made merely indicate how far mathematics has penetrated the sphere of music since the age in which it was first recognized that the music of the spheres could be reduced to mathematics.

Of course the mathematical analysis of musical sounds is of great practical importance. One illustration perhaps will be sufficient to convince us of this fact. The telephone seeks to reproduce sounds faithfully. In view of the variety of sounds, this goal appeared at one time to be almost unattainable with simple physical devices. But Fourier's theorem tells us that all vocal sounds are merely combinations of simple sounds of different frequencies. Hence the problem is simplified at least to the extent of reproducing simple sounds. Further analysis of the graphs of actual human sounds by means of Fourier's theorem shows that for intelligible audibility only the simple sounds with frequencies from 400 to 3,000 per second are needed. The design of the telephone was directed therefore towards the reproduction of simple sounds with frequencies lying in the range just mentioned. Considerable improvement in the quality of the reproduction was achieved.

The musical sounds of instruments have also been considerably improved by the application of mathematics. The analysis of vibrating strings has yielded knowledge useful in the design of pianos; the analysis of vibrating membranes has been applied to the design of drums; and similar studies of vibrating columns of air have made possible extensive improvements in the design of organs. The harmonic analysis of musical sounds is also used by the piano manufacturer when he positions the hammers so that they will suppress undesirable harmonics. Mathematics not only aids in the design of these instruments but, in some quarters at least, mathematics rather than the ear is the arbiter of perfect design. Many instrument manufacturers convert the sounds of their instruments into graphs by devices analogous to the phonodeik. They then judge the quality of their product by how closely these graphs correspond to ideal graphs for the sounds of these instruments.

It is no doubt true, nevertheless, that in so far as the design of musical instruments is concerned, experience has contributed more than mathematics. The reverse, however, is definitely the case in the design of reproducing instruments such as radios, phonographs, talking movies, and loud-speaker systems. The engineering of practically all the components of these complex instruments relies heavily on Fourier's analysis of musical sounds. Even the layman who becomes a high-fidelity enthusiast soon learns to speak Fourier's language. In view of the many contributions of mathematics to the production and reproduction of musical ideas the modern music lover evidently owes as much to Fourier as to Beethoven.

There are philosophical overtones to Fourier's work. The essence of beautiful music is, no doubt, more than what mathematical analysis furnishes. Nevertheless, through Fourier's theorem this major art lends itself perfectly to mathematical description. Hence the most abstract of the arts can be transcribed into the most abstract of the sciences, and the most reasoned of the arts is clearly recognized to be akin to the music of reason.

XX

Mastery of the Ether Waves

Mystery is in the air.

Anonymous

IN the discovery of the planet Neptune the nineteenth century witnessed a considerable addition to our material universe. It has already been related that the planet was observed after the mathematicians Adams and Leverrier had predicted its existence and location. But this addition to our universe, a planet many times larger than the size of our Earth, caused hardly a ripple in the daily affairs of mankind. The heavenly spirits of Copernicus, Kepler, and Newton merely smiled indulgently and murmured, 'I told you so.'

Not many years later the nineteenth century witnessed another addition to our physical universe. Like the discovery of Neptune this one, too, could hardly have been made without the aid of mathematics. But unlike Neptune, this addition was decidedly insubstantial. It weighed nothing, and it could not be seen, touched, tasted, or smelled; it was and is physically unknown to man. And unlike Neptune this shadowy 'substance' had manifest and even revolutionary effects on the daily lives of nearly every man, woman, and child in Western civilization. It whisked communications round the world in the flicker of an eyelid; it extended the political community from the street corner to the planet Earth; it quickened the tempo of life, promoted the spread of education, created new arts and industries, and revolutionized warfare. Indeed hardly a phase of human life was unaffected.

The central character in this second tale of discovery is a Scotchman, James Clerk Maxwell, who was born in Edinburgh in 1831 and was both student and professor at Cambridge. Although even as a youth Maxwell displayed an aptness for the abstract – his mathematical work at school was brilliant and he published his first paper at the age of 15 – he always wanted most to understand the physical workings of natural phenomena and mechanical devices. As a boy he inquired constantly, 'What's the go of that?' For his own satisfaction his theoretical analysis of the structure of Saturn's rings, an early piece of work, had to be supplemented by the construction of a model. It was hardly to be expected that a person so insistent on physical explanations should achieve his pre-eminence with purely mathematical reasoning about a most mysterious and physically inexplicable phenomenon.

In order to appreciate fully the problem that Maxwell faced we must go back into history a bit. Several thousand years ago a Cretan shepherd, Magnes by name, noticed that the iron nails in his sandals and the iron tip of his staff were attracted to a particular type of rock in the earth. The shepherd had discovered the lodestone or natural magnet and he had observed the fact that it attracts iron. In Europe during the twelfth century it was learned from the Chinese that a piece of lodestone can act as a compass, but the phenomenon of magnetism was not studied extensively until the Court Physician to Queen Elizabeth, William Gilbert, investigated its properties. Gilbert should be remembered especially for establishing the fact that the Earth itself is a magnet, and thereby accounting for the behaviour of the compass needle. With all his efforts, Gilbert made little progress in the direction of understanding the real nature of the attraction exerted by magnets, and his work had no influence on the superstitious attitudes towards the subject. Before and even after his time people believed the behaviour of magnets to be magical; they supposed that this magic power could cure almost every disease and even reconcile husbands to their wives. The phenomenon of magnetic attraction is 'explained' today by saying that the magnet sets up a field about it and that iron coming within the field is acted on by the field.

A very similar and related discovery was made by the Greek scientist, Thales. Thales noticed that a rubbed piece of polished amber attracts light objects such as pieces of straw and dry leaves. Apparently the rubbed amber, like a magnet, sets up a field which pulls certain objects falling within the field towards the amber.

For a long time the phenomena associated with amber and with lodestone were regarded as the same. It remained for Gilbert to point out differences; to distinguish between the two he called the attracting power of rubbed amber electric, which is Greek for amber.

In the late eighteenth century Professor Luigi Galvani of Italy noticed that a frog's leg twitches when the ends of an arc of wire formed by linking two unlike metals touch the ends of a nerve. The significance of this discovery was appreciated and used by another Italian, Alessandro Volta. Volta realized that the two unlike metals were producing a force, now called electromotive force, at the ends of the wire and he worked out a more effective combination of metals, that is, a battery. By replacing the frog's nerve with a wire and by attaching the ends of the wire to the terminals of his battery, Volta showed that the force could be utilized to make minute particles of matter flow in the wire. This flow of particles, which were later identified as electrons, is an electric current. Though neither Galvani nor Volta realized it, electrons are precisely what appear on rubbed amber, and it is these electrons that attract particles in other objects. Volta's battery made these electrons flow instead of leaving them bunched up and stationary as they are on rubbed amber.

A most important relationship between electricity and magnetism was discovered in 1829 by the Danish physicist, Hans Christian Oersted, who was working at the University of Copenhagen. Using Volta's new battery to force electric current through a wire, Oersted found that the wire acted as a magnet while the current passed through it, that is, the electric current set up a magnetic field about the wire. Such a field attracts or repels other magnets as does the natural lodestone. This discovery was really an accident, but as Pasteur once wrote, 'Chance favours only the prepared mind.' Oersted was worthy of this favouritism and he was able to explore his discovery fully. The French physicist, André-Marie Ampère, then showed that two parallel wires carrying currents behave like two magnets. If the currents are in the same direction the wires attract each other, and if in opposite directions they repel each other.

It remained for a self-educated, ex-bookbinder's apprentice,

Michael Faraday, who was working in England, and a school-master, Joseph Henry, of the Albany Academy in New York, to discover the other essential link between electricity and magnetism and thereby set the stage for the dramatic entrance of Maxwell. If a wire carrying current sets up a magnetic field, will not a magnetic field induce current in a wire? The answer, as these men showed about a hundred years ago, is yes, provided the wire is moved in the field of the magnet so as to vary the field about the wire.

Let us examine more closely the essence of Faraday's and Henry's discovery. Suppose that a rectangular frame of wire (fig. 64) is rigidly attached to a rod R and that the frame and rod are

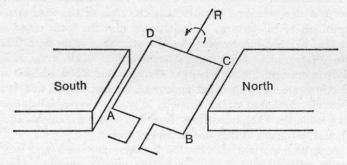

Figure 64. The principle of a generator of electricity

then placed in the field of a magnet. When the rod is made to rotate, by the use of water power or a steam engine, say, the frame of wire will also rotate. Suppose, too, that the rod rotates at a constant speed in a counter-clockwise direction and that the wire BC starts from its lowest position. As BC goes from this position towards a horizontal position on the right a flow of electric current takes place in the wire in the direction from C to B. This flow increases in strength as BC approaches the horizontal position and reaches a maximum at that position. As BC continues upwards, the flow decreases in quantity and vanishes when BC is at the highest position. As BC continues to rotate, a current again appears in the wire, this time in the direction from B to C. Again the flow increases in quantity as the wire rotates and reaches a maximum value for the new direction of flow when BC is again horizontal. As BC returns to the lowest position of its path, the flow of current diminishes and finally disappears. This cycle of changes repeats itself with each complete rotation of the rod. The appearance and flow of current in a wire that is moved in the field of a magnet is the phenomenon of electromagnetic induction.

The current generated, like the current caused by a battery, is a flow of billions of minute, invisible particles of matter called electrons. This electronic flow is caused by a force that appears in the wire simultaneously with the current and that goes through the same variations the current does; that is, it rises and falls and then reverses itself to rise and fall in the new direction. This force can be compared to the pressure that causes water to flow in a pipe. The electric current itself can then be compared to the flow of water.

Both the amount and force of the flow created by electromagnetic induction vary with time and, since we are dealing with measurable quantities, we can discover the functional relation involved. The relation between current and time is certainly periodic since the sequence of variations repeats itself with each complete rotation of the wire frame. It may be too much to expect that in this periodic phenomenon, as in those we encountered in the study of musical sounds, the function *sine x* should serve. But nature never ceases to accommodate itself to man's mathematics. The relation between current I and time t is of the form

$$I = a \text{ sine } bt,$$

where the amplitude a depends on such factors as the strength of the magnet, and the frequency b depends on how fast the frame rotates. If it makes 60 rotations in one second then the value of b, in view of our discussions of frequency in the preceding chapter, is 60×360 or 21,600. The current that furnishes electricity to most homes goes through 60 complete sinusoidal cycles of change in one second; for this reason it is called 60-cycle alternating current.

Electric current, then, can be thought of as a flow of electrons and it can be represented by a mathematical formula. But how does the process of electromagnetic induction produce electric currents? This phenomenon is replete with mystery. Somehow the mere motion of a wire in a magnetic field induces an electromotive force in the wire, and this force causes a current to flow. No one knows, however, how the magnetic field creates its effect, or, for that matter, how a magnet attracts iron or steel. No material causal agent can be detected in either phenomenon. In view of our profound ignorance about the physical nature of fields, an explanation of electromagnetic induction seems farther away from man's reach than the distant stars.

Fortunately, what may be beyond man's physical reach is nevertheless within his mathematical grasp. By Maxwell's time, the physicists of the nineteenth century had succeeded in formulating mathematically the quantitative aspects of various electrical and magnetic phenomena that had been studied over the preceding centuries. The behaviour of fields associated with fixed electric charges, such as those that appear on rubbed amber, and the behaviour of the fields that surround magnets, were expressed by two laws, known today as the laws of static electricity and magnetism. The phenomenon of electromagnetic induction, first observed by Faraday and Henry, was expressed in a third law, now called Faraday's law. Finally, the behaviour of magnetic fields that surround wires carrying current, the study of which had been conducted by Oersted and Ampère, was expressed in a fourth law named after Ampère. These last two laws are called laws of electrodynamics because they describe the behaviour of currents or magnetic fields in motion. All four take the form of differential equations, which are, unfortunately, too complicated to discuss here. We can, however, consider what Maxwell did with them.

While working with these laws of electromagnetism Maxwell made a deduction showing that the laws were inconsistent with another law of mathematical physics known as the equation of continuity. To a mathematician a contradiction is intolerable and Maxwell sought a resolution of the difficulty. He noticed that the addition of a new term to Ampère's law would secure the consistency of the laws of electromagnetism and therefore decided to add it.

Never one to be satisfied with the mathematics alone, Maxwell sought the physical significance of what he had done. He soon saw that the new term, which represented a changing electric field, had mathematical properties similar to that term in Ampère's law which represented the flow of current in a wire. Boldly Maxwell interpreted the quantity he had added. Its properties were those of a current. On the other hand, the changing electric field with which it dealt existed in space whereas the previously known currents flowed in wires. Maxwell thereupon decided that the new term represented a current or wave flowing through space. Unlike the current in wires this space wave appeared to have no material content nor was the manner in which it travelled physically clear to him. Nevertheless, convinced by the mathematics, Maxwell affirmed its existence and coined the term *displacement current* for it. Further reasoning showed that such a changing electric field, like electric currents in wires, must have an accompanying magnetic field. The combined fields are now known as an electromagnetic field.

Solution of the corrected differential equations of electromagnetism showed Maxwell that the electric and magnetic fields, when properly generated, travel in space, much as sound waves do; at any one point in space the strength of each field varies sinusoidally as time changes. The travelling electric and magnetic fields may each be compared to the wave that moves along a horizontally extended rope when one end is moved rapidly up and down. Thus Maxwell made the first of his great discoveries, the existence of electromagnetic waves.

His next discovery was probably a reward for his audacity. He observed that his corrected equations describing the behaviour of electromagnetic waves in space were the same as the equations previously obtained by other scientists for the motion of light. Moreover, his electromagnetic waves possessed the same velocity as light waves. Maxwell unhesitatingly drew the apparent inference. Electromagnetic waves are identical in nature with light waves. The identity obviously works both ways. Light waves must be electromagnetic waves. Hence the mathematical and physical knowledge already obtained about electromagnetic waves must be applicable to light. Conversely, knowledge about light could be applied to the study of electromagnetic phenomena. In other words, two formerly independent branches of physics were identified and the stock of knowledge about each was practically doubled.

To complete the physical interpretation of his mathematics, Maxwell had yet to explain what medium carried his newly found waves. In his day scientists accepted the fact that light waves moved in a medium called ether, a 'substance' which, though never detected experimentally in any manner, was believed to permeate all space and all material bodies. In view of the relationship he himself had established between electromagnetic waves and light, Maxwell assumed that his space waves, too, were propagated by motions of the ether. So many tasks had already been ascribed to the much abused ether that one more hardly mattered.

Maxwell's declaration of the existence of a new physical phenomenon which had never before been suggested and which could not be experimentally detected by the scientists of his time was indeed an audacious step. The most distinguished mathematical physicists of his day, Hermann von Helmholtz and Lord Kelvin, refused to believe in displacement currents. But, by definition, genius is not easily daunted. Convinced of the physical reality of his electromagnetic space waves, Maxwell went further and suggested the apparatus that could produce them. Twenty-three years after Maxwell had deduced the existence of space waves and ten years after his death, the German physicist, Heinrich Hertz, demonstrated the existence of these waves by generating and detecting them in the very manner Maxwell had proposed.

Hertz reasoned that Maxwell's displacement current or changing electric field should be identical in nature with the fields surrounding stationary electric charges or electrons. He therefore devised a way of making electric charges move back and forth on a wire so that the field associated with them was also set in motion. When the frequency of the alternating motion of the charges was high enough an appreciable part of the field moved off into space, just as waves move out along a rope when one end is moved up and down sufficiently fast. Some distance away the field acted on stationary electrons in another wire and caused these to move back and forth. Thus a current, which Hertz detected, was induced in the second wire. The wires Hertz used are the original form of modern antennas – the transmitting antennas high up on towers of the broadcasting stations and the receiving antennas that used to be on top of roofs but are now in the backs of radio sets. Wireless telegraphy, which involves merely long and short interruptions in the sending of electromagnetic waves, was just around the corner.

The wireless transmission of voice and music, however, presented another problem. The mathematical analysis of musical sounds, discussed in the preceding chapter, had shown the nineteenth-century scientists that these sounds consist of sinusoidal air waves with frequencies from a few to many thousand per second. Work on the telephone had demonstrated that these sound waves could be converted into electric currents possessing exactly the same mathematical properties as the sound waves have. Could these electric currents representing musical sounds be converted directly into electromagnetic waves and thus be transmitted through space? This is theoretically possible, but for reasons familiar to the radio engineer it is easier to radiate high-frequency currents of the order of millions of cycles per second than to radiate the low frequencies that correspond to vocal and instrumental sounds. Some scheme was needed whereby low-frequency currents could be converted into or attached to high frequencies.

Several such schemes were developed. In use at present is the system known as amplitude-modulation. The amplitude of a sinusoidal high-frequency current, which can be radiated into space easily, is made to vary above and below its normal value exactly as does the amplitude of the sound wave to be sent out. This is done with suitable equipment at every radio broadcasting station. The resulting amplitude-modulated, high-frequency current, or carrier (fig. 65) is then radiated into space, through which it travels hundreds and thousands of miles to receiving sets. Each receiving set 'removes' the carrier, that is converts the

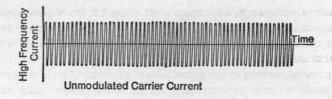

Unmodulated Carrier Current

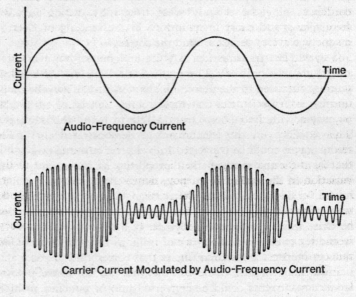

Audio-Frequency Current

Carrier Current Modulated by Audio-Frequency Current

Figure 65. An amplitude-modulated carrier wave

amplitude variations in the carrier to wire-borne low-frequency electric currents which vary with time precisely as does the amplitude of the high-frequency current. The low-frequency currents then operate a loud-speaker whose vibrations create sound waves. By these processes the very sounds uttered or played in the radio studio are reproduced in the home just a fraction of a second later despite intermediate transformations which defy the wildest imaginations.

Actual carrier frequencies for the amplitude-modulated radio waves of ordinary broadcasting stations range from 500,000 to 1,500,000 cycles per second. The man who 'tunes' his radio set to a particular station is adjusting it to receive the carrier frequency of that station.

In recent years another system of transmitting voice and music by radio has been explored and put to use, namely, frequency-modulation. In this system the *frequency* rather than the amplitude of the sinusoidal high-frequency current is varied in accordance with the sound to be transmitted. Suppose the frequency of the carrier or radio wave that propagates in space is 90,000,000 cycles per second and the sound to be transmitted is a 100 cycle per second note of amplitude 1. Were the carrier un-modulated, it would of course continue to oscillate at the rate of 90,000,000 cycles per second. But now suppose that this frequency is varied from 90,000,000 to 90,002,000, back to 90,000,000, then to 89,998,000 and then back to 90,000,000. This sequence of changes in frequency, or the modulation of the frequency, is made to occur at the rate of 100 times per second, that is, at the frequency of the musical sound. The extent of the variation in the carrier frequency, namely 2000 cycles, is determined by the amplitude of the musical note. If this amplitude were 2 instead of 1, the variation in the carrier frequency would be twice as much or 4000 cycles, so that the carrier frequency would range 4000 cycles above and below 90,000,000 again at the rate of 100 times per second (fig. 66).

Even higher frequencies than those used for frequency-modulation broadcasting are employed in radar sets. The electromagnetic waves sent out into space vary sinusoidally in strength at frequencies as high as 10 billion times per second. Such waves are sent out in short bursts lasting about one-millionth of a second each (fig. 67). If these bursts or pulses hit a metallic surface such as an airplane or ship they are reflected to the sender, who detects thereby the presence of the reflecting surface.

Incredible and staggering as these frequencies are, they hardly begin to tax the human imagination when compared with the frequencies found in light waves. It was believed even before Maxwell's time that light was some sort of wave motion. Maxwell's mathematical proof that light is electromagnetic in character made it clear that the essential difference between light and radio waves is the frequency of variation of the ether's motion.

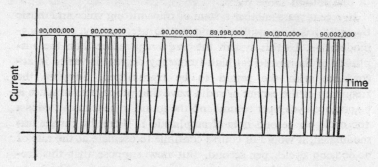

Figure 66. A frequency-modulated carrier wave

Light-wave frequencies are of the order of 1 followed by 14 zeros per second. Specifically, all waves whose frequencies range from 4 times 10^{14} to 7 times 10^{14} are visible waves, our eyes responding to these different frequencies by registering different colours. As the light received varies from the smallest frequency to the largest in the range above, our sensations of colour, a contribution of the nerves and brain, change gradually from red to yellow, to green, to blue, and, finally, to violet. Colour in light is

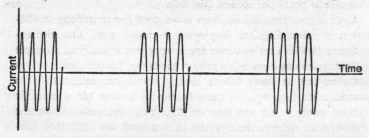

Figure 67. Radar pulses

thus analogous to pitch in sound. And just as we combine simple sounds to produce complex ones, so we can combine simple colours to produce new ones. White light itself, for example, is not a simple colour 'tone', but a light 'chord', a composite effect of many colours. Thus sunlight contains all colours from red to violet, the composite effect of which is white light.

More and more pieces of the electromagnetic jigsaw puzzle were soon filled in. Ultra-violet and infra-red rays, the former radiation detectable by its blackening of photographic film and the latter by its heating effect, were seen to be electromagnetic waves, with frequencies above and below those of light waves. X-rays, first detected in the last part of the nineteenth century, were also identified as electromagnetic waves with frequencies even higher than those of ultra-violet rays. And finally, gamma rays, which issue from radioactive substances, are also electromagnetic waves with frequencies still higher than those of X-rays.

The affinity among these various types of electromagnetic waves, which Maxwell's work uncovered, is now continually utilized. The electric lamps in our homes, for example, convert 60-cycle waves, which travel along wires, into light waves, which travel in space. The essential identity of the many types of waves is used most strikingly in the newest miracle of science now invading the American home – television. The variations of light in a scene to be broadcast are transformed into electric currents, which are in turn impressed upon a high-frequency radio wave and radiated into space. The receiving set in the home converts the radio wave into electric currents and the currents into light waves so that the eye sees precisely the original scene. Thus one form of electromagnetic wave is converted into another and that into a third; then the sequence of transformations is reversed. Every time we go to the movies we witness one type of electromagnetic wave being converted to another. Light passing through the sound track of film of varying shades of blackness strikes a photoelectric cell; this device converts the transmitted light into a varying electric current and the current in turn activates a loudspeaker. Thereby the mellifluous words of ardent males addressed to incomparable beauties waft us into romantic realms.

These practical attainments are indeed spectacular and make miracles a commonplace. They also have vast and immeasurable social consequences, some of which were stated at the beginning of this chapter. The use of the radio for political speeches should suffice to punctuate any remarks about the social import of the science of electromagnetism.

But there are values in Maxwell's contribution that dwarf its incalculable effects on society and on the daily business of living. Man does not live by bread and political pull alone. He wants to understand nature and his relation to nature; he means to satisfy his curiosity about ever-present phenomena such as sound and light; and he wishes to bring order out of the diverse impressions which a multitude of events cast upon his senses. Such values are obtained from mathematical accounts of physical phenomena.

Maxwell's electromagnetic theory surpasses even Newton's gravitational theory in embracing a variety of seemingly diverse phenomena in one comprehensive set of mathematical laws. The behaviour of the grain of sand and the heaviest star can be described and predicted with Newton's laws of motion. The invisible electron and the light of the sun can be described and harnessed with Maxwell's electromagnetic laws. Electric currents, magnetic effects, radio waves, infra-red waves, light waves, ultra-violet waves, X-rays, and gamma rays, sinusoidal waves with frequencies as low as 60 per second and as high as 1 followed by 24 zeros are manifestations of one underlying mathematico-physical scheme. This theory, which is at once so profound and so comprehensive that it beggars the imagination, has revealed a plan and an order in nature which speaks more eloquently to man than nature herself. With it man's reason, his sole claim to distinction from the rest of the animal world and his only basis for belief in his own importance, has secured another victory. Once more man has grasped with his mind the reins that direct nature's prancings.

Electromagnetic theory affords us another illustration of the power of mathematics to unearth nature's secrets. It was possible to conceive of and even to visualize the submarine and the airplane long before technicians produced working models. The notion of a radio wave, on the other hand, would hardly occur even in a flight of fancy and, were it to occur, would be immediately dismissed as such. Radio waves, whose physical nature are still not understood, were discovered, it might almost be said invented, because *mathematical reasoning demanded their existence*. And science is now systematically exploring other vast regions of the electromagnetic world clearly delineated by Maxwell's broad theory.

It is especially significant that it was not just ordinary mathematical reasoning which led to the prediction of radio waves. It was, rather, an insistence on *exact* reasoning. The mathematician, valuing logical consistency in his equations above all else, does not pass by the slightest contradiction. Nor does he permit an inadequate physical understanding limited by fallible and finite sense perceptions to deter him from taking the necessary steps to remove that inconsistency. Imbued with the spirit of exact reasoning the mathematician regards no demand on behalf of exactness as an unnecessary extravagance. So-called practical men and even scientists and engineers who confuse mathematical rigour with pedantry would do well to ponder Maxwell's work.

There is much more we can learn from even this brief survey of electromagnetic theory. Granted that through it mathematics has mastered another segment of the physical world. Granted, too, that radios, motors, optical devices, and X-ray machines designed and operated in accordance with this theory leave no doubt that the mathematics is dealing with real phenomena. But where and what are the physical agents that produce the effects mathematics describes? What are electrons that flow in wires and cause lights to glow? What are electric and magnetic fields that attract and repel objects and interact on each other? In particular, what is this displacement current that travels through space and that is in the air all about us? What is the ether that carries electromagnetic waves? Though the greatest mathematicians and physicists have plagued themselves with these questions, there are no answers. The weirdest ghosts ever conceived are no less realizable and no less tangible than the physical accounts concocted for electromagnetic phenomena. Electrons, fields stationary and moving, and ether are but fictions, 'shadows of speculations'. Electromagnetic phenomena are as mysterious and as awe-inspiring as purported supernatural manifestations.

Even the man who was most gifted in constructing a physical picture of electromagnetic induction, a picture Maxwell himself used to advance his own thinking, confessed that he was baffled in his attempt to understand the entire phenomenon physically. In a letter to Maxwell written in 1857, Faraday asks whether Maxwell could not express the conclusions of his mathematical work 'in common language as fully, clearly, and definitely as in mathematical formulae? If so, would it not be a great boon to such as I to express them so? – translating them out of their hieroglyphics that we might also work upon them by experiment. ... If this be possible, would it not be a good thing if mathematicians, working on these subjects, were to give us the results in this popular, useful, working state, as well as in that which is their own and proper to them?' Unfortunately Faraday's request cannot be filled to this day.

In no case is our ignorance of the real world or of its ultimate nature more shocking than in the phenomenon of light itself. Undoubtedly something travels through space when light from a source such as the sun or an electric lamp strikes our eyes. But what is it? For three centuries now scientists have seriously and persistently investigated the nature of light. Experimental evidence supports two vague, contradictory theories: one, that light is a continuous wave motion in ether; the other, that light is a motion of minute, invisible particles or corpuscles. Frequent shifts of scientific opinion from one theory to the other have given rise to a standing joke: the wave theory prevails on the odd days of the month; the corpuscular theory, on the even days.

It is true that Maxwell insisted on mechanical models of every phenomenon he investigated. He pictured the flow of electricity, for example, as the flow of an imaginary fluid and even studied real fluids to derive mathematical laws that might be applicable to the flow of electricity. He invented mechanical models involving particles and gears in order to picture and study the propagation of electric and magnetic fields. But he never forgot that the fluids and the mechanical models were merely aids to thought and ultimately he discarded them, although he retained the mathematical equations they had suggested. When he presented his classic paper 'A Dynamical Theory of the Electromagnetic Field' to the Royal Society in 1864, the physical scaffolding he had used to erect the mathematical architecture was omitted. Many of Maxwell's successors did retain the physical models and installed them as true explanations, probably because they themselves were unable to dispense with the pictures in their own work. The necessity for thinking in terms of a medium that carries electromagnetic waves soon established to their satisfaction the 'reality and substantiality of the luminiferous ether'. These pictures, however, cannot be taken seriously, for they are inadequate and experimentally non-verifiable.

The inability to explain electromagnetic phenomena qualitatively or materially contrasts sharply with the exact quantitative description furnished by Maxwell and his co-workers. Just as Newton's laws of motion furnished scientists with the means for working with matter and force without explaining either, so Maxwell's equations have enabled scientists to accomplish wonders with electrical phenomena despite a woefully deficient understanding of their physical nature. The quantitative laws are all we have in the way of a unifying, intelligible account. The mathematical formulas are definite and comprehensive; the qualitative interpretation is vague and incomplete. Electrons, electric and magnetic fields, and ether waves merely provide names for the variables that appear in the formulas, or, as von Helmholtz stated the point, in Maxwell's theory an electric charge is but the recipient of a symbol. The definitive statement about the physical nature of electromagnetic phenomena was made by Heinrich Hertz: 'To the question, What is Maxwell's theory? I know of no shorter or more definite answer than the following: Maxwell's theory is Maxwell's system of equations.'

If physical understanding and the power to reason in physical terms about electromagnetic phenomena are lacking, what is the nature of man's grasp of that phase of reality? On what does he base his claim of mastery? Mathematical laws are the only means of probing and mastering this large region of the physical world; of such mysterious goings-on mathematical laws are the only knowledge man possesses. Though the answer to these questions is unsatisfactory to the layman uninitiated into these latter-day

Delphic Mysteries, the scientist by now has learned to accept it. Indeed faced with so many natural mysteries, the scientist is only too glad to bury them under a weight of mathematical symbols, bury them so thoroughly that many generations of workers fail to notice the concealment.

With Maxwell's work physics took a new turn. Before his time a mechanical view of nature was not only popular but indeed reasonably satisfactory in supplying a physical account of natural phenomena. For a long time even electricity and magnetism were pictured as the actions of fluids though scientists did not know that they actually were such. The ether was regarded as a highly elastic solid and thereby a mechanical account of the propagation of light was rendered. The introduction of electromagnetic waves and the identification of light with these waves destroyed the validity of these physical accounts, however. Scientists began to have grave misgivings about the whole mechanical philosophy of nature and reluctantly abandoned it.

Physics thereupon passed from a mechanistic to a mathematical foundation. Whereas mathematics served previously to represent, study, and advance the mechanical analysis of phenomena, today the mathematical account is fundamental. In fact, the mechanical one has been abandoned except perhaps in very limited areas. The essence of any modern physical theory is a body of mathematical equations. Thus differential equations which in Newton's day were the servant of physical thought have now become the master.

Though Maxwell's work subverted the mechanical philosophy of nature it strengthened the philosophy of determinism which had grown up along with the mechanical view. To the nineteenth-century scientists Maxwell's work was the crowning achievement of the project begun by Copernicus, Kepler, and Galileo. Such a vast number of new phenomena were now subsumed under exact, mathematical laws that the mathematical design of the universe could hardly be doubted. Indeed no cockier group of scientists ever existed. Fulfilment of all the goals set up by the confident, unboundedly optimistic eighteenth-century scientists was the boast of their nineteenth-century successors.

Maxwell himself was not taken in. He was too shrewd to

become a devotee of his own great achievements. A keener student of metaphysics than his co-workers he again proved his genius by resisting the belief in a deterministic universe held by almost everyone at that time. Maxwell had done some fundamental work on the motion of molecules in connection with the theory of gases and was disturbed by the thought that any ordinary body is made up of molecules, each of which moves with the velocity of a cannon ball and yet never departs to a visible extent from its mean position. He was led to make a distinction between stable and unstable phenomena. A rock rolled along level ground is a stable phenomenon because a little push on the rock will produce just a little movement. A rock resting on the top of a mountain peak is unstable, however, because a little push might start an avalanche. Similarly, the match that starts a forest fire, the little word that sets the world fighting, and the little gemmule that makes us philosophers or idiots are unstable phenomena. Such unstable factors, or singular points as Maxwell called them, were to him flaws in the deterministic world. Laws break down in these instances and effects negligible under other circumstances can be dominating.

Maxwell cautioned his fellow scientists about the implications in the existence of singular points: 'If, therefore, those cultivators of physical science ... are led in the pursuit of the arcana of science to study the singularities and instabilities, rather than the continuities and stabilities of things, the promotion of natural knowledge may tend to remove that prejudice in favour of determinism which seems to arise from assuming that the physical science of the future is a mere magnified image of that of the past.'

The leader of his own generation was actually the prophet of the next one. Some of Maxwell's own contributions to the theory of gases helped to prepare the way for the demise of determinism. The cracks or flaws he saw in this scheme of things soon widened and the deterministic world fell apart. But this catastrophe, with its momentous consequences, must await adequate discussion later. Most unfortunately Maxwell's own work, of unsurpassed quality in many branches of mathematical physics, was terminated by death when he was only forty-eight.

XXI

The Science of Human Nature

The Proper Study of Mankind is Man.

Alexander Pope

'THE most useful and the least advanced of all the sciences,' said Rousseau, 'is that of man.' This son of a labourer looked about him and saw nothing but a diseased and corrupt state of human society. Political injustices, the strong preying upon the weak, luxury for the few and untold misery for the many, vice, greed, wars, the enslavement of peoples by military conquests, and the betrayal of the masses by their leaders appalled him.

The affairs of man were in sharp contrast to the affairs of nature. In nature law and order were clearly evident. The planets kept their appointed paths and never deviated from them. Wherever the physical scientists probed they found regularity and mathematical laws that attested to design and to harmonious behaviour. Nature was orderly, lawful, rational, and predictable.

But man was an integral part of the natural order. Was he not, like the physical world, a creation of God? Did not the current materialistic philosophy teach that mind and body are part of the material world? There must then be universal, natural laws of human behaviour. Men, like the planets, must be subject to forces of attraction and repulsion, so that man's behaviour too should be but the mechanical resultant of the action of such forces. In like manner it should be possible to derive economic laws from the interaction of elementary economic forces. Man's abuse of his fellow man, the chaos of political affairs, the widespread want and misery – such evils seemed characteristic of human relationships only because man had not sought the natural laws of society. The true laws, once obtained, could surely point the way to a better life and to institutions that would be stable and just, because they would at last be in tune with the 'natural order'. And if society could be required or persuaded to obey these natural laws, the ills of civilization would disappear.

There must then be a science of man. Rousseau pointed out, however, that this science cannot be studied experimentally because it would take the greatest philosophers to think up the proper experiments and the greatest monarchs to carry them out. Happily such experiments are not necessary, for the truth may be won by reasoning deductively from first principles. Hobbes stated this thought in his usual forthright manner. Politics, economics, ethics, and psychology must be reduced to exact sciences. Mankind has relied only on experience as the source of social and ethical knowledge; by this means, however, we can acquire only prudence, useful as that may be. But by means of science, Hobbes continued, we may acquire sapience, which is infallible and which enables us to predict. Science, to Hobbes, meant just one subject: 'Geometry is the only science it hath pleased God hitherto to bestow on mankind.' Kant agreed that there was a need for a science of society and added that a Kepler or Newton was wanted to find the laws of civilization.

Thus men arrived at the belief that it was necessary to found the deductive science of human affairs. Accordingly, the social scientists set out to identify, isolate, and abstract the universal laws at work in human relations. Like the detective who confidently expects to unravel the most complicated mystery by finding *la femme*, these social scientists expected to resolve all their problems by finding a few fundamental laws. Fields of thought formerly considered totally alien and inaccessible to mathematical analysis were re-examined with the purpose of duplicating there the accomplishments achieved in the exact sciences. Wine, women, and song, along with the wealth necessary to their enjoyment, became the objects of mathematical investigations. It will be our concern in this chapter to trace the influence of mathematical thought on the course of these investigations.

Granted the existence of social laws, how could the social scientists expect to discover them? The example of mathematics supplied the answer. They must first find the basic axioms that thought and experience vouchsafe to be so self-evidently true of human nature that all scientists would accept them. From these axioms, theorems on human behaviour would be deduced by the rigorous, impeccable reasoning used in mathematics.

Then, just as the theorems of mathematics supplemented with the axioms of motion and gravitation produced mathematical astronomy, so the theorems of human behaviour combined with special axioms of ethics, politics, or economics should produce sciences in these fields. Conclusions in these new social sciences might even be quantitatively formulated and thus permit the application of algebraic techniques for the deduction of further truths.

The search for axiomatic truths upon which the science of human behaviour was to be constructed took on the appearance of a gold rush. In rather rapid succession there appeared a parade of great works that analysed human nature for the purpose of discovering basic principles. Seventeenth- and eighteenth-century classics on the subject included Locke's *Essay Concerning Human Understanding*, Berkeley's *Principles of Human Knowledge*, Hume's *Treatise of Human Nature* and *Inquiries Concerning the Human Understanding*, and Bentham's *Introduction to Principles of Morals and Legislation*. James Mill's *Analysis of the Human Mind*, published in 1829, carried the movement into the next century. In all these works the authors advanced what they believed to be the axioms of the science of human nature and, following the deductive method, derived the laws that govern the actions and thoughts of men.

Some of the axioms of human behaviour advocated in these works merit attention not only for their own sake but also because they indicate the basic assumptions and generative ideas of the age. It was affirmed that all men are created equal: that knowledge and beliefs come from sense data; that the enjoyment of pleasure and the avoidance of pain are basic forces determining human behaviour; that human nature responds in well-known and constant ways to cultural and environmental influences; and that men always act in accordance with self-interest. This last axiom was most often emphasized as basic and comparable in its universality to the law of gravitation. Though men of the twentieth century might fear that self-interest is a disruptive force in society, not so the men of the eighteenth.

> Thus God and Nature fixed the general frame.
> And bade self-love and social be the same.

Private vices were but public benefits. Of course not all of the axioms above were accepted and advocated by all the theorists, but the ones stated were the most popular.

It would be rather difficult to survey in a short space the various deductions made in the science of human nature proper. Fortunately this is not necessary for our purposes. It is sufficient to know that such a science was erected.

In order to obtain results in the specific fields of ethics, politics, and economics, the general programme called for adding to the broad science of human nature axioms peculiar to the specific fields. Of those ethical systems which were developed by the men suffused with the spirit of reason one in particular has had so much influence, both directly and indirectly, on our twentieth-century civilization that it warrants examination in some detail. This system, erected by Jeremy Bentham (1748–1832), was not merely rational and deductive; it dared to be quantitative.

If there is such a thing as a mathematical mind, Bentham possessed it. He was so extremely logical and exact in his thinking that he would suspend a whole work and begin a new one because a single proposition struck him as slightly doubtful. He continually sought to classify all knowledge, to arrange ideas in their proper logical relationship – for example by subsuming the particular under the general – and to analyse all ideas into their constituents. Bentham has been justly described as a codifying animal.

Even his deficiencies, notably in the field of romance, were those commonly associated with mathematicians. After fifty-seven years of remoteness from the society of women he decided to marry and carefully reasoned to his choice. He then proposed by letter to a woman he had not seen in sixteen years. He was refused. But the logic of his proposal remained the same, and so after twenty-two more years, during which he carefully re-examined its impeccability, he again offered himself to the same woman, hoping, possibly, that she had learned some mathematics in the meantime and would see the force of his case. Apparently she was equally sure of her logic, or intuition, for she again refused.

It was not only with women that Bentham had the courage of his logical convictions. In an age when the various religious organizations were still powerful he stated bluntly that all were deleterious and fought the alliance of Church and State. When he became convinced of the wisdom of democracy he dared to advocate universal suffrage and abolition of the monarchy and House of Peers. The privileged classes were attacked in his *Book of Fallacies*. Corrupt individuals, corrupt courts, and dishonest lawyers were also attacked in his writings; one pamphlet, *The Elements of the Art of Packing (as applied to Special Juries)*, was directed at the Crown itself for its practice of fixing juries.

Bentham's fundamental axiom of human nature, that pleasure and pain are the realities underlying and determining human action, has already been mentioned. Man continually pursues happiness and retreats from pain. The words pleasure and pain were, of course, used broadly. Malevolence gives pleasure to some people and so must be reckoned among the pleasures.

Now a system of ethics in accord with and indeed derived from the science of human nature must build upon the pleasure and pain motives. And so Bentham postulated for his ethics that those acts which increase the happiness of people are right and those which decrease it are wrong. Since a particular act may please some people and harm others, he added that, 'The greatest good of the greatest number is the measure of right and wrong.'

Thus far in his development of ethics Bentham echoed and aptly phrased a then currently prevailing thought. He thereupon proceeded to explore its consequences and to refine it through the introduction of mathematical concepts. His objective was to measure pleasure and pain and to 'maximize happiness'. Towards this end the Newton of the moral world developed the 'felicific calculus'.

First, he listed fourteen simple pleasures such as sense, wealth, skill, and power, and twelve simple pains, for example, privation and enmity. To each act that causes pleasure or pain a measure can be assigned. The mathematical value of such an act, said Bentham, depends on objective factors, namely, its duration, intensity, certainty, propinquity, purity (freedom from other pleasures and pains), and its fecundity (tendency to produce other pleasures and pains). Each of these factors contributes to the measure of the pleasure or pain produced by the act. One more factor must be considered in evaluating an act, however. A pleasure or a pain affects people, and people, being highly complex machines, differ in sensibilities. If, for example, two men have $1,000 each and $500 is taken from one and given to the other, less pleasure is gained by the act than pain is incurred, for the recipient's wealth is increased by one-third but the loser loses half. Thus wealth is a measure of sensibility to some acts. Similarly, education, race, sex, character, and other factors determine the sensibilities of people.

The value of an act can now be computed as follows. The objective measure of the pleasure it gives is multiplied by the various sensibilities of the persons involved and then these products are added. The number obtained is considered to be positive. Then the pain this same act may induce in people is calculated in the same manner and considered negative. The value of the act is the sum of these positive and negative numbers. With this 'calculus' we not only obtain the value of an act but are also able to compare two courses of conduct.

Practical applications were soon forthcoming. Bentham's moral arithmetic was applied by some to decide whether it would be right to require vaccination against smallpox. Since at that time some children died as a result of the vaccination, the procedure was not universally approved. The proponents of vaccination argued, however, that if 10 per cent of those inoculated died from the inoculation whereas 50 per cent of the group would otherwise die from the disease, then surely the inoculations were warranted, provided the survival of the larger number would be a good for all of society.

This kind of argument, as well as Bentham's entire algebraic approach to morality, may seem to us to carry mathematics into fields where it has no business. It is certainly true that the measures of value he projected cannot be readily computed. This deficiency must be overlooked. 'Strict logicians are licensed visionaries.' What matters is that he boldly carried the banner of Reason into realms of thought previously ruled by authoritarian traditionalism, and that he sought a rationalistic approach to a system of ethics which served the common man. Here was a science of ethics founded not on religious precepts or on rationalizations of existing social patterns but on the science of human nature. Not the will of God but the nature of man gave rise to the new ethics. Virtue, in particular, was no longer to be rewarded by Heaven but was to be its own reward. The application of Bentham's philosophy would be desirable even today.

The theorists on ethics, typified by Bentham, had succeeded in carrying out the basic plan; that is, they had erected logical systems of ethics by utilizing the laws of human nature and special axioms about man's behaviour towards his fellow man. The political theorists proceeded to do the same. Spurred on by David Hume's confidence that 'politics may be reduced to a science', they sought axioms for their particular science and, of course, different schools of thought chose different axioms. Some, like Hobbes, sought axioms that would justify absolute monarchy; others, like Voltaire, sought to insure enlightened despotism; and still others, notably Bentham, argued for democracy.

Of the various political theories that were developed, two, at least, are of incomparable importance for our times, those of Locke and Bentham. Locke undertook to ascertain the natural origins and *raison d'être* of governments, that is, the logical basis for the existence of governments; the actual history of their rise was irrelevant to his investigation. His argument began with a doctrine from his famous theory of knowledge. All men are born with blank minds. Their character and all their knowledge are acquired through experience. Since, therefore, the essential differences among men are due to environment, it is correct to say that all men are born equal. In the hypothetical, earliest state, which was called the state of nature in the eighteenth century, all men possessed natural and inalienable rights, such as liberty, and were guided by the laws of reason. In order to secure protection of life, liberty, and property men made a 'social contract', giving to a government the right to determine and punish offences against society. When they entered into this contract, men agreed to be guided by the will of the majority; the government was supposed to determine that will and act accordingly. Hence if the rulers, chiefly the legislators, should betray their constituents, revolt would be justified. Thus a reasoned examination of the nature of government answered such questions as why it existed, whence it derived its power, when it exceeded this power, and what could be done about tyranny.

Nowhere is Locke's philosophy of government and the rational approach to it so succinctly expressed as in a famous 'mathematical' document of the eighteenth century, which is well known to all of us and which actually quotes many of Locke's phrases:

We hold these truths to be self-evident, that all men are created equal, that they are endowed by their Creator with certain unalienable Rights, that among these are Life, Liberty and the pursuit of Happiness. – That to secure these rights, Governments are instituted among Men, deriving their just powers from the consent of the governed. – That whenever any Form of Government becomes destructive of these ends, it is the Right of the People to alter or to abolish it, and to institute new Government, laying its foundations on such principles and organizing its powers in such form, as to them shall seem most likely to effect their Safety and Happiness.

The argument begins, it will be noticed, with the statement of self-evident truths, equivalent to the self-evident axioms that are the foundation of any mathematical system. The document then proceeds to state facts showing that the king had failed to provide the people with those rights which, according to the above axioms, governments are supposed to secure. Hence, by another of these axioms, the people were justified in abolishing this government and in instituting a new one.

The personal view of the writer of the document above went further. Each generation, said Thomas Jefferson, should make its own social contract. He calculated that every eighteen years and eight months half of the people over twenty-one years of age die.

Hence every nineteen years there should be a new contract and a new constitution.

Far more important than the mathematical form of the Declaration of Independence is the political philosophy it sets forth. The opening sentence is most pointed.

When in the Course of human events, it becomes necessary for one people to dissolve the political bands, which have connected them with another, and to assume among the powers of the earth, the separate and equal station to which the Laws of Nature and of Nature's God entitle them, a decent respect to the opinions of mankind requires that they should declare the causes which impel them to the separation.

The key phrase is 'the Laws of Nature'. Here is an explicit expression of the eighteenth-century belief that the entire physical world, including man, is ordered by laws of nature. Of course, this belief was based upon the evidence of design uncovered by the mathematicians and scientists of the Newtonian era. Obviously, since such laws existed, they must determine the ideals, conduct, and institutions of men. A valid law of government must be a natural law.

Equally significant are the words 'of Nature's God'. God's will and God's backing have, of course, been invoked on behalf of many diverse and even opposing causes. Here, however, it is not God's will as known to man through revelation or through the Scriptures; it is the God who speaks through nature. Reason uncovers His will, for reason, being part of man, is part of nature. In fact, the eighteenth-century thinkers practically identified 'right reason' and nature.

The Declaration was written by a small group of political leaders who sought to justify revolt from Great Britain. The justification received the backing of the people because it expressed their beliefs. As Jefferson himself pointed out, he had invented no new ideas or sentiments; he had merely stated what everyone was thinking. It was this widely accepted political philosophy, rather than the Stamp Act or the tax on tea, that fostered the American Revolution. Indeed both the American and French Revolutions were widely regarded as triumphs of Nature and Reason over wrong.

Rationalism of the mathematical variety and the doctrine of natural rights, applied to politics, produced a new philosophy of government and infused men with a determination to revolt against injustice. But the doctrine of natural rights did not fare too well in the nineteenth century. Many of the leaders of the revolt, notably Hamilton, Madison, and John Adams, were concerned more with the protection of private property than with the rights of the masses. Moreover, many special pleaders identified natural rights with the interests of the rising merchant class, which wanted freedom from governmental interference to make more money, or qualified the doctrine to mean the natural rights of free men, thereby validating slavery. In England the natural right of labourers to education was denied on the grounds that it would make them unhappy with their lot, make them fractious, and enable them to read seditious pamphlets, vicious books, and publications against Christianity. In addition, because it had inspired the French Revolution, the doctrine of natural rights was charged with the ensuing evils, such as the Reign of Terror and the Napoleonic aggrandizements. For all these reasons, the doctrine lost prestige and backing. As a consequence, the principle of democracy, that governments derive their just powers from the consent of the governed, lost its theoretical foundation and the practice of democracy might indeed have suffered. Fortunately, the philosophy of modern democracy was refounded by Bentham, who felt the cogency of reason even more than Locke. The new philosophy is called Utilitarianism.

Bentham had expounded his views on human nature and his system of ethics in his *Introduction to the Principles of Morals and Legislation* (1789). This same book treated the science of government and indeed made political science, as distinguished from statecraft, a branch of moral philosophy. Bentham discarded natural rights and God's will and sought a purely rational basis for government. To him the primary truth or fundamental axiom in the political field was that a government should seek the greatest happiness for the greatest number of people. From this basic principle he deduced many conclusions. Justice is not an end in itself; it is rather the means to increase the total amount of happiness. Law must consider the consequences of acts, not the motives, since only the effect of acts on the happiness of society is important. In penology Bentham contributed the deduction that the law must discourage by penalties acts that diminish happiness. Since punishment means pain, however, it is to be inflicted only when it prevents greater pain.

Then Bentham pondered this apparent paradox: rulers naturally seek their own happiness, but government should seek the greatest happiness of the greatest number. How can these opposing interests be reconciled? Only by securing an identity of interests of governors and governed. This can be accomplished by putting power in the hands of all. Democracy, then, is the preferred form of government. To clinch the argument, Bentham appealed to the 'uninterrupted and most notorious experience of the United States'. In that country, he asserted, there was no corruption, no useless expenditure, and none of the evils found in Great Britain.

Bentham's illustrious disciple, James Mill, took up the problem of who should constitute the electorate in a democracy. After eliminating voters whose interests are well protected by other voters as, for example, the wife's interests are protected – so Mill believed – by the husband, he concluded that only men over forty should vote.

Bentham may have been somewhat in error about the situation in the United States but his argument for democracy was very effective. The average American is a Utilitarian even if he has never heard the word. Bentham's greatest good for the greatest number and Locke's philosophy of natural rights and the social contract forged, and were fused in, the American democracy.

We need not consider further the course of political ideologies. Theorists were not able after all to found a science of government as successful as a mathematical theory of the heavens. Perhaps they did no more than justify and proclaim the political emergence of the common man. But by rational inquiry they did at least isolate and phrase the goals, ideals, and slogans of the democratic trend.

The full realization of democracy could not occur until there were changes in the philosophy and form of man's economic institutions, for the man who is politically free but economically a slave enjoys at best only an illusion of freedom. The great thinkers of the eighteenth century already at work at the task of reorganizing all knowledge were soon even more pressed to revamp economic thought by the approaching Industrial Revolution.

The new science of economics followed the rational, mathematically inspired lines of the ethical and political theorists. Basic to it was to be the science of human nature. To this the axioms of economics proper were to be added. The deduction of economic laws would follow readily.

The two leading eighteenth-century schools of economic thought, the Physiocrats under François Quesnay and the English classicists headed by Adam Smith and later John Stuart Mill, were in agreement about the existence of axiomatic economic truths. They agreed, also, that eternal and immutable laws rule in economic as well as natural phenomena. (The word 'physiocrat' means the rule of nature.) Hence it was possible to arrive at a natural science of wealth. The economist must ascertain the laws and proclaim them.

The axioms adopted by these schools of thought are familiar to all of us and are still, in the main, the dominant views. The individual acts in his own interest. Equally axiomatic are the rights to liberty, property, and security, and the proposition that land and (or) labour are the sole sources of wealth.

From such axioms it was not hard to deduce the theorems of free trade and unrestricted competition, doctrines incorporated in the phrases *laissez faire, laissez passer*. Any interference with man's normal and natural efforts to gain a livelihood was interference with God's design of the universe and therefore presumptuous. In particular, government must not interfere with business. Business must be left to businessmen whose enlightenment would insure the successful working of the economic system. The government need merely guarantee and protect contract rights. The Physiocrats, who believed land to be the sole source of wealth, advocated a single tax on land; Adam Smith, on the other hand, regarded labour as the sole source of wealth and therefore, though sympathetic with the problems of workers, preferred a single tax on incomes.

Axioms there were in these economic theories, as well as deductions in the mathematical spirit, but laws that corrected the economic ills of society there were not. These economists, however unconsciously, were special pleaders for the merchant and manufacturing classes. The theorists borrowed what they needed from the rationalistic attitude of the times merely to build a logical defence of the *laissez-faire* doctrine. Indeed as industrialization proceeded apace in the early nineteenth century, this doctrine failed miserably to mitigate the woes of the labouring classes. It merely justified the rich getting richer and the poor becoming penniless. So apparent were the inequalities and injustices that economists felt impelled to defend the existence of great masses working in factories at starvation wages. Their method was again to search for natural laws in order to establish that it was God's plan and an inevitability that women and little children should toil sixteen hours a day.

Thomas R. Malthus found the answer in the laws of population. The conclusions he sought were so readily discernible that he was able to write the *Essay on the Principles of Population* without having to cast his eyes on the world about him. The book established Malthus' reputation as a man of authority and earned for him a professorship of history and political economy. Malthus begins thus:

I think I may fairly make two postulata. [As usual the argument begins with the axioms.] First, That food is necessary to the existence of man. Second, That the passion between the sexes is necessary, and will remain nearly in its present state. . . . [In other words, sex is here to stay.] Assuming, then, my postulata as granted, I say, that the power of population is indefinitely greater than the power of the earth to produce subsistence for man.

In the language of John Adams: man has two wants, his dinner and his girl. But the second want is so intense that he forgets about the first and rushes into a rash marriage, from which come children. Hence the multiplication of population far transcends the multiplication of means of subsistence.

Perhaps to gain something of the authority of a mathematical demonstration Malthus states that population increases in geometrical progression, while the means of subsistence from a fixed area increases only in arithmetical progression. He estimates that the population doubles every twenty-five years. If other factors are not present it will be multiplied in two centuries by 256, whereas in the same time the food supply will increase by a factor of 9.

Malthus realized, however, that actual populations do not increase geometrically. Why not? The answer is that starvation, disease, vice, and wars check the increase of populations. These seeming evils are, in the long run, really beneficial; they are resorts of nature, dreadful but necessary. Since these happenings are part of the divine plan, no legislation can alleviate man's miserable lot. No society can exist in which all members can live in ease, happiness, and leisure. Malthus then emphasized the desirability of teaching restraint so that people will not have children they cannot support. He would in fact add an eleventh commandment: 'Thou shalt not marry until there is a fair prospect of supporting six children.'

The justification of wretched social conditions by appeal to natural laws did not end with Malthus. Another famous economist who took up the cause was David Ricardo. First, he segregated and labelled the factors that enter into economic life, namely, capital, labour, value, utility, rent, wages, profits, and so on. Everything in business, said Ricardo, followed inescapable, natural laws involving these factors, and the laws could be deduced from postulates. For example, it was self-evident that the price of a commodity was determined by supply and demand. This postulate, applied to the commodity of labour, implied that there was a natural price for labour. If wages were raised above this level, labourers would have larger families, thereby increasing the supply of labour and bringing about a reduction in wages. Hence it was pointless to raise wages. Ricardo summed up these considerations in his famous law of wages: 'The natural price of labour is that price which is necessary to enable the labourers, one with another, to subsist and perpetuate their race without either increase or diminution.' Thus it was natural to Ricardo, as well as to Malthus, that poverty, distress, and starvation should exist. It was natural, too, that labourer, landowner, and capitalist should be antagonistic to each other. All these laws and the conditions they bring about were the decrees of a far-seeing Providence.

As industrialization proceeded, the 'science' of economics failed more and more to treat the major problems of society. In fact, it worked against reform movements, against unions, against remedial legislation, and against charity so that instead of serving man the science served his enemies.

The rationalist movement in economics had not spent itself, however. The wonders of physical science were even more splendid and the power of mathematics even more evident in the nineteenth century than they were in the eighteenth. Yet economic theory, having adopted the methods of mathematics and science, was in greater confusion. The trouble was, reasoned some economists, that though they had used mathematical method and had sought natural laws, they had not used mathematics itself to any appreciable extent. Also, perhaps they had bitten off more than they could chew at any one time. It might be better to divide and conquer.

And so the economists attempted a quantitative, deductive approach to special phenomena instead of to entire fields, piecemeal investigations instead of wholesale ones. The first objective in each case was to find the fundamental formula or formulas that governed the particular phenomenon. The second was to use these formulas and mathematical techniques to deduce conclusions. In this more limited type of endeavour the economists were much more successful.

With the publication in 1838 of Cournot's *Researches into the Mathematical Principles of the Theory of Wealth*, there arose a new school in economic thought, the Mathematical School, which includes the work of Vilfredo Pareto in our own century. To illustrate its method of attack on specific problems we shall describe briefly the work done by two contemporary Americans, Raymond Pearl and Lowell J. Reed, on the very important problem of population growth.

For what follows we must keep in mind the fact that we are not concerned with the population of Middletown in 1947. We wish to study population changes in the large, to uncover the fundamental factors of growth rather than the incidental ones. In accordance with the mathematical approach to a problem Pearl and Reed start with reasonable assumptions:

(a) Physical conditions set an upper limit, denoted by L, to the population of a region or country.

(b) The rate of growth of the population is proportional to the existing population.

(c) The rate of growth of the population is also proportional to the possibility for population expansion, that is, to the difference between L and the existing population.

These axioms suggest to the mathematician a differential equation which can be solved readily. The result is a general

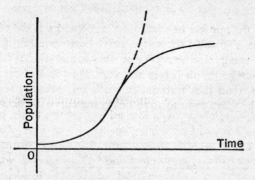

Figure 68. The curve of growth

formula for the growth of population. If y stands for the population of a country t years after a certain fixed date, then the formula is

$$(1) \qquad y = \frac{L}{1 + a(2 \cdot 718)^{kt}},$$

where a and k are numbers whose values depend on the region to which formula (1) is applied.

The reader need not worry too much about the details of formula (1). The shape of the curve corresponding to this formula is shown in figure 68. Known as the logistic curve, it represents in its entirety what is called a growth cycle. The broken line in figure 68 shows how population would have to change were it to increase indefinitely in geometrical progression as Malthus asserted.

Formula (1) states a general law of population growth that tells us how large masses of people ought to behave. But do they behave that way? With just a little algebra applied to the census figures of the United States from 1790 to 1910, the values of a and k in formula (1) were determined by Pearl and Reed. Accordingly, the general formula for population growth in the United States turned out to be

$$(2) \qquad\qquad y = \frac{917 \cdot 27}{1 + 67 \cdot 32(2 \cdot 718)^{-\cdot 0313t}},$$

where t is the number of years since 1780 and y is the population in millions. Figure 69 shows the curve corresponding to formula (2). The small circles represent the actual data; the broken portions of the graph before the year 1790 and after 1910 represent the trend that formula (2) calls for. We can see that the data for the years 1790 to 1910 lie on the curve of formula (2).

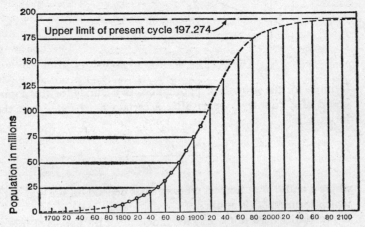

Figure 69. Growth of the population of the United States

How closely does the formula represent what happened after 1910? Well, according to the formula the population in 1930 should have been 122,397,000 and the census showed the population to be 122,775,000. For the year 1950 the formula predicts a population of 148,400,000; the census yielded 150,700,000. There seems to be very good agreement between the theory and the facts.

From formula (2) we can deduce several other interesting conclusions. It says that the upper limit to the population is 197,270,000 and that this figure will have almost been reached by the year 2100. Another deduction from the Pearl-Reed formula is that the United States passed the point of its most rapid rate of growth in 1914. The actual study of population growth shows, then, that a purely rational, theoretical approach to the problem has produced a formula or law that represents the facts, at least in their larger aspects.

Even the more restricted studies in mathematical economics typified by the work of Pearl and Reed have not always been productive, largely because correct premises have not been found. Too often the absence of any real contribution to a problem is hidden behind an immense mass of mathematical symbolism. There is no doubt, nevertheless, that the deductive mathematical approach to specific economic problems has produced some useful knowledge.

Unbounded optimism about the applicability and power of mathematics has led to some bizarre conclusions. One psychologist undertook to derive a formula for the strength of affections and quite naturally he began with love. He concluded that love between man and woman varies directly as the square of the time of association and inversely as the cube of the distance between them. In this 'law' we have a mathematical formulation of the dictum that distance makes the heart grow colder.

Another somewhat suspect mathematical formula was derived by the philosopher David Hartley. He offered a vest-pocket edition of his moral and religious philosophy in the formula $W = F^2/L$, where W is the love of the world, F is the fear of God, and L is the love of God. It is necessary to add only this, Hartley said, that as one grows older, L increases and indeed becomes infinite. It follows then that W, the love of the world, decreases and approaches zero. This is the sum and substance of moral truth.

We have been examining the influence of mathematics itself and of the rational spirit engendered by mathematics on the science of man. In so far as that spirit buoyantly and over-optimistically predicted the discovery of natural, universal laws of human behaviour and the consequent solution of all social problems it was, of course, wrong. Man has on the whole failed to understand and predict his own behaviour. His body, his emotions, and his desires apparently refuse to obey rigid laws or submit to mathematical regulation. At least no thinker has as yet built up a quantitative, deductive approach to an entire social science that would enable us to direct, control, and predict phenomena in that field. Especially in economics has success been signally absent.

Why is man his own Achilles' heel? One reason for the absence of any science of society was given by Hobbes long ago. 'For I doubt not, but if it had been a thing contrary to any man's right of dominion, that the three angles of a triangle should be equal to two angles of a square, that doctrine should have been, if not disputed, yet by the burning of all books of geometry suppressed, as far as he whom it concerned was able.'

Perhaps the severest criticism that may be levelled at the eighteenth- and nineteenth-century workers in the social sciences is that they were too mathematical and not sufficiently scientific. They wanted to find axioms or general principles from which the science of politics or economics would readily follow. But very few would, like Montesquieu, examine society itself, first to check the correctness of their axioms, and later to check their deductions.

Whatever may be the merits and shortcomings of the deductive approach to the social and psychological sciences, one value is pre-eminent. The concept itself of a science of ethics, politics, economics, or psychology and the stimulus to create such sciences stem directly from the fertilizing rationalism of the Newtonian age. Consequently, the clear light of reason has at least irradiated fields befogged by tradition, custom, and super-stition. In particular, the attempt to be reasonable about govern-ment instead of accepting established institutions opened men's eyes to inequalities, injustices, and cruelties. What Greek ration-alism did for mathematics, the mathematical spirit did in turn for these vague, ill-defined, confused domains of thought: it 'raised the edifice of Reason on the ruins of opinion'.

XXII

The Mathematical Theory of Ignorance: The Statistical Approach to the Study of Man

That which everywhere oppresses the practical man is the greater number of things and events which pass ceaselessly before him, and the flow of which he cannot arrest. What he requires is the grasp of large numbers.

Theodore Merz

A FAIRLY good rule of bridge, when there is little strength in the hand, is to lead from the weakest suit. As our tale will tell, this rule also works well for scientific 'hands'. It was applied with remarkable success by the social scientists when they realized that they did not hold trump cards.

The tactics of the successful mathematicians and physical scientists may be described briefly as *a priori* and deductive. By careful reflection on whatever knowledge of a phenomenon may be available they obtain broad fundamental principles which serve as axioms. Deductive reasoning then produces new conclusions and new knowledge. In this 'armchair' approach, observation and experimentation may help to arrive at first principles or to check the deductions but the mind rather than the senses is the effective agent.

The *a priori*, deductive approach has, on the whole, failed the social scientists for reasons that are most pertinent. Perhaps the principal one is that the phenomena they study are exceedingly complex. So many factors are involved even in relatively limited problems that it has been impossible to single out the dominant elements. How would we account, for example, for a period of national prosperity? Such a happy situation depends on natural resources, labour supply, available capital, foreign trade, war and peace, psychological considerations, and other variables. It is not surprising, then, that no one appears to have got to the core of this problem. If an economist should attempt to simplify the problem by making assumptions about some of the variables involved he is likely to make the problem so artificial that it no longer has any bearing on the real situation.

In many cases the *a priori*, deductive approach has not been feasible because there is practically no knowledge to work with. The treatment of some diseases cannot be prescribed because nothing is known about their causes and there is too little information about the factors favouring their spread. Vast phases of the chemistry of the body and of the operation of the brain are total mysteries to biologists. The mechanism of physical inheritance is almost a sealed book. In these fields analyses can hardly be begun.

In some problems failure to obtain fundamental laws through the classical method of deduction from axioms is due, paradoxically enough, to too much information. A gas consists of molecules which attract each other in accordance with the well-known force of gravity. In addition, the molecules are subject to the Newtonian laws of motion. If a given volume of gas contained only two or three molecules the behaviour of the gas could be predicted just as scientists can and do predict the behaviour of the planets. But a cubic centimetre of gas contains 6×10^{23} (6 followed by 23 zeros) molecules under ordinary conditions. Each molecule exerts an effect on all the others in accordance with the law of gravitation. Obviously, we cannot study the behaviour of this volume of gas by summing up the effects of all the individual molecules on each other. Some method is needed that permits treatment of a large number of molecules as one unit.

Another reason for dissatisfaction with the *a priori*, deductive approach to social problems was peculiar to the nineteenth century. The Industrial Revolution introduced large-scale factory production and led to increasing urban populations. From these developments there emerged a vast array of social problems connected with population changes, unemployment, quantity production and quantity consumption of commodities, insurance against the risks in large-scale enterprises, and diseases propagated by unhealthy living conditions in congested districts. These problems crowded upon the scientific workers so fast that, even had they been able to be solved by the *a priori*, deductive approach, more time would have been required than could then have been spared. This method, even when applied by such geniuses as Copernicus, Kepler, Galileo, and Newton, had required more than one hundred years to produce the laws of motion and gravitation. It could hardly have been expected to yield results more quickly in the social and medical fields.

For all these reasons, then, the *a priori*, deductive approach failed the social scientists and a new method of attack on their problems seemed necessary. If anyone stopped to think of how much was being demanded of a new method for obtaining scientific laws he could despair of ever finding it. It had to yield results quickly; it had to summarize the effects of many variables acting in one situation; it had to succeed where understanding was completely absent; it had to encompass the effects of uncountable millions of participants in a phenomenon; and it had to measure the effect of factors themselves unmeasurable. Despite these inordinate demands a new approach to scientific problems was created that met them all.

The new approach began with an analysis of the state of affairs. Here we have phenomena, argued the social scientists, whose essential nature we do not understand or, if we do understand it, as in the case of the motion of the molecules of a gas, the understanding does us no good, and so for all practical purposes we are ignorant. Hence we do not have the broad fundamental principles that could serve as the basis of a deductive approach. On the other hand, our weakness seems to be that we are confronted with a superabundance of undigested, bare facts which overwhelm us and underscore our ignorance.

It was at this point that the social scientists recalled the rules of bridge. Since they did not possess the fundamental principles that could serve as trump cards, they decided to lead through weakness. Said they, if we cannot understand *how* rainfall affects vegetation, let us nevertheless *measure* what it does. If we do not know *why* vaccination prevents deaths, let us *tabulate* the results of the practice. If we cannot *fathom* the complexities of national prosperity, let us establish a suitable index and *chart* its rise and fall. If we do not *understand* the mechanism of heredity in plants, animals, or human beings, let us reproduce the species and *record* what successive generations reveal. Let the world be our laboratory and let us gather statistics on what occurs therein.

The mere gathering of statistics was not a new thought, for statistics are found in the Bible and in older documents. What was new was the realization, which seems to have come first to a prosperous seventeenth-century English haberdasher, John Graunt, that statistics could serve as a major weapon in an attack on the problems of the social sciences. As a pastime Graunt had studied the death records of English cities and noticed the unvarying percentages of deaths from accidents, suicides, and various diseases. Thus occurrences which, on the surface, seemed to be the sport of chance were seen to possess surprising regularity. Graunt also discovered the excess of male over female births. On this statistic he based an argument: since men are subject to occupational hazards and war service, the number of men fit for marriage about equals the number of women, and so monogamy must be the natural form of marriage.

Graunt's work was supported and seconded by his friend Sir William Petty, professor of anatomy and music and, later, army physician. Though Petty made no observations as striking as Graunt's he is especially noteworthy because his point of view was broad. The social sciences, he insisted, must, like the physical sciences, be quantitative. Speaking of his writings on medical, mathematical, political, and economic subjects he said: 'The method I use is not yet very usual; for, instead of using only comparative and superlative words, and intellectual arguments, I have taken the course ... to express myself in terms of number, weight, and measure; to use only arguments of sense, and to consider only such causes as have visible foundations in nature.' He gave the infant science of statistics the name of 'Political Arithmetic' and defined it as 'the art of reasoning by figures upon things relating to the government.' In fact, he regarded all of political economy as just a branch of statistics.

When these alert, far-seeing Englishmen spoke for the potentialities in statistics and when a seventeenth-century priest used statistics to combat the superstition that the phases of the moon influence health, new foundations for science were conceived. The period of gestation lasted about a hundred years. During this time statistics came to mean quite generally the noteworthy quantitative information about a nation; that is, it was considered to be data for statesmen. Very little was done until the early part of the nineteenth century to follow up the suggestions contained in the work of Petty and Graunt, namely, to obtain laws on the basis of data. At this time an effective group of workers, conscious of the failure of the *a priori*, deductive approach to the social sciences, and equally conscious of the potentialities in statistics, started to tackle major problems.

Petty and Graunt were the discoverers of a vein of thought. In order to obtain pure gold, however, it is not sufficient merely to unearth the ore, back-breaking though this task may be. The ore must be sifted, strained, and refined. Similarly, the mere accumulation of statistics accomplishes little in itself, for only in the very simplest of problems will conclusions stand forth readily from data. The extraction of knowledge from large masses of data is accomplished by mathematics.

About the simplest mathematical device for the distillation of knowledge from data is an average. Suppose that the employees of some small business organization receive the following weekly salaries in dollars:

20, 30, 40, 50, 50, 50, 60, 70, 80, 90, 100, 1,000, 2,000

What is the average weekly salary? Usually we would take the sum of all these salaries and divide by the number of salaries. In this example the sum is 3,640 and the number of salaries is 13. Hence the average is 280. This type of average is called the *arithmetic mean*.

It is clear that the mean is not too informative. No one person actually earns this salary. Moreover, out of thirteen people only two earn as much or more. The others earn far less. In other words, the arithmetic mean is not a representative figure if some of the component quantities are very large compared to the others. In such cases other averages may be more informative.

Another frequently used average, called the *median,* is the datum for which there are an equal number of cases below and above it. In our example there are thirteen cases. The median salary, therefore, is 60 because there are six people earning less and six earning more.

The median does seem to be a more representative figure in this example but it, too, fails to tell the whole story. If the wages of the six people below the median were very much less than the above figures and the wages of the six above the median were very much more, the median would be the same. Such a gross disparity in earnings would not be reflected in the median figure of 60. Hence the median, too, often fails to be a representative figure.

Another average in common use is the *mode.* This is the figure in the data that appears most often. In our example the mode of the salaries is 50 because the largest number of people earn this salary. Though this average, like the others, gives some indication of the distribution of wages it, too, is inadequate. The range of the salaries above and below the mode is not reflected in this average.

What each of the averages fails to tell us in the distribution of the data above and below it. The mean does depend on all the values but we cannot infer the nature of the distribution from it. For example, if the two salaries of 1,000 and 2,000 were changed to 100 and 2,900 respectively the mean would still be the same but the nature of the distribution would be changed. What is needed is some measure of the dispersion of data about the average. For this purpose statisticians use a quantity called the *standard deviation*; it is denoted by σ (sigma). This quantity is computed as follows. First, the difference between any one datum and the arithmetic mean, that is the deviation of that datum from the mean, is calculated. To avoid negative numbers this deviation from the mean is squared. The squares of the deviations are then averaged by adding them and dividing by the number of data. The square root of this average is then taken to offset somewhat the squaring performed earlier. Briefly stated, the standard deviation of a set of data is the square root of the mean of the squares of the individual deviations from the mean of the data.

We could use the set of salaries above to illustrate the computation of a standard deviation; however, to avoid being lost in the arithmetic we shall use a simpler one. Let us compute the standard deviation of the data

$$1, 3, 4, 7, 10, 13, 18.$$

The mean of these data is 8. Hence, the deviations from the mean are

$$7, 5, 4, 1, 2, 5, 10.$$

The squares of these deviations are

$$49, 25, 16, 1, 4, 25, 100.$$

The sum of these squares is 220. The mean of these squares is therefore 220 divided by 7 or 31·4 approximately. The square root of this mean is about 5·6. Since the latter figure is large compared to the mean of 8, the dispersion of the data must also be large. Had we performed a similar calculation for the data on salaries, we should have obtained a standard deviation of 556. The mean, we recall, was 280. Again we would be justified in inferring that the dispersion of the salaries about the mean must be large.

Of course, even two representative figures such as the mean and standard deviation do not say as much as the data themselves but since the mind cannot carry and work with all the data, these figures are quite helpful.

An alternative to remembering the entire collection of data or to relying on just the two representative figures is the graph. Hardly a person who reads a daily newspaper has failed to observe that a graphical presentation of data makes facts stand out that would otherwise be far from obvious. Graphs of the rise and fall of the cost of living and of stock prices are common examples. The graphical approach to data, however, has produced far deeper and more significant conclusions than a mere display of rise and fall.

Suppose we measured the heights of all the men in a certain community. Corresponding to each height there would be the frequency with which this height appears. If we plotted the heights of the men as abscissas and the corresponding frequencies as ordinates we should obtain the graph of the distribution of these frequencies. A graph of actual data is shown in figure 70, wherein a smooth curve has been drawn through the data. There is no doubt that the graph gives a picture which is readily retained in the mind and displays at once a great deal of the information contained in the original data.

What is especially significant about the distribution of heights as well as of many other characteristics we shall discuss shortly is that the curve approximates an ideal distribution known to

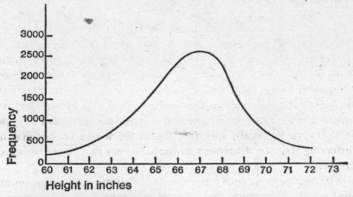

Figure 70. The heights of men in a certain community

mathematicians as the normal frequency curve (fig. 71). In fact, the larger the group whose heights are included the closer the curve comes to having the ideal shape, just as regular polygons with more and more sides approach the shape of the circle.

The normal frequency curve or normal distribution is so common and so important that we should notice its chief characteristics. The curve is symmetric about a vertical line that represents the largest frequency among the data. As we follow the curve to the right and to the left of this line the curve drops slowly at first, then very rapidly, and finally as it extends still further to the right and the left, it approaches but does not reach the horizontal axis. The shape has been likened to that of a bell and, in fact, the curve has been called the bell-shaped curve.

The abscissa corresponding to the largest ordinate or frequency in any normal distribution is certainly the mode of the distribution since it is the measured quantity that occurs most often. This mode must also be the median, for the symmetry of the graph tells us that as many cases occur to the left of this abscissa as to the right. It is almost obvious that the mode is also the mean because two abscissas, one on either side of the mode, and equally far from the mode, have the same frequency, and in

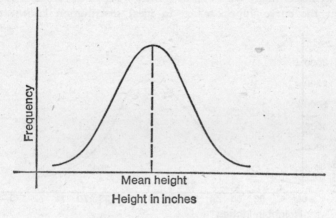

Figure 71. The normal frequency curve

the computation of the mean, the average of all these pairs of equally distant abscissas will be the middle one. Hence in a normal distribution the mode, median, and mean coincide.

The normal frequency curve has been familiar to astronomers and scientists since about 1800 because it occurs often in connection with measurements. Suppose a scientist is interested in the exact length of a piece of wire. Partly because the hand and the eye are not perfectly accurate and partly because surrounding conditions such as temperature may fluctuate, he measures this length not once but perhaps fifty times. These fifty measurements will differ from each other, sometimes perceptibly and sometimes imperceptibly. A graph of the various measurements against the number of times that each measurement appears among the fifty approximates the normal frequency curve. In fact, the more measurements made the more nearly will their frequency distribution follow this curve.

There is good reason to expect that a set of measurements carefully made should follow a normal curve. The errors in measurement should be due to random errors made by the eye or hand or to random variations in the instruments employed. These errors should distribute themselves on either side of the true value and cluster about this value, just as the hits of a rifleman on a target will, if he is a marksman, cluster about the bull's eye and become rarer with greater distance from the centre.

The fact that measurements follow a normal curve is very helpful to scientists. In a normal distribution the data cluster about the mean value, which, as just noted, should be the true value. Hence, the mean value of a large number of measurements, if they appear to follow a normal curve, should be a good approximation to the true measure. If, moreover, a large set of measurements does not appear to follow a normal curve some disturbing influence has crept into the measurements and should be eliminated. For example, if the length of a piece of metal were being measured in a room with a rising temperature, the measurements would undoubtedly increase steadily and fail to follow a normal curve. The mean of these measurements would be grossly in error and a graph of the measurements would reveal this disturbing factor at once.

The normal curve has been used thousands of times to determine astronomical distances; to measure mass, force, and velocity; and to fix solubility, boiling and freezing temperatures, and hundreds of other chemical quantities. Because of its use in the elimination of errors of measurement the normal curve has come to be known also as the 'error curve'. Its very existence affirms the seemingly paradoxical but none the less true conclusion that accidental errors in measurement do not occur haphazardly but always follow the curve above. Humans may not even err at will.

In the use of normal distributions it is important to know how many of the cases lie in any given range of the measured quantity. Consider, for example, the different heights of 100,000 American men. It has already been described how the frequencies of the various heights lie on a normal curve. Suppose that the mean and standard deviation of this distribution are 67 inches and 2 inches respectively. Then (fig. 72) ideally 68·2 per cent of the measured heights lie within 1 standard deviation or 2 inches of the mean; that is, 68·2 per cent of the men have heights between 65 inches and 69 inches. In addition, 95·4 per cent of the heights lie within 2 standard deviations or 4 inches of the mean; and 99·8

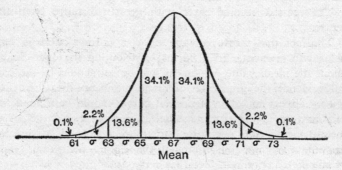

Figure 72. The percentages of cases falling in different regions of a normal frequency distribution

per cent of the heights lie within 3 standard deviations or 6 inches of the mean. The percentage lying within any given fractional number of standard deviations from the mean has also been computed and may be found in tables. Thus if a person is studying a normal distribution and computes the mean and standard deviation, he can obtain all the information he desires about the distribution from these two quantities.

About 1833 the Belgian astronomer, meteorologist, and statistician, L. A. J. Quételet, decided to study the distribution of human traits and abilities in the light of the normal frequency curve. Much of his data, incidentally, was taken from the thousands of measurements on parts of the human body made by the Renaissance artists, Alberti, Leonardo, Ghiberti, Dürer, Michelangelo, and others. Quételet found what hundreds of successors have since confirmed. Almost all mental and physical characteristics of human beings follow the normal frequency distribution. Height, the size of any one limb, head size, brain weight, intelligence (as measured by intelligence tests), the sensitivity of the eye to various frequencies of the visible portion of the electromagnetic spectrum – all are found to be normally distributed within one 'racial' or 'national' type. The same is true of animals, vegetables, and minerals. The size and weights of grapefruit of any one variety, the lengths of the ears of corn of any one species, and so on, are normally distributed.

The fact that human traits and abilities follow the same distribution curve as do errors of measurement was of the utmost significance to Quételet. He argued that all human beings, like loaves of bread, are made from one mould but differ only because of accidental variations arising in the process of creation. For this reason the law of errors applies. Nature aims at the ideal man but misses the mark and thus produces deviations on both sides. On the other hand, if there were no type to which men conformed, we could measure their characteristics – height, for example – and not find any particular significance in the graph or any definite numerical relation in the data.

The more measurements Quételet made the more he noticed that individual variations are effaced and that the central characteristics of mankind tend to be sharply defined. The mean of each of these characteristics identifies the ideal or 'mean man'. The mean man was, furthermore, the centre of gravity around whom society revolved. The central characteristics, he then declared, result from general causes, and therefore society exists and is preserved. More than that, the evidence of design and determinism appeared to be as clear in social phenomena as in physical phenomena.

We shall defer judgement of Quételet's philosophical inferences to a later chapter. Let us content ourselves, for the moment, with observing that the applicability of the normal curve to social and biological problems has led to knowledge in these fields and to laws. Indeed, conviction today that the distribution of any physical or mental ability must follow the normal curve is so firmly entrenched that any measurements on a large number of people that do not lead to this result are suspect. If, for example, a new examination is given to a large group and does not produce a normal distribution of grades, the conclusion about the distribution of intelligence is not challenged; rather, the test is declared invalid.

Graphical studies of distributions lead to some provocative questions. Mental and physical characteristics, we saw, are normally distributed. If we graph the distribution of incomes, however – that is, each of the various incomes against the number of people having that income – the graph would look much like figure 73. This curve says that most people get incomes that are at the lower end of the income scale. As a matter of fact, studies

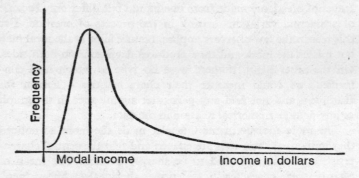

Figure 73. The frequency distribution of income

indicate that the most common income, the modal income, is at the 'wolf-point' or the point of mere subsistence. The curve also shows that there are many people who make much less than a subsistence income and only a few who make much more.

The graph makes immediately apparent, then, gross differences in income levels, and calls attention to the disparity between income, on the one hand, and physical and mental abilities on the other. This disparity almost demands an explanation. Why does the distribution of income differ so radically from the distributions of the abilities of the people who earn the incomes?

Valuable conclusions which are not only useful in special problems but also theoretically important, can obviously be drawn from data or from graphs based on data. But the cream of any scientific investigation, as judged by modern standards, is the mathematical formula. A conclusion ensconced in a formula is doubly valuable. Not only is the formula a compact and valuable result in itself but it permits the application of all the mathematical techniques of algebra, calculus, and other branches for the derivation of new conclusions. The point here can be understood by reference to an earlier illustration. The concept of universal gravitation is in itself a highly informative generalization. Because its behaviour can be stated as a formula, however, we may combine it with the laws of motion and derive the paths of the planets around the sun.

Now the compression of data into formulas is sometimes possible and when this is the case the process is fraught with meaning. For the moment we shall illustrate the process of representing data by a formula, and to do this we shall consider a somewhat specialized and slightly oversimplified problem.

Let us suppose that we were to set out to study the variation in food prices over a period of years. The level of food prices, as well as that of other commodities, is gauged by an 'index number', which is roughly an average price calculated by methods that do not matter here. The following table lists the index numbers (represented by y) of retail food prices in the United States for many years. In the table x stands for the number of years after 1900; that is, $x = 1$ corresponds to 1901, and so on.

x	1	3	5	7	9	11	13	15
y	71·5	75·0	76·4	82·0	89·0	92·0	100·0	101·3

Mere observation will not yield the formula relating x and y. The next step is to graph these pairs of x and y values, letting abscissas represent values of x and the ordinates the values of y (fig. 74). The points plotted seem to lie along a straight line. In fact, the line through the points (3, 75) and (9, 89) passes very close to each of the other points. Unavoidable errors in the determination of the index numbers could account for the fact that these other points do not lie exactly on the line. Thus far we have determined that the graph of our function is a straight line.

It is a simple problem in co-ordinate geometry to find the equation of this line. The result is the formula

$$y = \frac{7}{3} x + 68,$$

where y is the index number corresponding to any given year x. This formula fits the known data about as closely as the points in figure 74 come to the straight line.

The formula represents a considerable achievement. Without any knowledge of the factors affecting the rise and fall of food prices a law describing their course has been obtained. The law

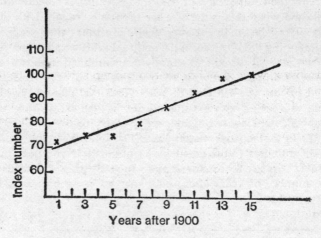

Figure 74. Graph of the data on food prices

certainly covers the period from 1900 to 1915 and, like other laws of science, can be used to predict – in this case, the level of food prices for some time after 1915.

There is a temptation to go further. Does the formula give the true law of the behaviour of food prices for all time? Certainly not. In fact a fundamental question exists as to whether food prices follow any unchangeable pattern. In any case, food prices do not rise continually and hence the formula above can at best represent the true law only approximately and for only a short period of time. It fails to be more representative partly because it is based on a limited amount of data and partly, perhaps, because the index of food prices may not be reliable.

While the particular problem of the level of food prices may not lead to any basic or universal law, the approach described above can lead to such laws where they exist, that is, where the data do follow a fixed pattern. The technique is to graph the data and to fit a formula to the graph. As might be expected, the process may involve some complicated mathematics when the graph does not happen to be a straight line.

A more significant example of a formula obtained from data and one purported to be a true economic law was furnished by the renowned student of political economy, Vilfredo Pareto. Pareto's study of income distribution in a given society led him to the formula $N = Ax^m$, where N represents the number of people having an income *equal to and higher than* any given quantity x, while A and m are two constants that must be determined from the data for any one country. Pareto also found that m had about the same value, approximately $-1·5$, in every country he tested. The invariability of this number from country to country and epoch to epoch appeared to Pareto and to many other economists to be profoundly significant.

Pareto himself attributed the existence of the same law of income distribution in many countries not to the economic structure of society but to the common distribution of certain natural qualities in men. He used the constancy his law revealed to refute Karl Marx, who argued that the trend of capitalistic society is to reduce the income of more and more people. And he used it further to argue that a country should not attempt to improve inequalities in income by legislation.

We may now raise the same questions for Pareto's study of income distribution that we raised for the behaviour of food prices. Is there a universal law of income distribution and, if so, does Pareto's formula state that law? There is more reason to expect the existence of such a law in the case of incomes than in the case of food prices. We can believe that the major factors affecting income would operate in about the same manner in all societies and at all times. At least the likelihood of this being the case is as good, on *a priori* grounds, as the likelihood that the planets should follow unvarying paths from year to year.

Actually there has been much dispute among the economists about whether or not Pareto's law is correct. It was first announced in 1895 and since then has been tested against data chosen from numerous countries. In many test areas, such as England during the nineteenth and early twentieth centuries, the formula fits the data very well. On the other hand, the failures do not necessarily disprove it for there is always some question about the reliability of the data.

Actually, we cannot be sure that any law obtained by fitting a formula to data is correct. After we graphed the table of index numbers and time, we selected a straight line that passed through as many points as possible and came very close to the other points. There is more than one straight line, however, that will go through some points and approximate the others. If the line is changed so is the formula that is derived from it later. Of course the difference may be negligible for practical purposes but this cannot always be ascertained in advance.

The formula may be even less accurate than the discussion above would indicate. The points on the graph of index numbers lay almost on a straight line; it was then *assumed* that the graph was truly a straight line, the discrepancies being attributed to errors in the process of gathering the data. The true situation, however, may be that the data are accurate and that the points do not lie on a straight line but on a curve that passes exactly through all of them. If this is the case the formula we found is certainly not the correct one though again perhaps sufficiently close to it for practical purposes.

What can we do about the errors that may creep into the process of fitting a formula to data? All we can really do is to make yesterday and today the guides for tomorrow. We predict by means of the formula obtained, and we check the predictions against what actually happens. If the prediction is incorrect we can use the new data together with the old to fit a formula to the enlarged set of data. Despite the uncertainties that underlie the derivation of formulas from data and of predictions based on such formulas, there is no question but that the formulas summarize and represent the known data in a most desirable form. Moreover, some of these formulas fitted to data have proved to be so constantly applicable that they seem to express the invariable behaviour of nature as much as do the Newtonian laws of motion and gravitation. The weighty implications of this fact will be discussed in a later chapter.

In some statistical studies, however, the very concept of a formula is not applicable and yet we may wish to glean knowledge from the data. Let us consider one of the problems investigated by Sir Francis Galton, a cousin of Darwin and founder of the science of eugenics. He tackled the problem of whether abnormal height is hereditary and his method was essentially as follows. He took a thousand fathers and recorded their heights and then the heights of their sons. Were a formula at all applicable it would have to relate two variables, the height of the fathers and the height of the sons. Moreover, for each value of one variable the formula would yield just *one* value of the second. For example, the formula $y = 3x$ yields one value of y for each value of x. Now to any one height of a father there corresponded several heights for the sons. Hence a formula was out of the question. What Galton did was to introduce the notion of correlation. The correlation between two variables is a measure of the relationship between them. This measure or number is obtained by substituting the individual values of the variables into a specially constructed expression known as the correlation coefficient which can take on values from -1 to $+1$.

A correlation of 1 indicates a direct relation; when one variable rises and falls the other does also; when one is large the other is. A correlation of -1 means that one variable behaves directly opposite to the other one; when the values of the first are high, the second is low in its range, and conversely. A correlation of 0 means that the behaviour of one variable has nothing to do with the behaviour of the other; they proceed independently of each other. A correlation of $\frac{3}{4}$, say, means that the behaviour of one variable is similar to that of the other though not exactly the same.

Galton found that there is a definite positive correlation between the heights of fathers and the heights of sons. Tall fathers in general have tall sons. Galton also found that the deviation of the sons from the mean of the race is smaller than that of the fathers – that is, the sons of tall fathers are not quite so tall. Their heights regress towards the mean of the race. Galton obtained analogous results in his study of the inheritance of intelligence. On the average, talent is inherited but the children are more mediocre than the parents. (This study should be read by parents who suffer from the intellectual condescension of their children.)

Like Quételet, Galton was very much impressed with what his studies revealed. After finding that the results he had obtained with respect to height and intelligence applied to many other human characteristics, he jumped to the conclusion that human physiology is stable and that all living organisms tend towards types.

The most valuable feature of Galton's work was the notion of correlation. This can be and has proved immensely useful. To make a study of the level of industrial production in this country requires the gathering of complicated data. If, however, there is a high correlation between industrial production and the number of shares traded on the stock exchange, the latter, easily accessible data could be used. If general intelligence correlates highly with ability in mathematics, then persons with good intelligence can expect to do well in mathematics. Knowledge of the correlation between success in high school and success in college or success in college and financial success later in life can be extremely valuable in predicting the future of groups of individuals.

There are difficulties involved in the use of statistical methods where care and judgement, and not mathematics, must supply the remedies. One such difficulty arises from the meaning of the terms used in a study. Suppose we wished to study unemployment in the United States. Who are the unemployed? Should the term include those people who do not have to work but would like to? Or people who are employed two days a week and would like full-time employment? Or the well-trained engineer who can find no job other than driving a cab? Or the man unfit for employment?

The interpretation of statistical conclusions is also fraught with difficulties. Statistics show that with each succeeding year more people die of cancer. Does this mean that modern life is more likely to produce cancer? Hardly. Many people died of cancer fifty years ago but the cause of the death was not recognized because medical techniques were not so advanced. Also people live longer today than fifty years ago, and since cancer is primarily an affliction of older people it occurs more frequently. Many of the people who died of tuberculosis years ago might have been subject to cancer had they lived longer. Finally, better records are being kept today. In other words, while cancer may now be more of a 'killer' than formerly, we cannot conclude that modern life is more likely to engender it or that people living today are more susceptible.

Unfortunately such difficulties in the use of statistics have often been deliberately concealed or glossed over by advertising men and propagandists in order to 'establish' their claims. Such misuse of statistics has induced unjustified mistrust and has provoked derogatory characterizations. Statisticians have been described as men who draw precise lines from indefinite hypotheses to foregone conclusions. There is also the undoubtedly familiar quip: there are lies, damned lies, and statistics.

The abuses of statistics should not blind us to their effectiveness in studies of population changes, stock-market operations, unemployment, wage scales, cost of living, birth and death rates, extent of drunkenness and crime, the distribution of physical characteristics and intelligence, and the incidence of diseases. Statistics are the basis of life insurance, social-security systems, medical treatment, governmental policies, and the numbers racket. Even the hard-headed businessman uses statistical methods to locate his best markets, to control manufacturing processes, to test the effectiveness of his advertising, and to gauge the interest in a new product. The statistical approach eliminates haphazard guesses and the captiousness of individual judgements and replaces them by highly useful conclusions.

Indeed it is an understatement to say that statistical methods have been successful in numerous problems. They have been decisive in making sciences out of speculative and backward fields, and they have become a way of approaching problems and thinking in all fields. The idea of measurement now pervades all activities of Western civilization. Long ago the famous Dr William Osler affirmed that medicine would become a science when doctors learned to count. And the importance of statistical studies led Anatole France to say, in effect, that people who don't count won't count. The mathematical conclusions drawn from the 'data for statesmen' are indeed shaping the courses of nations.

XXIII

Prediction and Probability

> It is remarkable that a science, which began with the consideration of games of chance, should be elevated to the rank of the most important subjects of human knowledge.
>
> *Pierre Simon Laplace*

OVER a period of forty years, Jerome Cardan, the Renaissance professor of mathematics and medicine, who was so rich in genius and devoid of principle, gambled daily. Early in his career he decided that if a person did not play for stakes there would be no compensation for the time lost, which could otherwise be spent in learning. Because he did not desire to waste his time in unprofitable pursuits, he studied seriously the probabilities of throwing sevens and of picking aces out of a deck of cards. To aid fellow gamblers he incorporated the results of his investigations in a manual called *Liber De Ludo Aleae*. This work presents the results not only of his thoughts on the subject but also of his practical experience. He points out, for example, that the chance of obtaining a particular card when cutting a deck is considerably increased by rubbing the card with soap. Thus was founded the branch of mathematics that is now fundamental in the theory of gases, the insurance business, and the physics of the atom.

About a hundred years later another gambler, the Chevalier de Méré, encountered a problem of probabilities and, not possessing the redeeming mathematical abilities of Cardan, sent it to that mathematical prodigy, Blaise Pascal. The alacrity with which Pascal undertook the solution is probably explained by his expectation that a theory of probability would resolve the fundamental and complex problems that throughout his life perplexed his mind, strained his body, and tormented his soul.

No behaviour was more riddled with contradictions than that of Pascal. Conflicting beliefs and desires produced strange vagaries of conduct and caused him to oscillate between the sacred and the profane. His literary efforts were divided between serious argumentation on theological controversies, such as that stylistic masterpiece, the *Provincial Letters*, and counsel on love, as in his *Discourse on the Passions of Love*. Deeply disturbed by the differences between the doctrines of the Bible and the dogma of the Roman Catholic Church, he nevertheless ignored both when he sought to rob his sister of her inheritance. He awarded to himself a prize he had offered for competition to the scientists of his times, and then complained of their lack of sincerity in the pursuit of knowledge. He advised people to restrict their love, even love of children, to a mental rather than an emotional act; yet, he himself did not hesitate to check experimentally his conclusions on the passions of love. Though he worried about the way to salvation he trangressed sufficiently to be in dire need of finding it. His fervid joy in his religious experiences equalled those of a saint, but his conduct towards people was marred by the excesses of a sinner. A renowned contributor to the most rational of man's activities – mathematics – he nevertheless maintained that truth comes from the heart. He was the believer in miracles whose probability he could show to be too small to warrant belief; and he was the defender of faith who helped found the Age of Reason.

Even Pascal's scientific life involved conflicts. Forbidden as a boy to study mathematics by a father who feared the strain on his young son's health, he finally at the age of twelve demanded to know what the subject was about. On learning the answer from his father he proceeded to devour its contents. Two years later he was admitted to the weekly scientific meetings of the great French mathematicians of the time. At the age of sixteen he proved the famous theorem that we discussed in our study of projective geometry. He had lived thirty-one of his thirty-nine years when the Chevalier presented him with the problem in probability. Pascal communicated with Fermat and in the interchange of letters that ensued, the two men produced basic results in the field.

The potential usefulness of a theory of probability should be apparent. Nothing about our future, even an hour from the present, is certain. The ground under us may be rent asunder in the next minute. Such possibile calamities do not disturb us, however, for we know that the probabilities of their occurrence are small. In other words it is the probability of whether or not an event will happen that determines our attitudes and actions in regard to the event.

In our daily use of the notion of probability we are satisfied merely to estimate whether it is high or low. Moreover, whatever numerical judgements of probability may be made are usually rough estimates. But estimates which may be wide of the mark do not suffice as a basis for decisions on large engineering, medical, and commercial ventures. It is necessary in such situations to know the exact numerical probabilities of particular events. And this mathematics supplies. Where we are uncertain mathematics tells us exactly how uncertain we are. Such numerical probabilities are reliable guides to action.

Let us see how these are obtained. For example, what is the probability of turning up a four on a single throw of a die? One way of solving this problem might be to cast the die 100,000 times and count the number of times a four appears. The ratio of this number of appearances to 100,000 is the answer or pretty close to it. But mathematicians are not likely to adopt such a procedure unless compelled to. They are essentially lazy and would prefer to sit still and think the matter through rather than weary their arms by throwing dice – unless perhaps, as in Cardan's case, more than an intellectual exercise is at stake.

Instead, Pascal and Fermat argued thus: A die has six faces; any one of these is equally likely to show up since nothing in the shape of the die or in the method of throwing it favours any one face; of these six equally likely possibilities only one, namely the appearance of the four, is favourable to us since that is the face we want to turn up. Hence the probability of a four is 1/6. If we were interested in either a four or a five turning up we should say that the probability is 2/6, since in this case two of the six possibilities would be favourable to us. If we were interested in not having a four or five turn up, there would be four favourable possibilities and the probability would be 4/6.

In general, the definition of a quantitative measure of probability is this: *If of* n *equally likely possibilities* m *of these are favourable to the happening of a certain event, the probability of the event happening is* m/n *and the probability of the event failing is* (n − m)/n. Under this general definition of probability, if no possibilities were favourable, that is, if the event were impossible, the probability of the event would be 0/*n* or 0, while if all *n* possibilities were favourable, that is, if the event were certain, the probability would be *n/n* or 1. Hence the numerical measure of probability can range from 0 to 1, from impossibility to certainty.

As another illustration of this definition consider the probability of selecting an ace, unsoaped, from the usual deck of 52 cards. Here there are 52 equally likely choices, of which 4 would be favourable. Hence the probability is 4/52 or 1/13.

There is often some question about the significance of the statement that the probability of drawing an ace from a deck of 52 cards is 1/13. Does it mean that if a person draws a card 13 times (each time replacing the card drawn) then one draw will be an ace? This is not so. He can draw a card 30 or 40 times and not obtain an ace. The more times he draws, however, the closer will the ratio of the number of aces drawn to the total number of draws approximate 1 to 13. This is a reasonable expectation because the larger the number of draws the more likely is it that each card will be drawn about the same number of times as any other card.

A related misconception is to suppose that if a person draws an ace, let us say on the very first draw, then the probability of drawing an ace on the next draw must be less than 1/13. Actually the probability is still the same and would be 1/13 even if three aces had turned up on three successive draws. A card or a coin has neither memory nor consciousness and what has already happened does not affect the future. The essential point about the probability of 1/13 is that it tells us what will happen in a *large* number of draws.

A term frequently used in connection with probability statements is 'odds'. The probability of throwing a four on a toss of a die is 1/6. The probability of not throwing a four is 5/6. The odds in favour of throwing a four is the ratio of the first probability to the second, that is, 1/6 to 5/6 or 1 to 5. The odds against throwing a four is the ratio 5/6 to 1/6 or 5 to 1. Again, the probability of 'heads' on a throw of a coin is 1/2; the probability of not throwing a head is 1/2. The odds in favour of a head, as well as the odds against a head, are 1 to 1. In this case the odds are said to be even.

The definition of probability we have been discussing is remarkably simple and apparently readily applicable. Suppose we were to argue, however, that the probability of a person crossing a street safely is 1/2 because there are two possibilities, crossing safely and not crossing safely, and of these two only one is favourable. If this argument were sound the reader would be wise not to bother finishing this page but to put his effects in order. The fallacy in the argument is that the two possibilities, crossing and not crossing safely, are *not equally likely*. And this is the fly in the ointment. The definition of Fermat and Pascal can be applied only if the situation can be analysed into equally likely possibilities.

Since it is so important to the application of the definition of probability that the possibilities be equally likely perhaps we should reconsider whether the chances of the various faces on a die turning up are equally likely. This is exactly what *some* of the dice throwers we occasionally read about are doing, namely, checking the number of times the various faces turn up.

But if we must throw dice to verify conclusions about dice arrived at by the mathematics of probability, we may as well dispense with the theory. As a matter of fact, in the case of a die thrown up into the air we can be fairly sure, even without testing, that the possibilities are equally likely. It is logically an assumption, of course, but one about as strongly supported by our knowledge of cubes – if not of dice directly – as the axioms of plane geometry are supported by experience. And where we are sure that the possibilities are equally likely we do employ the approach of Pascal and Fermat described above.

Let us apply it to the problem of coin tossing. Suppose two coins are tossed up into the air. What are the probabilities of (*a*) two heads, (*b*) one head and one tail, and (*c*) two tails? To calculate these probabilities we must notice first that there are *four* different, equally likely ways in which these coins can fall. These are: two heads, two tails, a head on the first coin and a tail on the second, and a tail on the first coin with a head on the second one. These last two possibilities are sometimes mistakenly counted as only one since both yield one head and one tail. If we consider two coins such as a penny and a nickel, however, it is clear that the case of a head on the penny and tail on the nickel is different from and equally likely as a tail on the penny and a head on the nickel. Of the four possibilities, then, only one is favourable to securing two heads. Hence the probability of two heads is 1/4. Similarly, the probability of two tails is 1/4. However, the probability of one head and one tail is 2/4 because two of the four ways in which the coins can fall produce this result.

If a person pursues the problem of coin tossing to the case of three coins, he must first analyse the equally likely possibilities. Again it is easier if he thinks in terms of three different coins, say a penny, a nickel, and a dime. There is, of course, only one possibility of three heads. There are three possibilities of two heads and a tail, however, since the tail can occur on any one of the coins while the other two are heads. Also there are three possibilities of one head and two tails, and one possibility of three tails. The total number of possibilities is eight. The probabilities of the various occurrences, therefore, are these; three heads, 1/8; two heads and a tail, 3/8; two tails and a head, 3/8; three tails, 1/8.

We could now consider, purely as an intellectual pastime for the moment, the probabilities involved in throwing four coins, five coins, and so on. Unfortunately, the possibilities become much more numerous as the number of coins is increased. At this point Pascal came to the aid of the mathematicians with a very interesting 'triangle' now named after him. Let us consider the following triangular array of numbers:

```
                    1
                 1     1
              1     2     1
           1     3     3     1
        1     4     6     4     1
     1     5    10    10     5     1
  1     6    15    20    15     6     1
  .   .    .    .    .    .    .    .    .
```

Each number in this 'triangle' is the sum of the two numbers immediately above it (o must be supplied where one of these two numbers is missing). Thus 4 in the fifth row down is the sum of 1 and 3; 6 is the sum of 3 and 3; and so forth. Hence we could construct row after row by mere arithmetic.

The really interesting feature of the Pascal triangle is that it gives at once the probabilities involved in coin tossing. For example, the numbers in the *fourth* row, namely, 1, 3, 3, 1, add up to 8 and this is the number of ways in which *three* coins can fall. Moreover, if we put each of the numbers in this row over 8, thus 1/8, 3/8, 3/8, 1/8, we obtain the probabilities of the various different possibilities, that is three heads, two heads and a tail, and so forth. If we wished to know the various probabilities involved in throwing five coins, we should use the sixth row. The sum of the numbers in this row is 32. This is the total number of ways in which the five coins can fall. If we now form the fractions 1/32, 5/32, 10/32 ... we obtain the probabilities of five heads, four heads and a tail, three heads and two tails, and so on. The number 1 at the very peak of the triangle should evidently be associated in some way with the throwing of zero coins. It does in fact yield the probability of retaining our money if we bet on the fall of zero coins.

Historically, the theory of probability was initiated as an aid to gamblers. The present widespread interest in the subject, however, is not evidence of tremendous gambling activity. Rather, the permeation of statistical methods into problems of industry, economics, insurance, medicine, sociology, and psychology raised questions that had never arisen in previous applications of mathematics and that could be answered only by a theory of probability. In order to appreciate the present scope of the subject let us examine a few uses of the theory.

One of the most original and most impressive applications was made by Gregor Mendel, abbot of a monastery, who in 1865 founded the science of heredity with his beautifully precise experiments on hybrid peas. Suppose there are two types of purely bred peas, green and yellow. If these peas are cross-fertilized, the second generation will be either all green or all yellow. Mendel explains this by saying that one of these colours dominates the other.

Let us suppose green is the dominant colour. These green peas of the second generation are not like those of the first; the first generation is pure while the second is hybrid. If we cross-fertilize the peas of the second generation we might expect mixtures of genes, the supposed carriers of hereditary characteristics, as follows. In the mixture of genes from two hybrid peas green can mix with green, yellow with yellow, yellow with green, and green with yellow. *These are precisely the possible associations of heads and tails in throwing two coins.* Hence 1/4 of the third generation should be a green-green mixture; 1/4 should be a yellow-yellow; and 1/2 should be a green-yellow and yellow-green mixture. Because green is the dominant colour, all those peas of the third generation that contain at least some green genes should be coloured green while the others will appear yellow. Hence 3/4 will appear green and 1/4 yellow. This proportion predicted by the theory of probability was actually obtained by Mendel and, much later, by many other experimenters. The statement of this proportion is Mendel's first law of inheritance of characteristics.

Mendel went on to consider the proportions that should appear in later generations from the interbreeding of various types of the third generation, as well as the proportions that should appear when several independent characteristics are cross-bred at the same time. In each case the mathematical theory of probability predicts what actually takes place.

This knowledge is now used with excellent practical results by specialists in horticulture and animal husbandry who create new fruits and flowers, breed more productive cows, improve strains of plants and animals, grow wheat free of the disease of rust, perfect the stringless string bean, and produce turkeys that have plenty of white meat and are small enough to fit the home refrigerator.

The use of the theory of probability in the study of human heredity is especially valuable. Scientists cannot control the mating of men and women; nor, if they could, would experimental results be so quickly and easily obtained. Hence they must deduce the facts of heredity from just such considerations as were illustrated above. Also because individuals may be biased in their judgement of human characteristics, the objectiveness of the mathematical approach is far more essential there than in the study of plants and animals.

The theory of probability also decides practically every move made by the biggest business in the United States – insurance. Consider the problem an insurance company faces in connection with John Jones. On his payment of an annual premium the company agrees to pay $1,000 either at the end of twenty years or on his death if it occurs before that time. What should the company ask Mr Jones to pay as a yearly premium? Obviously this depends on how long Mr Jones may be expected to live.

To determine this probability the company could list the various possible causes of death – cancer, heart disease, diabetes, automobile accidents, falls, and others. It could then attempt to decide how soon these causes would affect John Jones. To answer this question the company would have to study the family background, personal history, and daily activities of Jones; it would also have to study the condition of all the organs of his body. With this information it could start to calculate the answer. After several days of calculation only one fact would be sure to emerge, namely, that the calculations are best consigned to the wastebasket. No analysis of Mr Jones as an individual will ever enable the company to decide how soon the various causes of death may act on him.

The solution of the problem is obtained in quite a different way. John Jones is just one of the hundreds of thousands of people with whom the company deals. If the company merely knew what is most likely to happen to the average man within a small numerical margin of error, it would be safe, for what it may lose on Jones it may make up on Smith and in the end come out, where it would like to – ahead of the game.

What the insurance companies did do was to study the death records of a random collection of 100,000 people who were alive at the age of ten. Now these records say, for example, that at the age of forty, 78,106 people out of the 100,000 were alive. Hence the companies decided to take 78,106/100,000 as the *probability* that *any* person aged ten will live to be forty. In like manner, to obtain the probability of a person aged forty living to be sixty the companies took the number of those alive at age sixty and divided this by the number alive at age forty.

The approach to probability exemplified by the procedure of the insurance companies is a basic one. Essentially, it is an appeal to experience for the primary data to which mathematical reasoning is then applied. This use of experience to obtain probabilities is, strictly, outside the domain of mathematics. Mathematics begins after the probabilities are known and is concerned with reasoning about the numbers so obtained. For example, if an insurance company wishes to issue a thirty-year policy in which a husband and wife are involved, it is important to know the probability that both will live thirty years from the commencement of the policy. Let us suppose that both are forty years old. Now the probability of one person of age forty living to be seventy is about ·50 because of 78,106 people alive at age forty, 38,569 were alive at age seventy. This is the probability of throwing a head on a throw of a single coin. Hence, the probability of *both* living to be seventy is the probability of throwing two heads in a throw of two coins; therefore, the probability that both people will live to be seventy is ·25. The foregoing is a simple, run-of-the-mill problem. As might be expected, mathematics is employed for the solution of far more complicated probability problems arising in insurance.

The use of experience to obtain the basic probabilities is unavoidable in medical problems. Suppose, for example, that it is known from the records of a great many cases that 50 per cent of the people afflicted with a certain disease die from it. The probability of death from the disease is then taken to be 1/2. The probability may be applied now to a practical problem. A doctor who believes he has a new treatment tries it on four patients and all recover. Does this result mean that the new treatment is effective and should be applied to all cases?

At first blush it does appear that the treatment is remarkable. Whereas we would expect two deaths, none occurred. The theory of probability may be called in to decide the matter. In a particular collection of four people it is not true that two must die. In such a group all or none or any number between zero and four may die. Only in a very large number of cases would 50 per cent succumb. The situation is mathematically equivalent to the throwing of coins. The chance of any one person recovering from the disease is the chance of throwing a head on a throw of a single coin. The probability of four people recovering is the probability of throwing four heads in a throw of four coins. If we consult the fifth line of Pascal's triangle we find that the probability of throwing four heads in a throw of four coins is 1/16. This number then is also the probability that the doctor would strike a group of four people all of whom would recover from the disease even without his treatment. This probability means that if we were to pick very many groups of four people afflicted with the disease then, on the average, one out of sixteen groups would contain four people who would recover. Now the doctor who administered his treatment to a group of four may have struck the one group in which all would recover. Since this is by no means very unlikely – many a 100 to 1 shot wins a horse race – it is not safe to conclude that the new treatment is effective. It should be tried on many more cases before any conclusion is drawn.

The problems we have thus far considered have involved situations wherein only a few possibilities could occur. When a person throws a die, for example, the possibilities in regard to the outcome are just six. In the case of mortality the possibilities are only two. In many problems of probability, however, the possible outcomes are either infinite or so large that it is mathematically convenient to treat them as infinite. Suppose, for example, that measurements of a length are made. These measurements are just a few of the infinite number of different measurements that could be made. Hence, the calculation of the probability that the mean of the measurements, let us say, is correct, must take into account the infinite number of possibilities. Similarly, the output of a machine that produces hundreds of thousands of units of an item is not uniform; the variations from unit to unit, though slight perhaps, are so numerous that the entire collection is treated as though it were part of an infinite collection.

The theory required to treat problems in which the number of possible outcomes is infinite – the theory of continuous probability – was created by the peasant, aristocrat, politician, and

superb mathematician, Pierre Simon Laplace (1749–1827).
Cardan, Pascal, and Fermat were attracted to probability by
problems of gambling. Laplace's interests, equally impractical,
were in the heavens. He used the theory of probability to obtain a
measure of the reliability of numerical results derived from data
and to determine the likelihood that certain astronomical
phenomena were due to definite causes rather than pure chance.
It is perhaps no longer a surprise to us that a mathematical
theory intended to serve the astronomer should prove useful in a
thousand walks of life. Nevertheless, we shall examine a few of
these uses to see once more how broad the reach of mathematics
is.

Where the possibilities in a particular phenomenon are
infinite, the frequency distribution of these various possibilities is
most often and very fortunately normal. Hence it is possible to
apply the knowledge acquired about such distributions to the
problems of continuous probability. It is necessary merely to
make a slight transformation in the facts bearing on the normal
curve to use it for this purpose. We may recall that a normal
distribution is characterized or fixed by the mean and standard
deviation. In addition, 68·2 per cent of the cases fall within one
standard deviation or one σ of the mean; 27·2 per cent of the
cases fall in the interval between σ and 2 σ away from the mean;
4·4 per cent of the cases fall in the interval between 2 σ and 3 σ
away from the mean; and the balance of the cases, namely ·2 per
cent, lie more than 3 σ away from the mean. These statements
need merely be translated into probabilities. For example, the
probability that any one datum falls within one σ of the mean
must be ·682 since 68·2 per cent of all cases fall in this interval.
Another way of stating this fact is that on the average 682 out of
1,000 cases will fall within one standard deviation of the mean.
Similar translations should, of course, be made for the percentages occurring in the other intervals. Because the normal
frequency distribution curve can be reinterpreted in the manner
just described it is often referred to as the normal probability
curve (fig. 75).

Let us consider one or two examples of the use of the normal
probability curve. The frequencies of the heights of all American

men practically fill out a normal distribution with a mean of about 67″ and a standard deviation of about 2″. What is the probability, therefore, that any American man chosen at random has a height between 65″ and 69″? Since all heights between 65″ and 69″ lie within one σ of the mean and since 68·2 per cent of the cases have heights in this range, the probability is ·682. In like manner, the probability that a man chosen at random has a height between 67″ and 71″ is ·477 because the range 67″ to 71″ is 2 σ to the right of the mean, and 47·7 per cent of the cases have heights within this range.

It should be noticed that we do not ask the question, what is the probability that a man chosen at random has a height of exactly 68″? The answer to this question is zero because this is the probability of any one possibility out of an infinite number. Such

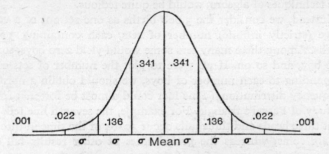

.341   .341

.001   .022   .136   .136   .022   .001

σ   σ   σ Mean σ   σ   σ

Figure 75. The normal probability curve

a question is not too significant, however. All measurements are approximate. If the error in measuring heights were say ·1″, then it would be more significant to ask, what is the probability that a man chosen at random has a height between 67·9″ and 68·1″? This question can be answered by referring to data on the normal curve just as we did in answering the questions in the preceding paragraph.

A more interesting problem in probability arises when we try to determine from a limited number of cases, that is a sample, whether boy and girl babies are equally likely. Statistics in one community showed 1,890 male to 1,710 female births in 3,600 cases. Does this departure from a 50–50 ratio indicate that boys

and girls are not equally likely? Not necessarily, for to say that
boys and girls are equally likely, or that the probability of a boy is
1/2, means only that in a very large number of cases there will be
about as many boys as girls. What can we conclude, then, from
the data on the 3,600 cases?

We could approach this problem by assuming that boy and girl
babies are equally likely and then ask for the probability of obtaining 1,890 boys in 3,600 births. Now in 3,600 births the possible outcomes are finite, namely, zero boys, one boy, two boys,
and so forth up to 3,600 boys. Since the probability of a boy, like
the probability of a head on a coin, is assumed to be 1/2, we
could resort to the thirty-six hundred and first line of Pascal's
triangle to obtain the probability of 1,890 boys. The calculation
of the required terms of this triangle, however, even if speeded by
the techniques of algebra, would be quite tedious.

Instead, we consider the 3,600 births as one set out of a very
large (strictly infinite) number of sets, each containing 3,600
births. Among these many sets some would yield zero boys, some
one boy, and so on. If we were to plot the number of sets corresponding to each number of boys, we should obtain a normal
frequency distribution. (This fact could almost be foreseen from
a study of Pascal's triangle. For example, the seventh line tells us
that three heads and three tails occur 20/64 of the time in throws
of six coins, whereas the probabilities of other results fall off
symmetrically on either side of this outcome.) On the *assumption*
that boys and girls are equally likely, the largest number of sets
would contain 1800 boys and 1800 girls. This number of boys,
1800, is, then, the mean number of boys. We must now appeal to
a formula of statistics, which will not be given here, to obtain the
standard deviation of this frequency distribution. In this case
$\sigma = 30$. This means that 68·2 per cent of the sets would contain a
number of boys between 1,770 and 1,830. In the set of 3,600
actually observed the number of boys was 1,890. This number
would lie 3 σ to the right of the mean. Now the probability of
an occurrence which is 3 σ or more to the right of the mean
is only ·001 or one chance in a thousand. Since this probability
is indeed small, our assumption that boy and girl babies are
equally likely must be wrong. As a matter of fact, the records of

many thousands of births show that the ratio of male to female births is 51 to 49, which may be evidence of God's good judgement or of the fact that a girl is worth slightly more than a boy.

The problem we have just examined amounts to asking for the probability that a particular happening, the birth of 1,890 boys in 3,600 babies, should fall within a specified interval of the entire range of possibilities. This question was readily answered by reference to the normal probability curve. A somewhat different question is posed by the following type of problem.

A manufacturer of cord sells his product in balls weighing one pound each on the average. He claims that practically no ball of cord leaves his factory which deviates by more than ·1 of a pound from the one-pound standard. A retailer buys 2,500 balls of this cord, weighs the 2,500 balls, and finds that they weigh 2,450 pounds, that is they average ·98 of a pound. The average weight of the balls is then well within the ·1 of a pound limit to which the manufacturer claims he holds. On the other hand, he could be deliberately producing his cord to weigh ·98 of a pound and thereby be making a hidden profit. Is the manufacturer honest? That is, is it likely that a random selection of 2,500 balls from the product the manufacturer offers should just happen to weigh ·02 of a pound less on the average?

The question concerns the behaviour of the means of samples. Just how close should the mean of a sample be to the mean of the entire population – that is, the factory output – for us to believe that it is a sample of that output? This question can be answered by a study of the frequency distribution of the *means of all possible samples of 2,500 units each*. We cannot develop here the theory of the distribution of means. It must suffice to say that these means form a normal distribution; moreover, the mean of this frequency distribution of *means* could be shown to be 1, the mean of the entire output, and the standard deviation of this distribution of means could be shown to be ·0006. Now the particular sample that the customer received had a mean of ·98. This means differs by ·02 from the mean of 1 and is therefore about 30 times ·0006 or about 30 σ to the left of the mean of the samples. The probability of a datum falling as much as 30 σ away from the mean is so small as to be negligible. Hence the likelihood that the

2,500 balls received by the customer was indeed a sample of the manufacturer's purported output of one-pound balls is not to be credited. We should be fully justified in concluding that the manufacturer was deliberately producing balls of cord averaging less than one pound in weight.

A recent and most interesting application of the mathematical theory of probability was to 'prove' the existence of extra-sensory perception. Here again the proof hinges upon the relation between the sample and the complete population. The existence of extra-sensory perception has been stoutly maintained by Professor J. B. Rhine and others because certain individuals are able to predict the numbers and colours of cards drawn from a deck in a greater percentage of cases than the mathematical probability would call for. That is, if the probability of correct prediction in a given case is $1/5$, say, the subject should be able to guess correctly *about* $1/5$ of the time. But suppose that in 800 trials the subject guesses correctly 207 times instead of the expected 160. Is the excess over the expected 160 an accident in this particular set of 800 trials or is it significant? Such an unexpectedly large number of correct guesses is interpreted by Rhine to mean an unusual mental faculty to read hidden cards by extra-sensory perception. Whether the additional 47 correct guesses is sufficient evidence for such a belief is subject to dispute. Rhine has calculated that the probability of getting the additional 47 correct guesses in a particular set of 800 trials is $1/250,000$. This probability is so small that Rhine does not attribute the extra guesses to mere chance.

In all of the applications we have thus far examined, the theory of probability has served to measure the likelihood of some event or possibility. Not content to perform this humble service to science and industry, the theory became a tyrannical master. The problem that incited this step is one we mentioned earlier. The molecules of a gas attract each other in accordance with the Newtonian law of gravitation. Any attempt to predict the motion, expansion, contraction, or temperature variation of a gas on the basis of this Newtonian law becomes hopelessly complicated, however, because of the large number of molecules. Mathematics cannot solve exactly the problem of the motion of even one molecule subject to the attractive forces of just a few others.

The successful attack on this problem was made by Clerk Maxwell, and the method was provided by the theory of probability. The endless number of molecules in a volume of gas is replaced by one ideal or representative molecule whose size is the most probable size of all those in the gas, whose velocity is the most probable velocity, whose separation from other molecules is the most probable one, and whose other properties are always the most probable ones. The most probable behaviour of this ideal molecule is then taken to be the behaviour of the gas itself. It is astonishing, but none the less true, that the laws obtained in this manner describe and predict the behaviour of gases as the laws of astronomy predict the motions of the planets. In essence, the most probable behaviour of the gas turns out to be the actual behaviour.

The explosive implications of this application of the theory of probability will be considered later. For the moment it is enough to notice that probability theory thus emerged from its role as an evaluator of data and hypotheses to demand respect as a primary method for obtaining laws.

The vital role that the theory of probability has come to assume in scientific work and in philosophic thought was foreshadowed in the work of Pascal. He began by applying the theory to gambling; he ended by applying it to God. Pascal stood at a turning point in history, at the time when the new science had begun to challenge vigorously the old faith. Like every thinking man of his century, he was impelled to take part in the conflict and to seek some resolving philosophy. Intensely religious by nature and yet a notable contributor to science and mathematics, Pascal felt the conflict more poignantly than any other man. Because he saw both sides so well his mind became a battleground and in a most appealing passage he openly declared his bewilderment:

This is what I see that troubles me. I look on all sides and I find everywhere nothing but obscurity. Nature offers nothing which is not a subject of doubt and disquietude; if I saw nowhere any sign of a Deity I should decide in the negative; if I saw everywhere the signs of a Creator, I should rest in peace in my faith; but, seeing too much to deny and too little confidently to affirm, I am in a pitiable state, and I have longed a hundred times that, if a God sustained nature, nature should show it without ambiguity, or that, if the signs of a God are fallacious, nature should suppress them altogether: Let her say the whole truth or nothing, so that I may see what side I ought to take.

But God refused to reveal himself. Pascal then bethought himself of his early work on probability and of the problems in gambling he had solved thereby. Did the theory have any message for the problem of religious belief? The answer came to him in the form now known as Pascal's wager.

The value of a ticket in a lottery is the product of the probability of winning and the prize at stake. Even though the probability may be small, if the prize is very great, the value of the ticket is great. So, reasoned Pascal, though the probability that God exists and that the Christian faith be true is indeed small, the reward for belief is an eternity of bliss. The value of this ticket to heaven is, then, indeed great. On the other hand, if the Christian doctrine is false, the value lost by adherence is at most the enjoyment of a brief life. Let us then wager on the existence of God.

Pascal's wager was not a flippant remark. It was a cry of despair. The problem he faced has reappeared in but slightly altered guise. It was reopened in recent times by the theory he created.

XXIV

Our Disorderly Universe: The Statistical View of Nature

> Here then we rest: The Universal Cause
> Acts to one end, but acts by various laws.
>
> *Alexander Pope*

Is there law and order in this universe or is its behaviour merely the working of chance and caprice? Will the Earth and the other planets continue their motions around the sun or will some unknown body, coming from great distances, rush through our planetary system and alter the course of every planet? Cannot the sun some day explode, as other suns are doing daily, and burn us all to a crisp? Was man deliberately planted on a planet especially prepared for his existence or is he merely an insignificant concomitant of accidental cosmic circumstances?

The thinking person would like, more than anything else, to know the answers to such questions. Insignificant by comparison are his grandiose plans for a United Nations, his pressing monetary concerns, and the irritations of daily life. His irrepressible desire for answers is man's ennobling quality and his ceaseless search for knowledge about himself, about the wonders of nature, the structure of the universe, and the forces that keep all the activities of the universe going gives point to lives which would otherwise spend themselves in orgies of meaninglessness. The answers may never be fully known, but thanks to the great mathematicians man does have many significant clues. Unfortunately, there is more than one interpretation of these clues.

One of the interpretations is already familiar to us. Arguing from the existence of the mathematical laws uncovered during the Newtonian era, the eighteenth-century thinkers erected the most comprehensive and influential philosophical system of modern times. It propounded a world that is designed and orderly and that functions according to plan. Mathematical laws made manifest that design, and the unfailing fulfilment of scientific predictions gave proof that the design was being adhered to. Of course, laws that regulated the motions of planets and other inanimate objects did not make clear where man fitted into the scheme of things. But could it be doubted, since evidence of design was unmistakable, that man was included?

This philosophy of determinism still dominates our thoughts and beliefs and guides our actions. Unfortunately, the order of nature, supremely simple and harmonious to the founders of modern science, now appears to be breaking up in the maelstrom of statistics and probabilities which the nineteenth and twentieth centuries have used so effectively.

Mathematicians themselves, needless to say, were proud of the new ideas and techniques they had introduced for handling statistical data. They were pleased, too, with their conversion of the intuitive notion of probability into a highly useful tool for the guidance of man's actions. But as members of the intellectual community in which they worked, their joy was short-lived, for it was the very success of statistical methods and of the theory of probability that caused the orderly structure of nature to crumble about their heads.

If the formulas and laws obtained by the new procedures had been inaccurate they would have been dismissed as unreliable substitutes to be used only when the sure procedure of deducing conclusions from thoroughly acceptable mathematical and scientific axioms failed. And had they been merely rough approximations, no undue philosophical significance would have been attached to the new methods. But such was not the case. They were, in fact, surprisingly accurate and effective, and thereby hangs the tale.

Let us approach the heart of the problem and examine the challenge to the philosophy of determinism posed by the advent of statistical methods. We shall allow ourselves the privilege of borrowing the Platonic technique of dialogue, so that the arguments pro and con will be stated by Mr Determinist and Mr Probability of High Degree. The latter, the youthful protagonist, will open the discussion with a fuller statement of the problem.

The most disturbing consideration, he points out, is that statistical methods and the theory of probability produce thoroughly reliable laws where we have no reason to expect any at all. Consider, for example, the distribution of intelligence. Select any large random group of people and measure their intelligence by a well-designed test; the distribution will approximate the normal frequency curve. The larger the group on which the test is tried the closer to a perfect normal distribution will the curve be. The apparently unplanned bestowal of the varied and inexplicable qualities that determine intelligence hardly bespeaks a law; yet the distribution of intelligence follows a curve that expresses regularity and an invariable relationship.

Again, consider the phenomenon of inheritance. The chromosomes of the parents mingle freely in the fertilized egg and endless transformations take place from conception to maturity; yet the transmission of hereditary characteristics can be accurately predicted by the theory of probability.

Now make numerous measurements of a length and graph the various measures against the frequency with which each occurs. The crudity of the eye and the hand should produce considerable irregularity in these measurements; yet the curve will show almost a normal distribution and the larger the number of measurements the closer does the curve approach a normal distribution. Even man's errors follow a law. In short, concludes Mr P., we have the surprising and disturbing result that laws describe phenomena which, to all appearances, should be unlawful.

But why be disturbed, old Mr D. asks, if there are laws covering phenomena for which none was expected? Why not be grateful for having more laws? Do they not strengthen the argument for determinism? Design apparently exists everywhere, even where you did not expect it.

That is precisely why I am concerned, replies Mr P. Not only have we no reason to expect laws in these situations but we have every reason not to expect them. Since we do have laws governing such situations, how much significance can we attach to the existence of the mathematical laws produced by Newtonian science? Why infer design and determinism from the existence of those laws?

Not so fast, rejoins Mr D. Suppose we put it this way. We appear to have mathematical laws describing phenomena which to the best of our knowledge appear to be haphazard and disorderly, and for this reason you question the significance of laws we have always regarded as proving the design of the universe. May it not be, however, that the seemingly disorderly phenomena do follow physical laws but, because these phenomena are so complex, they appear to our limited intelligences to be the results of chance.

Your argument sounds reasonable enough, answers Mr P., who is merely softening up his opponent with a few kind words. Under close inspection the motions of the molecules of a gas appear completely irregular; yet physicists believe that each molecule follows the same physical laws that the Earth follows in its path around the sun. Similarly, it could be argued that the distribution of qualities that make for intelligence and the process of heredity follow orderly physical procedures which determine precisely the state of each individual, but that these procedures are too intricate to be grasped by our understandings. The same could be said of economic phenomena, the incidence of death, and other seemingly unlawful affairs. Thus phenomena that appear to be disorderly may be completely determined, and the mathematical laws obtained from statistical studies may merely reflect the existence of these underlying orderly physical processes.

Mr D. is now complacently off his guard, while Mr P., who knows his theory of probability, prepares to strike.

But now consider the following facts, Mr D. When six coins are tossed up simultaneously any number of heads from zero to six may result. We have no way of telling what the exact number of heads will be because too many known and unknown factors determine the outcome: the strength of the wind, the force imparted to the coins by the hand, the shape of the floor on which they fall, and other factors. We assume then that the result of tossing the coins is a matter of chance. Moreover, the greater the number of times the coins are tossed the more chance is permitted to play a role. And yet if these six coins are tossed up a great number of times the theory of probability enables us to calculate in advance about how many times zero heads will show up, about how many times one head will show up, and so on to the last possibility. The larger the number of throws the closer do the results agree with the predictions of the theory. Hence, regardless of whether or not the fall of a coin is determined by some series of inviolable rules, the assumption that chance alone decides the outcome yields mathematical laws that predict the outcome.

As a matter of fact, continues Mr P., you know that nineteenth-century physicists obtained some very famous laws on the behaviour of gases by just such a procedure as I sketched a moment ago for the fall of coins. They manoeuvred around the difficulties in studying the action of billions upon billions of molecules in a gas by working with an ideal, fictitious molecule whose mass, velocity, and other properties have the *most probable* values that can occur among the various mass and velocity values of the molecules in the gas. Yet the laws built up by reasoning with this ideal molecule are as applicable as any that mathematics and science have produced, despite the fact that they state only the most probable behaviour of a gas rather than the necessary behaviour. Hence, the belief that individual molecules follow a pre-assigned pattern is not supported at all by the lawful behaviour of masses of molecules. Indeed the belief is irrelevant.

Mr D. is far from ready to retreat from his position.

You agree, Mr P., that the motion of the molecules in a gas and the fall of a coin may be following definite, inescapable laws but as a convenience you assume that each coin falls haphazardly and that the molecules of a gas have most probable characteristics. Just because this assumption of chance behaviour and your mathematics of probability predict successfully, we should not lose sight of the overwhelmingly important existence of underlying, fundamental laws. Although the use of probability arguments for complex phenomena is convenient and fruitful, it does not in itself discredit the underlying laws. In fact, it is only because these laws do hold that probability arguments yield sensible and useful conclusions.

Mr D., you do not appreciate as yet the full force of my argument. Indeed you will see that you are mistaken in believing in any necessary laws. Consider, for example, the fall of a coin and consider, in particular, its weight, which is involved in any conceivable Newtonian law describing its motion.

During the time the coin is falling its weight is not even constant. The coin consists of an enormously large but *continually changing* number of molecules, for every solid object is continually gaining and losing molecules. The wind that blows on the coin while it is falling consists of billions and billions of molecules, set into motion we know not how, all of which dance around the coin in quite different manners. The surface of the floor on which the coin falls is not fixed in shape. As molecules of wood leave or join it, the shape changes and so the angle at which the coin hits the floor is not precise. Neither is the distance the coin falls. Suppose we try to measure the distance from the centre of the coin to the surface of the floor. Where is the centre of a coin whose shape is continually changing? Where does the surface of the floor start since its molecular layers are completely irregular? Shall we use a ruler to measure the distance? After all, even the ruler is not constant in length any more than any mass is. The molecules at its ends leave and return and continually alter its length.

Now that we see the complexities in the structure of matter, Mr P. hurries on, aren't we audacious to speak at all of scientific laws? All such laws deal with matter, with masses, surfaces, lengths, pressures, densities, and other properties that are never constant for any object. Only the crudity of our hands, eyes, and measuring instruments deceives us into believing that there are such things as fixed lengths and masses and that we can speak of exact scientific laws. Laws can involve mass, length, volume, weight, and other qualities only in so far as they use average figures for these quantities. The laws can therefore be no more than convenient summaries of irregular physical states wherein the variations cluster around some average numbers. To sum up, Mr D., our examination of the fact that some laws cover apparently chaotic phenomena has led us to the conclusion that *all* scientific laws do. What shall we say now about the significance of scientific laws for the existence of an orderly nature?

The point of your argument as I understand it, Mr P., is that when we examine the structure of matter itself we find that seemingly constant quantities are actually continually changing. Hence, you ask, how can we speak of definite, invariable scientific laws when all they can be are convenient statements about average effects just as a statement about the mean income of workers is an average. But consider, for a moment, Mr P. Why discredit broad, revealing, well-verified laws merely because of some microscopic irregularities that don't affect at all the major events covered by the laws?

What you say might be true, Mr D., if the situation were not much worse than I have thus far described it. Let us go a little further into the nature of matter itself and let us consider the molecules themselves. They are, you know, made up of atoms, and these in turn are made up of free electrons and a nucleus with a very complicated structure of its own. And now, Mr D., hold tight while I tell you a thing or two about the nucleus and the electrons. You probably think of these particles as little chunks of matter each existing in a definite place at any one time. Well, so did scientists years ago. Today we no longer can say that. We must say that each electron and each constituent of the nucleus exist *everywhere* but with more or less probability depending on the place. In effect, modern atomic theory says that you are not seated in a chair in a corner of this room. You exist everywhere to a degree of probability that varies from place to place and that is greatest for the corner in which you think you sit. A fantastic theory of matter, you say? As fantastic as the medieval concept of hell? Maybe, for it is just this theory that has brought the hell of atomic bombs into our world.

Now, Mr D., where is the good, old-fashioned, solid matter that obeys precise, compelling mathematical laws? The stone that Dr Johnson once kicked to demonstrate the reality of matter has become dissipated in a diffuse distribution of mathematical probabilities. The ladder that Descartes, Galileo, Newton, and Leibniz erected in order to scale the heavens rests upon a continually shifting, unstable foundation.

I fail to see the point, Mr P. You are simply telling me that the structure of the atom is so complex, in the light of our present understanding, that scientists have resorted to probability arguments to master it. What does that prove? You have merely shifted the argument from the falling of coins to the structure of the atom. I do not doubt that atomic structure is complex nor do I question the wisdom of using probability theory in studying the structure. Yet the very existence of laws for the atom, like the existence of laws for the distribution of intelligence or for the inheritance of characteristics, does not in itself deny the possibility of underlying determined behaviour. Dr Einstein has said apropos of this very point, 'I shall never believe that God plays dice with the world.'

Maybe, Mr D., but my point, which you claim not to see, is that you cannot conclude that the acquisition of workable laws does, in itself, *necessarily* establish design, invariable order of nature, and causality – in short, determinism. You must, I believe, concede that much. I know, however, that you still feel that you have a few facts up your sleeve. You would argue that the very existence of laws describing the behaviour of matter, despite the complex structure of matter, is even *added* proof that these laws imply design.

Mr P. has, of course, stated Mr D.'s argument because he found it hard to stop talking even when his own point was made and because, with the usual overconfidence of youth, he thought he could state Mr D.'s case better than Mr D. himself. According to parliamentary procedure, the floor is still Mr P.'s to take up his side of the argument.

Let me obtain a law for you, Mr D., and let's see how happy you will be about it. List for me the data on national prosperity for the past fifty years and the strength of the sunspots during those years. You know, of course, the statistical process of fitting a formula to data. This process will give me a formula, a mathematical law that relates national prosperity and strength of sunspots. What conclusion should we draw about the existence of some inevitable connection between the two variables? None at all, wouldn't you say? And yet how does this formula differ from so many scientific formulas which, you say, proclaim laws of the universe?

Mr D. here rises with some emotion from his chair (which exists everywhere with varying degrees of probability).

The answer is apparent, Mr P. The scientific laws will continue to hold indefinitely whereas the formula fitted to the data on sunspots and national prosperity will not. Take Kepler's laws, for example. All the observations of the past four hundred years support them. Is it not significant that the Earth has followed the same laws for so long a period of time?

I am glad you picked Kepler's laws as an example, Mr D. First, let me remind you that Kepler's laws were originally obtained by fitting formulas to data. After many years of supreme effort and after trying some fifty different types of curves, Kepler found that the path of Mars is an ellipse. All the observations of Copernicus and Brahe supported him. Luckily for Kepler and the history of science those observations were not too good. Today we know from theory and more accurate observations that the true path is not an ellipse but is distorted by all sorts of perturbations arising from the gravitational attraction of the other planets. Kepler's laws, then, happen to be descriptions of the *average* behaviour of the planets. *Strictly, they do not hold today.*

Moreover, the fate of Kepler's laws has been the fate of all scientific laws. They hold for a time and then some refinement is shown to be necessary by the general increase in scientific knowledge. Kepler's laws were in themselves refinements on Copernican theory and Copernicus, as we know, improved on Ptolemy. Because Kepler had the advantage of building on earlier theory his laws proved to be good descriptions. But we see that even his work is not the last word.

Maybe not, Mr D. hastens to counter, pausing for a moment in his nervous pacing up and down. But you agree that the history of those laws shows more and more refinement. Refinement leading to what? To the true laws, no doubt; and Kepler's laws, if not the final word, come very close. But how could we come closer to true laws if there were no true laws to aim at?

The answer, Mr D., is that if the Earth in its motions adheres exactly to one pattern, which Kepler's laws approximate so well, that pattern may still be only the most likely behaviour; no necessity that we know of obliges the Earth to continue to do the most likely thing any more than necessity tells coins how frequently they must turn up heads. Tomorrow the Earth may crash into the sun. In other words, Mr D., and if you stopped that nervous walking you might concentrate better, *we are not questioning the existence of working laws but rather the significance to be attached to them.*

I am sorry if my pacing has been upsetting you, Mr P. Let me present a major argument in favour of Kepler's and other laws that does not hold for your statistical laws, your laws obtained by fitting formulas to data. Let us remember that Galileo and Newton successfully analysed the phenomena of motion. As a result, we have a *physical* explanation of the behaviour of the planets in terms of a force of gravitation. This force keeps the planets in their paths and keeps them obeying Kepler's laws. Indeed these laws are a mathematical consequence of the law of gravitation. Even the perturbations in the paths of the planets are now explained by the action of gravitation.

Mr D., I am embarrassed for you. Fie on your explanations! A fig for your theory of gravitation! You know very well that it is no more than a fiction. What is this force of gravity that keeps the planets in their paths? No fantasy of literature strains our intellects so much as the attempt to understand how the sun exerts its pull on the Earth. Far more reasonable connections could be devised to relate sunspots and national prosperity. All we really have are formulas and we have no more reason to attach philosophical importance to the existence of these formulas than to the existence of a formula relating sunspots and national prosperity.

Mr D. once more sought the reassurance of his comfortable chair whose real existence he began to doubt. Mr P. continued to hold forth and decided to drive home his advantage.

Let us look back for a moment, he resumes. Doesn't the argument between us boil down to one point? On the knowledge of a few laws of nature you have based a philosophy of nature. The introduction of statistical methods and the theory of probability now compel us to appreciate how little is really implied by the discovery, or shall I say manufacture, of a few laws.

Mr D. was hardly listening; he had become lost in thought. The various arguments of his loquacious opponent had evidently opened his mind to the existence of an underlying irregularity and disorder even in phenomena formerly considered lawful. The development of atomic theory by chemists and physicists revealed new problems and uncertainties in that domain and certainly made patent the fact that matter is far more complex than had been supposed. The development of the kinetic theory of heat, which explained this phenomenon in terms of the rapidity of motion of molecules, made it clear that the flow of heat and cold is no more than a mass effect of irregular motions of billions of molecules. The constant pressure of a liquid, instead of being a definite single force, was simply the mass effect of an irregular bombardment of the walls of the container by the individual molecules of the liquid. A smooth mirror surface was really only a collection of molecules, each behaving differently even though the entire collection gives the net effect of reflecting light steadily and in accordance with mathematical laws. The sounds of human voices and of musical instruments, reproduced day in and day out with almost complete fidelity and so well represented by mathematical formulas, were but the average effects of irregular mass movements of air molecules. Galton's use of statistical methods to find the laws of heredity – after his failure to find or understand the mechanism – made that phenomenon also appear to be the sport of chance. The forms and varieties of plants, animals, and even human beings were limitless. The weather was more contrary than obliging. Man could not predict, much less control, droughts, hurricanes, and cloudbursts. The very forces of nature that had been admired for their simplicity, order, and invariability included unexpected and unexplainable tidal waves, volcanic eruptions, and earthquakes. Nature suddenly appeared unpredictable, perverse, and capricious.

Thus the very same world which the eighteenth century regarded as rigidly determined and designed in accordance with immutable mathematical laws seemingly had to be viewed now as chaotic, lawless, and unpredictable. Reality appeared totally void of purpose, a 'tale told by an idiot, full of sound and fury, signifying nothing'. Man, in particular, was but an accident of the blind, fortuitous concourse of events. *The mathematical laws of science amounted to no more than convenient, usable summaries of the average effect of disorderly occurrences.* This attitude towards nature and its laws, which affirms that nature is chaotic and unpredictable and that its laws are no more than convenient, impermanent descriptions of average effects, is known as the *statistical view of nature.*

This statistical view and the deterministic view are unalterably opposed. Although they both agree on the existence and applicability of scientific laws, they differ widely on the interpretation of these facts. Determinism asserts that scientific laws are statements of the necessary, invariable, universal behaviour of natural objects. The statistical view regards laws as statements possessing merely a high degree of probability. The determinist believes in an essential connection between objects related by law as the Earth and sun are in the laws of Kepler. The statistical theorist maintains that the law is merely an observation of a temporary situation, an accidental juxtaposition as significant as my wearing a brown necktie and my neighbour's smoking a cigar at the same time. Determinism asserts that the present state of nature determines the future unalterably. If I throw a ball into the air it must follow a parabolic path right down to Earth again. The statistical view says that not only may it fail in any one case to follow a parabolic law, but it may travel directly to the sun.

An example or two may further clarify the difference between the two points of view. Suppose a bat hits a ball. Under the deterministic point of view, the forces at work when the bat contacts the ball compel the ball to trace a definite pattern of flight which can be predetermined and which is described by the mathematical laws of motion. Given a few quantitative facts, the motion of the ball could be predicted with certainty. According to the statistical view, we could say that the billions of molecules in the bat, when brought close to the billions in the ball, would be very likely during their random motions to hit many of this second group of molecules and impart to them their own velocities. Since so many of the molecules in the ball would be affected, the ball itself would probably be set moving in that direction in which most of its molecules are sent by contact with those in the bat. The probability that the ball will move in a definite direction is so large that we could hardly expect to encounter any departure from this expected behaviour, though a radical departure is at least possible. The needle does exist in the haystack even though the probability of finding it is very, very small.

One other example may further clarify the distinction between the deterministic and statistical views. In normal times a nation regarded as an entity displays a continuous, regular behaviour pattern. People go to work; they eat; men and women marry and raise families; old and young enjoy their respective amusements; elections are held and the winners take office. If we did not know any more about the nation than just these facts, and if such behaviour could be deduced from very reasonable axioms about human beings, we should be tempted to assert that the behaviour of nations and even life itself was designed and determined by some super-being and constrained to follow this invariable design. The pleader for the statistical view, however, would urge us to look closer. What do you discover when you examine the behaviour of individuals themselves? Many people *don't* go to work; they beg, borrow, and steal. Some people *don't* eat; they starve. Quite a few *don't* marry, or marry and have no children. At election time only a fraction of the people vote; of the rest, some don't care to and others are not allowed to. In view of these facts what shall we say of the behaviour of the people as a group? Do they follow invariable, predetermined laws? Are not statements about group behaviour merely descriptions of general, mass effects which conceal all sorts of opposing actions, irregularities, and even disorder? The statistical view recognizes the variability and even haphazardness of individual actions. It expects, however, that the over-all effect of multifarious acts, though differing from one individual to another, will nevertheless produce an average result in the whole nation. But it specifically allows for the possibility that the mass effect may sometimes be revolution and a thorough alteration in the average behaviour of the people.

The question, whether determinism or the statistical view of nature is correct, is not an academic one. In a designed and orderly universe, life has meaning and purpose. Assurance of this design gives man courage and reason to live and build. It also reinforces his faith in a Supreme Being, for the strongest rational argument for the existence of God is the argument based on a designed universe. A thinking, superhuman Providence or Grand Designer is almost a necessary antecedent of a mathematically guided natural world. The existence of a God, in turn, gives substance to vast areas of religion and ethics. On the other hand, if the statistical view of nature is correct, the physical world and man's role in it are irrational. Occurrences obviously serve no purpose and head nowhere since they are merely accidental, chance happenings. The whole cosmos may even be destroyed tomorrow by some universal cataclysm. Life offers nothing but the meaningless pleasures and pains of the moment.

Undoubtedly it was because so much hung in the balance that the determinists returned to the fray. New reasons were found for reading design, causality, and determinism into the laws of Kepler, Galileo, and Newton. Let our 'bloody but unbowed' Mr D., who all this time has been marshalling new arguments, speak for himself.

There is, he declares, an essential distinction between statistical laws and the formulas of the Newtonian class. The former are based on tables of data or on probability arguments; the latter are deduced from unquestionable mathematical and scientific axioms which surely are true of nature, despite the fact that the underlying structure of matter is complex and largely unknown. For this reason we can be certain that the Newtonian laws are also exact truths and, therefore, genuine laws that nature must follow.

Mr P. was, of course, also prepared to take up the cudgels again and he began to speak with an assurance that the discussion must soon come to a satisfactory conclusion.

The burden of your argument, Mr D., rests on the truth of the axioms. Now do axioms describe facts inherent in the universe or are they merely fitted to experience in essentially the same way as a law of retail food prices is fitted to actual prices? Consider, for example, Newton's axiom about the force of gravity. It says that the force with which one mass attracts another equals the product of the masses divided by the square of the distance between them. This axiom has proved itself to be fairly accurate time after time in that deductions based on it have led to numerical results in agreement with observations to within the limits of accuracy of these observations. Yet the applicability of this axiom to the motion of the Earth around the sun, or the moon around the Earth, is ascertained only after many observations of the heavens and after many measurements of masses, distances, and time intervals. Hence the axiom may be no more than a good but approximate description of the average behaviour of nature. As a matter of fact, before Newton decided on his formula other formulas very much like his had been tried and rejected because they did not give such accurate results. Why should Newton's axiom be the last word? Evidently a person cannot be any surer of the truth of such scientific axioms than he can be of a law of food prices.

Mr D. may have been expecting some such argument, for he was prepared to reply at once.

Very well, Mr P., you may doubt the absolute truth of scientific laws in so far as they depend on axioms such as Newton's law of gravity. You must grant, however, that the theorems of pure mathematics itself are unchallengeable, for these rest on axioms that are entirely self-evident. Moreover these axioms do not involve measurement at all. Would you challenge the axiom that the whole is greater than any of its parts or the theorem that the sum of the angles of a triangle is 180°? Surely axioms and therefore the theorems of pure mathematics are absolute truths about nature and constitute definite laws. The existence of these laws in the structure of the universe makes the existence of others very likely.

The argument seemed incontrovertible, but Mr P. was not at all dismayed. Having recently completed his formal education, he had learned that new, non-Euclidean geometries had been created which apply to physical space as well as Euclidean geometry does. Hence, he confidently set about deflating his opponent's contentions.

An excellent point, Mr D., but unfortunately a hundred years behind the times. You have heard, no doubt, of non-Euclidean geometry. On some other occasion* we shall look into that subject a bit. Let me assure you here that the axioms and theorems of non-Euclidean geometry, which contradict Euclid's, are at least as good a description of physical space as Euclid's. We have not the slightest evidence for the *truth* of Euclidean geometry. Not the slightest.

As we might well imagine, Mr D. was indeed baffled. Every argument he had set forth had been swiftly demolished. But suddenly a gleam of cunning came into his eyes and a mild excitement shone forth. He began cautiously and with a slight touch of irony in his voice.

You have no doubt heard of the theory of probability, Mr P.? Do you grant too that the Keplerian and Newtonian laws, which we both admit to be broadly applicable, are very simple laws? Now what is the probability that the laws of a disorderly universe arising in a haphazard fashion should be simple? And compare that probability with the probability of finding simple laws in a universe operating in accordance with design. By which probability would you be guided?

Mr P. appreciated the force of the argument only too well. The probabilities were against him. He reflected carefully and then slowly unwound the opposing argument, apparently working it out as he talked.

After thousands of observations of the planet Mars, he resumes, Kepler found that its path is a simple ellipse. This does not mean that the observations he possessed all lay exactly on an ellipse; the small differences were charged to errors in measurement and ignored. Kepler, who believed that God used mathematics to construct the universe, was well pleased with the ellipse because it supplied a simple law of motion. But mathematicians would argue that all Kepler had done was to pick one of many curves that fit the data within the range of experimental error. Had he been willing to take a more complicated curve he might have found one fitting his measurements even more closely than the ellipse. Was Kepler correct in picking the simpler curve and charging the departure from this curve to errors in measure-

* See Chapter XXVI.

ment? Evidently we cannot be sure. Since no measurements will ever be exact this uncertainty will never be removed. The argument for design based on the simplicity of scientific laws can be reduced to this: from among the many formulas man can find that describe a natural phenomenon within errors chargeable to measurement, he selects the simplest one. Viewed this way the argument of simplicity reflects a preference of man's mind rather than the state of nature.

Though Mr D. by this time was secretly preparing to abandon his ship, he advanced timidly, though not without some hope, one more argument.

I see, he states, at least one more weighty argument in behalf of the truth and necessity of scientific laws, namely, their universal application in practical engineering. Bridges, buildings, dams, engines, and power plants constructed with the indispensable aid of these laws hold up. Bridge spans do not collapse; engines do the job they were designed to do. If there were not a large measure of truth in the laws, if nature were not constrained to obey them, why should they apply so universally and so well?

Sir, your argument has more emotional than logical force. For thousands of years people worked under the hypothesis – conviction to them – that the Earth was flat. Within the restricted geographical areas inhabited during those years, this hypothesis was good enough to give results in accord with experience. The hypothesis was, of course, incorrect. Similarly, since the time of Newton, scientists have utilized his quantitative law of gravitation and every engineering project has depended on its extensive use. Today, as a result of the creation of the theory of relativity, we know that Newton's law is not accurate. More than that, the new theory dispenses with the force of gravity entirely. Yet for over two hundred years the law of gravitation has been a scientific dogma. It is still used because it gives results good enough for most purposes in man's work-a-day world. Hence the applicability of a formula or a theory has little to do with truth or with the existence of design in the universe. You have made the mistake, Mr D., a very common mistake, of believing that a theory which works for many years must be true, whereas its actual status is no better than a working hypothesis. The error was made with Ptolemaic theory, the flatness of the Earth, Euclidean geometry, and the concept of gravitation. Actually man has just been stumbling from one description of nature to another; only we discover our errors so slowly and correct them ever so much more slowly that for long periods of time we relax in the delusion that we have discovered laws of nature. Fortunately men like Copernicus, Newton, and Einstein prevent us from losing ourselves irrevocably in false beliefs.

Of course, parries Mr D., I might abandon my argument for determinism based on the universal applicability of all scientific laws and base an even better one on the applicability of just one simple theorem of mathematics. Although my mathematics is apparently not up to date, Mr P., I do know that in mathematics we build long chains of pure reasoning that are absolutely independent of experience. The conclusions we arrive at are often far removed from the axioms. For example, the proposition of Euclid asserting that a tangent of a circle is perpendicular to the radius drawn to the point of contact is hundreds of steps removed from the axioms on which it ultimately rests. Yet the theorem is as much in accord with experience as the axiom is. Why should the result of so many steps of *pure reasoning* accord thus with experience? Is it not because nature herself is rationally designed and lawful? Nature does not permit contradictions any more than man's mind does.

Since you do hold such naïve views, Mr D., I must ask you how you know that the long chains of reasoning will continue to produce theorems in accord with nature. May not human reason behave at times like an automobile that keeps to the road for many miles but then edges imperceptibly to the side and finally lands in a ditch? The ditch for the sweet chariot of reasoning may be just ahead and when the vehicle finally lands there it will be fit for the academic junk yard next to that one-horse shay – the argument based on the orderliness of nature.

Possibly such a frightful accident will occur, Mr P., but until it does the existence of logical, intricate, and extensive mathematical developments whose theorems apply as broadly and effectively to nature as do the axioms must, under any philosophy other than determinism, be labelled a miracle.

Not really, Mr D. This miracle is readily explained. How did man obtain the principles of reasoning which enabled him to deduce the hundreds of theorems you refer to? Suppose, for example, I were to argue that because all fallible beings are human and because all mathematicians are human, all mathematicians are fallible. How would you determine whether I have used correct logic? Would you not check the principles involved against your experience with familiar classes of objects? In other words, Mr D., man learns to reason by studying nature's behaviour. He then finds that the conclusions of his logical processes are in accord with nature if his axioms are. Is there anything startling in this concordance? What you call a principle of logic is no more than an abstract formulation of nature's apparent behaviour.

Mr D.'s defences appeared to be completely shattered. In desperation he decided to attack.

As a proponent of the statistical view of nature, Mr P., how do you account for one element in the organization of nature which seems irreconcilable with your position? The tendency of energy is to dissipate itself so that it cannot be harnessed for man's needs; for example, after water falls from a height it levels off and can no longer be used for power. Since energy is available to man in the form of the sun's heat, coal, oil, atomic processes, and waterfalls, it does appear that the energy was especially created in usable form rather than that it came about as a result of a haphazard arrangement of molecules. In fact, the probability of an arrangement such as exists on our Earth arising by chance is smaller than the probability that an arbitrary selection of one million people will produce a group all of the same height.

Mr P. was on home ground and he therefore felt able to counter with confidence.

The strength of your argument, Mr D., appears to be in the fact that the particular organization of energy found on our planet is a highly improbable one. It is indeed highly improbable. But now consider a lottery in which 100,000 tickets are sold, one of which is the winner. The odds against holding a winning ticket are 99,999 to 1. Yet one person with those odds against him does win. Agreed then that the conditions on our Earth do constitute a highly improbable state, this state is *possible* and *has occurred*. Conscious design need not have been responsible for it. Further, there appear to be millions of planets in the sky on which the particular organization of energy found on the Earth is not encountered. Hence its occurrence on one planet is all the less surprising.

In spite of the irrefutable finality of this answer, old Mr D. now felt that he had won at least a moral victory. He had forced Mr P. to concede that our Earth was a highly improbable one. Deeming it best to terminate the discussion at the most favourable position he had been able to achieve throughout the long conversation, he excused himself on the ground that he had to complete a proof of a new law of electromagnetism.

Perhaps we can rejoin them if they re-open their discussion at some future time. Let us mention, before leaving the subject, that between the two views, the world as an orderly, determined organization and the world as chaos in which pure chance reigns, there are many intermediate points of view. One such view asserts that nature is neither lawful *nor* chaotic. The human mind thinks in such terms and unconsciously imparts these qualities to nature just as man makes God in his own image. The mind possesses within itself the desire to organize experience in the form of mathematical laws. It also possesses concepts such as exact quantitative laws and exact geometrical forms and applies these concepts to experience in order to comprehend it. The laws that result do not exist in the universe at all. They are but natural projections of our desire, reflecting the essential nature of mind and perhaps also its limitations, much as the lover's description of the loved one reflects the lover.

It is not our intention to explore all the schools of thought on the subject of law in nature. They run the gamut from absolute determinism to complete chaos. We must be content here to conclude with a restatement of the main theme, namely, that the development of mathematical ideas and methods has determined the dominant attitudes towards nature and, as a consequence, towards religion and society.

XXV

The Paradoxes of the Infinite

> We admit, in geometry, not only infinite magnitudes,
> that is to say, magnitudes greater than any assignable
> magnitude, but infinite magnitudes infinitely greater,
> the one than the other. This astonishes our dimension
> of brains, which is only about six inches long, five
> broad, and six in depth, in the largest heads.
>
> *Voltaire*

TRISTRAM SHANDY was hopelessly perplexed. He had begun to write his autobiography and found that he could record only half a day's experiences in one day of writing. Consequently, even if he were to start writing at birth and even if he were to live for ever, he could not record his whole life, for at any time only half of his life would be recorded. And yet if he did live on indefinitely he ought to be able to record his whole life, for the experiences of his first ten years would be recorded by the end of his twentieth year; the experiences of his first twenty years by the end of his fortieth year; and so on. Thereby every year of his life would be reached at some time. Hence, depending on which way he reasoned, he could or could not complete his autobiography. The longer Tristram puzzled over this paradox the more confused he became and the further he appeared to be from a decision.

Tristram's inability to resolve the paradox was really to be expected, for his problem involved an infinitude of time. The greatest mathematicians and philosophers from Greek times on had been plagued with problems involving infinite quantities and had not fared any better. For example, Galileo recognized that the number of whole numbers is infinite; that is, the number of these whole numbers is greater than any finite number that can be named. He recognized, also, that the number of even whole numbers is also infinite. Which of these two infinite sets, he asked, is the larger one? On the one hand, it seemed as though the first should be, for it contains all the numbers of the second set and more besides. On the other hand, to each number in the first set there corresponds exactly one number in the second, as 5 corresponds to 10. Also, to each number in the second set there corresponds exactly one in the first, as 10 corresponds to 5. In view of this *one-to-one correspondence* between the two sets there should be as many in the first as in the second. Galileo concluded that it was impossible to compare infinite quantities and abandoned further thought on the subject. He says 'infinity and indivisibility are in their very nature incomprehensible to us'. Leibniz, too, considered the very same question and concluded that the notion of the number of whole numbers is self-contradictory and should be rejected.

Not many years before a successful attack was finally made on the problems of the infinite, the supreme nineteenth-century mathematician, Karl Friedrich Gauss, expressed a horror of infinite quantities: 'I protest against the use of infinite magnitude . . . which is never permissible in mathematics.'

However much mathematicians recoiled from or repudiated the thought of infinite quantities, by the middle of the nineteenth century mathematics could no longer dispense with the concept. During the period from 1600 to 1850 mathematics had made gigantic strides. In this heroic age great intellectual adventurers had dared to leap over chasms of difficulties to goals envisaged by their genius and far-sightedness. These trail blazers expected that others would build the bridges to support the circumscribed steps of the more cautious thinkers who were to follow.

But the bridges were not easily built. The attempts to fill the gaps left during the heroic age were frustrated by paradoxes, contradictions, and more paradoxes. There developed an imperative need for critical thinkers with imagination and daring of another kind, the kind that would be able to dispense with and even override intuition and 'common sense'. This need was finally met. Neither the more circumspect workers, however, nor the trail blazers could have anticipated the astonishing and profound disclosures which the critical efforts brought forth.

The first successful attack on the problems of the infinite was made by Georg Cantor. His father had urged him to study engineering, a more profitable pursuit than teaching, and Cantor had started to follow this down-to-earth career; he ended by contributing to the most abstract regions of mathematics. His work got the reception that innovation and originality usually encounter – neglect, ridicule, and even abuse. One fellow-mathematician, Leopold Kronecker, attacked it viciously. Somewhat milder and more typical of the response to it was the remark made in 1908 by Henri Poincaré, the most famous of the late nineteenth-century mathematicians: 'Later generations will regard [Cantor's] *Mengenlehre* as a disease from which one has recovered.' Mathematicians, let it be known, are often no less illogical, no less closed-minded, and no less predatory than most men. Like other closed minds they shield their obtuseness behind the curtain of established ways of thinking while they hurl charges of madness against the men who would tear apart the fabric. So severe were the attacks on his work that Cantor began to doubt himself, became depressed, and suffered mental breakdowns.

Towards the end of his life (he died in 1918), the uncommon sense of his logic finally gained some recognition from a few colleagues. It is comforting to contrast Poincaré's statement above with one made just a little later by David Hilbert, the greatest mathematician of this century: 'No one shall expel us from the paradise which Cantor has created for us.' Today Cantor's work is so widely and completely accepted that many profound mathematicians are quite willing to devote themselves to the solution of further problems which the acceptance of Cantor's work brought in its train.

And now let us see how Cantor attacked the problem of infinite quantities. The most familiar examples of infinite collections are the collection of whole numbers, the collection of fractions, and the collection of all real numbers, that is, whole numbers, fractions, and irrational numbers such as $\sqrt{2}$, $\sqrt{3}$, and π. To obtain the number of objects in such collections by counting is impossible because the process is endless. On the other hand, describing them as infinite sheds very little light on them, for all this word says is that they are not finite. Such a description is about as informative as the statement that *Pithecanthropus erectus* is not a cow. We must substitute, if possible, a positive answer to the question of how many objects there are in infinite collections.

Cantor recognized, of course, that the number of objects in an infinite collection or class cannot be obtained by counting. He also recognized the deeper significance of another seemingly superficial observation. Suppose we have two classes of objects such that to each object in the first class there corresponds one and only one object in the second and conversely. For example, if a squad of soldiers, each carrying a gun, were to pass before us there would be just such a correspondence between soldiers and guns. Technically, the relation between the two classes, soldiers and guns, is described by the phrase *one-to-one correspondence*. Obviously, two classes that are in one-to-one correspondence must contain the same number of objects. Moreover, it is not necessary to count the classes to reach this conclusion.

Cantor's greatness lies in his perception of the importance of the one-to-one correspondence principle and in his courage to pursue its consequences. *If two infinite classes can be put into one-to-one correspondence then,* according to Cantor, *they have the same number of objects in them.* For example, the class of positive whole numbers

$$1 \quad 2 \quad 3 \quad 4 \quad 5 \quad 6 \quad \ldots$$

and the class of reciprocals of these numbers

$$1 \quad \tfrac{1}{2} \quad \tfrac{1}{3} \quad \tfrac{1}{4} \quad \tfrac{1}{5} \quad \tfrac{1}{6} \quad \ldots$$

are in the one-to-one correspondence whereby each number in the first class corresponds to one and only one number in the second, namely, its reciprocal. In like manner, to each number in the second class there corresponds one and only one in the first. Hence, these two classes have the same number of objects in them. The number that represents the quantity of objects in these particular classes Cantor designated by $\aleph_0$ (aleph-null). It is called a transfinite number.

To say that the number of positive integers, as well as the

number of any set of objects in one-to-one correspondence with the positive integers, is $\aleph_0$ does not seem to answer the fundamental question of how many there are in each set. The reader may say that $\aleph_0$ is a stranger to him and gives him no information on the number of positive integers. The objection is not a valid one. This number is as informative as the number one billion billion. The latter number is no more than a symbol representing the quantity of objects in a particular collection just as $\aleph_0$ represents the number of positive integers.

Of course the reader may retort that he can count the objects in a collection of one billion billion whereas he cannot count $\aleph_0$ objects. Hence, the former number means something to him whereas the latter does not. The distinction is correct but insignificant. Who has counted a billion billion objects? Theoretically it is possible to do so but theoretically it is also possible to assign numbers to infinite collections of objects. And just as the knowledge that two different collections each have a billion billion objects is definite and valuable, so the knowledge that two infinite collections contain the same number of objects, a fact indicated by the use of the same number to represent both collections, is definite and valuable – indeed, as we shall see, perhaps more valuable than knowledge of the number one billion billion.

The argument for Cantor's definition is even stronger. $\aleph_0$ is as meaningful as the number three itself. This number means something to us because we can now readily call to mind a group of objects for which this number denotes the quantity. To a child learning to count, however, the number three is meaningless. But just as the child grasps the meaning of three by associating it with three fingers or three blocks, so the man may grasp the meaning of $\aleph_0$ by becoming familiar with collections having $\aleph_0$ objects in them. Cantor's theory permits him to decide which these collections are.

With Cantor's definition in mind, let us reconsider the difficulty that bewildered Galileo and blocked his thinking about infinite quantities. Galileo, we recall, recognized the one-to-one correspondence between the collection of positive integers and the collection of positive *even* integers and was unable to reconcile this fact with the fact that the first collection contained all the numbers in the second collection and more besides.

Cantor's solution to the dilemma is that the collection of positive integers and the collection of positive even integers *both* contain $\aleph_0$ objects, despite the fact that the second collection is contained in the first one. The number of whole numbers and the number of even whole numbers *is* the same because the two collections of numbers are in one-to-one correspondence.

Is it not absurd that the collection of positive integers should have as many numbers as the subcollection of positive *even* integers? Yet if we accept one-to-one correspondence as a basis for deciding the numerical equality of infinite collections we must agree to this seeming absurdity. Apparently we are being led into contradictions that make nonsense of all our reasoning. Here we must swallow hard and face a surprising fact. There are no logical difficulties in Cantor's concept of infinite numbers. Our belief that it is absurd to have as many positive even integers as positive integers is merely a habit of thought formed while working with finite collections of objects. This mode of thought, however, which serves for *finite* collections, is no reliable guide to an understanding of infinite collections. Once again in the history of mathematics we face a conflict between logic and traditional thinking. And once again we face a parting of the ways. It was the failure of mathematicians before Cantor's time to understand that they must abandon some habitual ways of thinking about quantity that kept them from developing the subject of infinite numbers. But the critical thinkers of the nineteenth century were not so easily deterred.

In fact they took the bull by the horns. Following a suggestion of Bernard Bolzano, a professor of philosophy and a notable predecessor of Cantor in the development of the theory of infinite classes, an infinite set was defined to be one that can be put into one-to-one correspondence with a part of itself whereas a finite set cannot be. Thus the set of positive integers is infinite because there is a one-to-one correspondence between the whole class and the even numbers, which are only a part of that class.

Can every infinite collection be put into one-to-one correspondence with the positive integers? By no means. The set of *all* numbers between 0 and 1, a collection that includes whole numbers, fractions, and irrationals, cannot be put into one-to-one correspondence with the positive integers. The proof is readily made by showing that any supposed one-to-one correspondence between the positive integers and the set of all numbers between 0 and 1 leads to a contradiction. We shall omit the details, however.

Since the infinite collection of all numbers between 0 and 1 cannot be put into a one-to-one correspondence with the positive integers, the two collections cannot be equal in number. The number of numbers between 0 and 1 is represented by the transfinite number C. Accordingly, any collection of objects in one-to-one correspondence with all the numbers between 0 and 1 must also contain C objects.

An example of a set of C objects is furnished by the points on a line segment. Consider a line and a fixed point O on that line. Let us attach to each point on the line the number that expresses the distance of that point from O, with the added condition that distances to the right of O are to be positive and those to the left, negative. There is, then, a one-to-one correspondence between the numbers from 0 to 1 and the points on the line to which the numbers are attached. This implies that the number of these points is C.

We have defined C as the number of real numbers between 0 and 1. This set is in one-to-one correspondence with *all* the positive real numbers. We shall prove this fact geometrically. The set of real numbers is itself in one-to-one correspondence with the points on a line, such as the X-axis used in co-ordinate geometry. Hence let the points on the line L (fig. 76) and to the right of O represent all the positive real numbers, and let OA be the unit segment so that its points are in one-to-one correspondence with the real numbers between 0 and 1. We construct a rectangle such as $OABC$ and draw the diagonal OB. Now let P be any point to the right of O. Draw CP and let it intersect OB in Q. From Q drop a perpendicular to L thus obtaining P'. Under the correspondence determined by the construction just described any point P anywhere on L and to the right of O corresponds to one and only one point P' in OA. Conversely, if we start with any point P' in OA and construct a perpendicular to OA at P', this perpendicular will cut OB in Q, say. We then draw CQ and where CQ cuts L we have the point P corresponding to P'. Since the points on OA are in one-to-one correspondence with all the points to the right of O and on L, the number of points in OA as well as on the entire half-line is C. Stated arithmetically, the set of positive real numbers is in one-to-one correspondence with the real numbers between 0 and 1, and hence the number of positive real numbers is C.

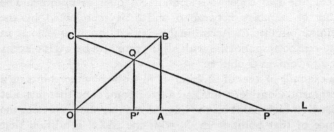

Figure 76. The one-to-one correspondence between the points of a unit segment and the points on a half-line

The number of points on a line segment and the number of points on an entire half-line are the same despite the fact that one is infinite in length and the other is just one unit long. Actually OA could have been two units in length or any other finite length and our result would have been the same. Hence the number of points on *any* line segment is always C.

This conclusion, like others established above, seems to violate our intuition. What right have we, however, to expect more points on the larger of two line segments? What precise knowledge about points and lines supports such an expectation? Euclidean geometry does require that any line segment contain an infinite number of points since any line segment however small can be bisected, but this geometry says nothing about the number of points on a segment. Cantor's theory does, and it informs us that *any two line segments, regardless of their lengths, possess the same number of points.* This conclusion is not only logically sound but it also permits us to dispose of some perplexing questions about the nature of space, time, and motion that had bothered philosophers for over two thousand years.

Our intuitions of space and time suggest that any length and any interval of time, no matter how small, may be further subdivided. The mathematical formulation of these concepts takes into account this property. For example, any line segment may be bisected by a precise Euclidean construction. The mathematical line contains additional properties. Any length consists of points, each of which has no length; moreover, these points are related to each other as are the numbers of the number system. Now between any two numbers there is an infinite number of other numbers; for example, between 1 and 2 there are 1½, 1¼, 1⅛, and so on. Hence, between any two points on a line there is an infinite number of other points. Similarly, the mathematical concept of time regards time as consisting of instants, each with no duration, which follow each other as do the numbers of the number system. Thus twelve o'clock is an instant and there is an instant corresponding to any number of seconds after twelve o'clock that we can name. It is true, then, for instants as for points on a line that there is an infinite number of instants between any two.

There are difficulties in these mathematical concepts of length and time which were first pointed out by the Greek philosopher Zeno, but which can now be resolved by use of the theory of infinite classes. Let us consider a formulation by Bertrand Russell of Zeno's Achilles and tortoise paradox.

Achilles and the tortoise run a race in which the slow tortoise is allowed to start from a position that is ahead of Achilles' starting point. It is agreed that the race is to end when Achilles overtakes the tortoise. At each instant during the race Achilles and the tortoise are at some point of their paths, and neither is twice at the same point. Then, since they run for the same number of instants, the tortoise runs through as many distinct points as does Achilles. On the other hand, if Achilles is to catch up with the tortoise he must run through more points than the tortoise does since he has to travel a greater distance. Hence, Achilles can never overtake the tortoise.

Part of this argument is sound. We must agree that from the start of the race to the end the tortoise passes through as many points as Achilles does, because at each instant of time during which they run each occupies exactly one position. Hence there is a one-to-one correspondence between the infinite set of points run through by the tortoise and the infinite set of points run through by Achilles. The assertion that because he must travel a greater distance to win the race Achilles will have to pass through more points than the tortoise is not correct, however, because, as we now know, the number of points on the line segment Achilles must traverse to win the race is the same as the number of points on the segment the tortoise traverses. Again we must notice that the number of points on a line segment has nothing to do with its length. In other words, it is Cantor's theory of infinite classes that solves the problem and saves our mathematical theory of space and time.

In his fight against the infinite divisibility of space and time Zeno proposed other paradoxes that confounded his adversaries and that can be answered satisfactorily only in terms of the modern mathematical conceptions of space and time and the theory of infinite classes. Consider an arrow in its flight. At any instant it is in a definite position. *At the very next instant,* says Zeno, it is in another position. When does the arrow *go* from one position to the other?

How does the arrow manage to get to a new position by the very next instant? The answer is that there is no *next* instant, whereas the argument assumes that there is. Instants follow each other as do the numbers of the number system, and just as there is no next larger number after 2 or 2½ there is no next instant after a given one. Between any two instants an infinite number of others intervene.

But this explanation merely exchanges one difficulty for another. Before an arrow can get from one position to any near-by position it must pass through an infinite number of intermediate positions, one position corresponding to each of the infinite intermediate instants. How does it ever manage to get to that near-by position if it has to pass through an infinite number of intermediate ones? This too is no difficulty. To traverse one unit of length an object must pass through an infinite number of positions but the time required to do this may be no more than one second, for even one second contains an infinite number of instants.

There is, however, a greater difficulty about the motion of the arrow. At *each* instant of its flight the tip of the arrow occupies a definite position. At that instant, the arrow cannot move, for an instant has no duration. Hence at each instant the arrow is at rest. Since this is true at each instant, *the moving arrow is always at rest.* This paradox is almost startling: it appears to defy logic itself.

The modern theory of infinite sets makes possible an equally startling solution. Motion *is* a series of rests. Motion is nothing more than a correspondence between positions and instants of time, the positions and the instants each forming an infinite set. At each instant of the interval during which an object is in 'motion' it occupies a definite position and may be said to be at rest.

Does this mathematical concept of motion satisfy our conception of the physical phenomenon of motion? Does not our intuition suggest that motion is something more than an object's being in different positions at different instants of time? Here again our intuition cannot be trusted too much. A 'motion' picture is no more than a series of stills flashed on the screen at the rate of sixteen per second. That is, it consists of motionless pictures presented to the eye at a rate rapid enough to give the illusion of motion. This motion, then, is no more than a series of rests. The mathematical theory of motion should be more satisfying to our intuition for it allows for an infinite number of 'rests' in any interval of time. Since this concept of motion also resolves paradoxes it should be thoroughly acceptable.

The algebra of transfinite numbers also possesses some surprising features that aid us in solving other difficulties in our ideas of time and space. Consider the two classes of objects (a) and (b):

(a) 1 2 3 4 5 6 7 ...
(b) 6 7 8 9 10 11 12 ...

The two classes are obviously in one-to-one correspondence for each number in class (a) corresponds to the one below it in class

(b) and vice versa. Hence the two classes have the same number of objects in them. This number is $\aleph_0$ since that is the number of positive integers. The second class, however, contains 5 numbers less than the first class does. That is,

(1) $$\aleph_0 - 5 = \aleph_0.$$

The curious fact represented by equation (1), namely, that if we subtract a finite number from an infinite quantity we still have the same infinite quantity, was expressed more dramatically if less tersely by the Roman poet Lucretius about A.D. 100.

You may complete as many generations as you please; nor the less, however, will that everlasting death await you; and for no less long a time will he be no more in being, who beginning with today has ended his life, than the man who has died many months and years ago!

Since the collection of positive integers can be put into one-to-one correspondence with the collection of positive even integers, and since there are as many positive even integers as positive odd ones, the number of these odd integers, as well as the number of the even integers, is $\aleph_0$. However, the collection of all positive integers is exactly the same as that of the odd and even positive integers together. The latter contains $\aleph_0 + \aleph_0$ or $2\aleph_0$ objects while the collection of positive integers contains $\aleph_0$. Hence.

(2) $$\aleph_0 = 2\aleph_0.$$

Were we seriously concerned, we could use the fact expressed in equation (2) to solve the dilemma of Tristram Shandy presented at the beginning of this chapter. Tristram was puzzled because he could record only half a day's experiences in one day, so that even if he were to live an infinite number of years he apparently could record only half of his life. On the other hand, it was equally clear to him that if he were to live forever, every year of his life would be recorded at some time. The mathematical theory of infinite quantities supports the latter argument. If he were to live $2\aleph_0$ years he could record $\aleph_0$ years of his life. But to live $2\aleph_0$ years is to live $\aleph_0$ years, and so Tristram could favour posterity with his completed autobiography.

Equations such as (1) and (2) involving $\aleph_0$ seem incorrect to us because we are accustomed to thinking in terms of what holds for finite numbers. Yet there is nothing illogical here. Properties that hold for finite numbers need not hold for transfinite numbers, nor does the reverse need to be the case. The logic of this statement is no different from the logic of saying that though cats and dogs are both four-legged animals there are statements true about cats that are not true about dogs.

Our brief examination of Cantor's contribution to the study of infinite quantities has shown some of the valuable results to which his theory has led. There are, however, entries on the other side of the ledger and these too warrant some attention.

The basic concept in the study of infinite quantities is that of a collection, a class, or a set of objects, as, for example, a set of numbers, a set of points on a line, and a set of instants in time. Unfortunately, this seemingly simple and fundamental concept is beset with difficulties we have not as yet considered. Let us support this statement with a few examples.

Our first one is classic. In different forms it appears in much ancient literature including the New Testament. Paul in his Epistle to Titus says of the Cretans, 'One of themselves, even a prophet of their own, said, The Cretans are always liars, evil beasts, slow bellies. This witness is true.' This libel against the Cretans is more commonly phrased as, 'Epimenides the Cretan affirms that the Cretans always lie.' If Epimenides is correct, however, he is recording a truth and so it is not true that Cretans *always* lie. On the other hand, according to his own statement, he, as a Cretan, is a liar and so his statement that all Cretans are liars is a lie. In either case, Epimenides contradicts himself. Apparently he cannot logically make the statement that all Cretans are liars even though the fact may very well be so. His mouth is taped with logic.

Consider next the dilemma of the honest village barber whose advertisement proudly proclaimed that though he did not shave those people who shaved themselves he did shave all those people who did not shave themselves. One day, while lathering his face, it suddenly occurred to him to ask whether he should shave himself. If he did, he would then be one of those people who shaved themselves; he should therefore, in accordance with his own advertisement, not shave himself. On the other hand, if he did not shave himself, his own advertisement boasted that he did. In brief, if he shaved himself he shouldn't; if he didn't, he should. The poor barber had defined a class of people that did and did not include himself. Unfortunately, we shall have to leave our barber, with his face lathered and razor poised, to make his own way out of his predicament.

A related difficulty may be found in the following rather amusing example. The word 'monosyllabic' is not monosyllabic whereas the word 'polysyllabic' is polysyllabic. The first of the two words is not a description of itself whereas the second is. Let us agree to call all words such as monosyllabic, which cannot be applied as descriptions of themselves, heterological. Hence we may say that any word x is heterological if x is not itself x. But suppose x is the word heterological. Then we are saying that the word heterological is heterological if heterological is not itself heterological. In other words, we are saying that something is something if it is not that something. About all that can be said at this point is that something is wrong.

In all of these paradoxes a distinct class of objects is involved, the class of Cretans, the class of people to be shaved, and the class of heterological words in the last example. Analysis shows that the statements about these classes are self-contradictory. Yet just such difficulties were introduced into mathematics by Cantor's use of the class concept. It is no wonder, then, that his work aroused a storm of criticism and became the subject of fierce controversies.

It is painful to relate that the difficulties have not been cleared up. Because they involve problems on the borderline between logic and mathematics, several different approaches to the two subjects have been advanced, each of which claims to be the correct one, though no one approach has as yet proved satisfactory. Mathematicians are now divided into schools of thought, each advocating its own philosophy of the foundations of mathematics.

It should be added that not all of mathematics has been cast in doubt. Moreover, even those portions that are the subject of controversy need not be discarded even temporarily. There is fortunately a pragmatic sanction for these portions. Just as the soundness of the calculus was debated all the while it was being used to produce majestic laws, so today the debatable theorems are being applied and are proving highly useful. The history of the calculus is encouraging, too, because just as the difficulties there were finally resolved so we may expect solution of the current ones.

The doubts have at least given mathematicians the opportunity to spoof their own work. Recognition of the fact that each age has the problem of rigorizing what it creates led E. H. Moore, a prominent American mathematician, to remark, 'Sufficient unto the day is the rigour thereof.' Other mathematicians have expressed more cynicism. A proof, runs one quip, tells us where to concentrate our doubts. Logic, says another, is the art of going wrong with confidence.

Despite the paradoxes to which Cantor's work led and which still remain to be cleared up in a thoroughly satisfactory manner, many mathematicians have come to see that he made the only real progress man is capable of making. Mathematicians create by acts of insight and intuition. Logic then sanctions the conquests of intuition. It is the hygiene that mathematics practices to keep its ideas healthy and strong. Moreover, the whole structure rests fundamentally on uncertain ground, the intuitions of man. Here and there an intuition is scooped out and replaced by a firmly built pillar of thought; however, this pillar is based on some deeper, perhaps less clearly defined intuition. Though the process of replacing intuitions by precise thoughts does not change the nature of the ground on which mathematics ultimately rests, it does add strength and height to the structure.

We feel obliged to conclude this chapter with a caution. Puzzles and paradoxes have been so much to the fore that the reader may regard the theory of infinite numbers as a mathematical *divertissement*. This is far from the correct evaluation. We should see rather how exact thinking has been applied to the shadow of one of the vaguest and most intangible intuitions. In rendering precise the notion of quantity as applied to infinite sets of objects, Cantor disposed of reams of philosophic disputes which had taken place from Aristotle's time right up to modern times.

The theory of infinite numbers is only one of the creations of the nineteenth-century critical thinkers. Almost bizarre in its contents it is nevertheless both logical and useful. The next mathematical creation we shall examine will strike the uninitiated as even more fantastic; yet it proved to be sound enough to revolutionize mathematical, scientific, and philosophical thought. It would seem as though the nineteenth-century mathematicians were being forced farther and farther from normal channels of thought in order to restore to mathematics the rigour which the Greeks first injected but which the seventeenth century lost sight of in its haste to keep pace with scientific activity.

XXVI

New Geometries, New Worlds

> I have made such wonderful discoveries that I am
> myself lost in astonishment: Out of nothing I have cre-
> ated a new and another world.
>
> *John Bolyai*

THE first man to challenge Euclid was Euclid himself. The creator of the most widely and most completely accepted system of thought – the abode of truth, and the birthplace of philosophies and sciences – doubted his results even before he issued them to the world. Euclid's questioning of himself marked the beginning of a two-thousand-year 'behind the scenes' attack on the obvious.

It is well known that Euclidean geometry is founded on ten axioms whose truth appears so self-evident that no 'sane' man would dare question them. From this sound basis impeccable logic produced more 'truths' just as appealing and as immediately acceptable as the axioms. Two millenniums of application climaxed by the successes of the Newtonian era added practically incontrovertible evidence of the soundness and reliability of these truths. Century after century buttressed logic with experience and common sense with tradition, until Euclid's system acquired inviolate sanctity. By 1800 educated people were far more likely to swear by the theorems of Euclid than by any statement in the Bible.

Whether a person appealed to experience, accepted the Kantian philosophy, or inclined to the obvious, the inescapable conclusion appeared to be that Euclid was truth and truth, Euclid. Despite this enviable position, which Euclidean geometry possessed from the outset and which time enhanced, a few thinkers, including Euclid, were not at ease. They were disturbed by two apparently innocent axioms.

The first of these says that a line segment can be extended *as far as one pleases* in either direction. The second one is the axiom on parallels, which says that through a point P not on a line L there passes one and only one line M (in the plane of P and L) that does not meet L *no matter how far M and L are extended* (fig. 77). If the axioms of Euclidean geometry are accepted because experience with physical space warrants our doing so, then these axioms are open to some doubt. No man has had direct experience with what happens in space more than a few miles beyond the Earth. All we can really say is that these axioms appear to be true in the limited regions in which we actually move. And even here we cannot be too sure of our assertions, for it was pointed out in the chapter on projective geometry that we never *see* parallel lines even in the portion of

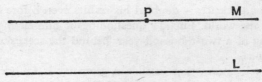

Figure 77. Euclid's parallel axiom

space immediately around us. As we look into the distance along lines Euclid refers to as parallel, we find that they seem to meet.

Euclid reveals his concern about these axioms by the manner in which he uses them. He does not use the parallel axiom, which is the more questionable of the two, until he has proved as many theorems as he can without it. He is equally cautious about the unlimited extensibility of the straight line. An examination of the theorems of his geometry shows that he uses line segments (portions of lines between two points) but never supposes that he has an infinite straight line to start with. When necessary he extends a segment in either direction only as far as the theorem requires. It should not be inferred that Euclid doubted the truth of these axioms; rather, because of their seemingly weighty implications, he would have preferred, no doubt, to derive their contents as consequences of simpler axioms.

A few hypercritical thinkers in every age following that of Euclid also hesitated to use as axioms statements on which any hard-headed businessman would bet his bottom dollar. To eliminate any lingering qualms these men from Missouri all tried the same thing. They concentrated on the parallel axiom and sought either to deduce it from the other axioms or to find a more acceptable substitute. Several hundred such worthy attempts on the part of the best mathematicians ended in failure. By 1800 the parallel axiom had come to be labelled the scandal of geometry.

It is hardly desirable, nor would it be too profitable, to review most of the efforts. The work of one man, however, the Jesuit priest Girolamo Saccheri, a professor of mathematics at the University of Pavia and a keen student of logic, merits our attention. Saccheri had a brand new idea. His novel attack on the problem of the parallel axiom was to argue in effect: Given a line L and a point P, then either (a) there is exactly one parallel to L through P, or (b) there are no parallels to L through P, or (c) there are at least two parallels to L through P. Alternative (a) was Euclid's parallel axiom. Suppose it were replaced by alternative (b), and the latter together with the other nine axioms of Euclid were shown to lead to contradictory theorems. Then surely alternative (b) could not be correct. Similarly, if the use of alternative (c) and the other nine Euclidean axioms led to contradictory theorems, then alternative (c) could not be correct. Then it would follow that Euclid's parallel axiom is the only possible one.

By using alternative (b) together with the other nine Euclidean axioms Saccheri did deduce theorems that contradicted each other. But he failed to deduce contradictions from the nine Euclidean axioms and the alternative axiom postulating the existence of at least two parallels. Though his efforts were determined and extensive, and though some of his deductions were indeed strange when compared with analogous results in Euclidean geometry, contradiction there was not.

Saccheri was on the threshold of an epoch-making discovery but he refused to step over it. For the moment we shall leave it to the reader to determine the conclusion Saccheri should have drawn from his failure to deduce inconsistencies. As for Saccheri himself, he was so unprepared for the strange theorems he established from his set of axioms that he decided Euclid's parallel axiom must be right. Accordingly, in 1733 he published his results in a book titled *Euclid Vindicated from All Defects*. Apparently when one man sets out to vindicate another he will most likely do so regardless of the facts.

One explanation for Saccheri's failure as well as the many others is that great as the mathematicians were who approached the problem presented by the parallel axiom, none was discerning enough to recognize and reject a two-thousand-year-old habit of thought. But within the mathematical world of the early nineteenth century a change in the intellectual milieu took place, bringing with it a sweeping, critical re-examination of fundamental beliefs. Undoubtedly this change accounts for the fact that three men, Gauss, Lobatchevsky, and Bolyai, with no knowledge of each other's thoughts, discovered the correct interpretation of Saccheri's work at about the same time; Lobatchevsky and Bolyai published their results within a few years of each other.

Of these three the greatest, and one who ranks with Newton and Archimedes, was Karl Friedrich Gauss. Karl showed unbelievable precocity in many fields and a particular predilection for mathematics. When as a young man he proved that the regular polygon of 17 sides could be constructed with straightedge and compass, he was so delighted that he abandoned his intention of becoming a philologist in order to study mathematics. He soon contributed masterful work to many branches of the subject and also achieved note as an inventor and experimentalist. Though his contributions were no less numerous and no less profound than those of other mathematicians, Gauss was extremely modest. He said, 'If others would but reflect on mathematical truths as deeply and as continuously as I have, they would make my discoveries.' Those who believe that genius is 99 per cent perspiration as well as those who despair of their mathematical abilities may find comfort in Gauss's statement.

Gauss was still a youth when the problem of the parallel axiom first came to his attention. At first he worked hard to replace the parallel axiom by a simpler one, and he failed. He then followed

Saccheri's line of thought by adopting a parallel axiom contradicting Euclid's – essentially Saccheri's third alternative – and by deducing consequences from this new axiom and the other nine of Euclid's. Like Saccheri he arrived at strange theorems. Instead of allowing the strangeness to frighten him Gauss fought fire with fire. He drew the brand new, astounding conclusion which the great and near-great had failed to consider. He decided that *there can be other geometries as valid as Euclid's.*

Gauss had the intellectual courage to create non-Euclidean geometry but not the moral courage to face the mobs who would have called the creator mad, for the scientists of the early nineteenth century lived in the shadow of Kant whose pronouncement that there could be no geometry other than Euclidean geometry ruled the intellectual world. Gauss's work on non-Euclidean geometry was found among his papers after his death.

Of the two other men who deserve honour for the creation of non-Euclidean geometry the first was the gifted Nicholas Lobatchevsky. Born in 1793 to a poor Russian family, he studied at the University of Kazan and at the age of twenty-three became a full professor there. Lobatchevsky, too, was attracted to the problem of the parallel axiom. He says that he was struck by the fact that two thousand years of effort by the greatest mathematicians had failed to produce a better axiom. And so, like Saccheri and Gauss, he built a new geometry on the basis of a parallel axiom contradicting Euclid's. The almost unbelievable theorems to which he was led did not discourage him any more than they had Gauss. Sound reasoning had led to them and sound reasoning was the unquestionable guide. And so Lobatchevsky, too, affirmed the radical but inescapable conclusion: There are geometries different from Euclid's and just as valid.

The man who shares honours with Lobatchevsky for the discovery and the courage to publish his work on non-Euclidean geometry is the Hungarian, John Bolyai. Like the other two he was blessed by the gods and, in addition, was encouraged and developed by his father Wolfgang, also a mathematician. Wolfgang had himself been bitten by the bug of the parallel axiom problem and had spent many vain years in work upon it. He bequeathed it to his son who, in 1825, at the age of twenty-three, suddenly saw the light. There are axioms contradicting Euclid's, he contended, that can nevertheless serve as the basis for new geometries. John proceeded to build one. On the urging of his father, he published his work in 1833 as an appendix to his father's text.

How were the epoch-making documents of Lobatchevsky and Bolyai received? How did scientists react to the startling news that Euclidean geometry now had rivals? How did the very rational philosophers greet the most thoroughgoing refutation of the leading philosophy of the times? The works of Lobatchevsky and Bolyai were completely neglected. Moreover, in 1847 Lobatchevsky was dismissed by the University, despite brilliant contributions and unselfish devotion to his work. If Bolyai had been a professor rather than an Austrian army officer, he might have suffered the same fate.

About thirty years after Lobatchevsky and Bolyai published their monumental work, Gauss's correspondence on non-Euclidean geometry was published posthumously along with other papers. His name attracted attention to the subject and shortly thereafter the mathematical world began to read Lobatchevsky and Bolyai.

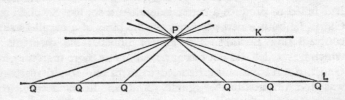

Figure 78. Euclid's parallel as a unique limiting line

To appreciate their work on the problem of the parallel axiom we must go back for a moment. Consider any straight line L (fig. 78) and any point P not on L. Euclid's parallel axiom asserts that there is one and only one line K through P which does not meet L. Now let Q be any point on L. As Q moves to the right, the line PQ revolves counterclockwise about P and seems to approach the line *K*. In like manner, as *Q* moves to the left along *L*, the line *PQ* rotates clockwise about *P* and again approaches *K*. In each case, then, *PQ* approaches one and the same limiting line *K*.

Bolyai and Lobatchevsky, however, assumed that the two limiting positions of *PQ* are not the same line *K* but two different lines through *P*, and that these limiting lines, *M* and *N* (fig. 79)

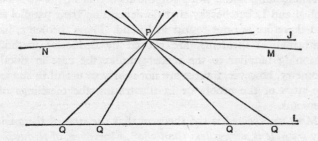

Figure 79. The parallel axiom of Lobatchevsky and Bolyai

do not meet *L*. Moreover, they assumed that *every* line through *P* and between *M* and *N* such as *J*, does not meet *L*. Hence Bolyai's and Lobatchevsky's parallel axiom affirms the existence of an infinite set of parallels to *L* through *P*. (These men reserved the word parallel for just the limiting lines *M* and *N*, but we shall use it to denote any line through *P* that does not meet *L*.)

The reader will perhaps feel, along with the mathematicians of Bolyai's and Lobatchevsky's time, that this is a ridiculous assumption to make. The diagram suggests that *M* and *N* will meet *L* if all three lines are extended far enough. Let us remember, however, that Bolyai and Lobatchevsky were interested in picking an axiom which, regardless of whether or not it describes the space we believe we live in, would be a logical alternative to Euclid's. And since the theorems to be derived from this axiom and the remaining Euclidean axioms would depend only on reasoning and not at all on accord with diagrams, the failure of the axiom to correspond with visual sensations is irrelevant.

What theorems were Bolyai and Lobatchevsky able to prove with their axioms? Of course, all the theorems of Euclid's geometry proved without the use of his parallel axiom are, automatically, theorems in Bolyai's and Lobatchevsky's geometry, since these men retained the other axioms of Euclid. As examples of such theorems we might mention these: Vertical angles are equal; from a point *P* at most one perpendicular can be drawn to a straight line; and, in a triangle with equal sides, the angles opposite these sides are equal.

Astonishing is the word for the theorems in the geometry of Bolyai and Lobatchevsky that do depend on their parallel axiom and therefore are not found in Euclid. These theorems, like all mathematical theorems, are proved by deductive methods of reasoning familiar to the reader; unlike the case in Euclidean geometry, however, figures are not nearly so useful in suggesting the steps of the proofs or in illustrating the meanings of the theorems.

Most unexpected is the theorem that *the sum of the angles of any triangle is always less than 180°*. Moreover, *of two triangles, the one with a larger area has a smaller angle sum.* Even more surprising is the fact that the new geometry wipes out a vital concept of Euclidean geometry, that is, that two geometric figures may have the same shape but different sizes. We say, in such a case, that the figures are similar but not congruent. In the new geometry *two similar triangles must also be congruent.* As a final example of a new theorem, we mention this one: *The distance between two parallel lines approaches zero in one direction along the lines and becomes infinite in the other direction.*

Bolyai and Lobatchevsky had succeeded in erecting a new geometry with many surprising theorems. But was their work any more than an exercise in logic? Let us realize, first, that hundreds of deductions in the new geometry had produced no theorems contradicting each other. This meant that the old parallel axiom could not be deduced from the other Euclidean axioms; otherwise the assumption of the new one would surely have led to contradictions within the system. It was not exactly news that the Euclidean parallel axiom could not be deduced from other Euclidean axioms. This fact had been suspected before.

The second implication in the work of Bolyai and Lobatchevsky was not anticipated. It is that we could not hope to establish the incontrovertible truth of the Euclidean parallel axiom by showing that any alternative produced contradictions. It was clear, therefore, that both of the schemes used by earlier mathematicians to vindicate the parallel axiom would never have been successful.

But the greatest significance of the new geometry was completely unexpected. Though the logical exercise was over the conclusion lingered on in people's minds: *There are geometries different from Euclid's.* A mathematician who possessed this knowledge was like a boy with an air rifle in his hands. The temptation to use it was too strong to resist. Euclidean geometry was known to be an accurate description of physical space. The non-Euclidean geometry of Bolyai and Lobatchevsky, on the other hand, did not seem and was not intended to apply to the physical world – but could it?

First reactions to this question are generally negative. If Euclidean geometry is correct, how can this new, conflicting geometry also be correct? Moreover, how can such absurd theorems apply to our familiar world? A little thought indicates that first reactions may be too hasty. What guarantee do we have that Euclidean geometry is correct? True it has been used for thousands of years. It also has the favour of long-established habits of thought. But let us recall Euclid's own reason for being concerned about the parallel axiom. Was it not that it makes an affirmation about regions of space far removed from man's daily experience, a space so vast that the accessible regions are in comparison only a dot on the surface of the Earth? Who of us knows the geometry of the universe in the vicinity of Mars, or for that matter even ten miles above the surface of the Earth? By what right do we assume that it must be the same as what seems to apply on Earth? Euclidean geometry may be no better than the hundreds of scientific laws that served well enough in their day but ultimately had to be discarded.

After carefully considering just this problem Gauss suggested a criterion for determining the truth of Euclidean geometry. The sum of the angles of a triangle equals 180° in this geometry but is less than 180° in the new geometry. Hence measuring the angles of a triangle should decide which geometry fits the physical world. For two reasons a very large triangle had to be chosen. In the first place, the error in sighting is greater in a smaller triangle. In the second place, it is a theorem of Lobatchevsky's and Bolyai's geometry that the angle sum of a triangle approaches 180° as the triangle shrinks in size. For a small triangle the sum might be so close to 180° that measuring instruments might not be sensitive to the difference.

Gauss, himself, performed the experiment. He stationed an observer on each of three mountain peaks. Each observer measured the angle formed by his lines of sight to the other two observers. The sum of the angles of the triangle turned out to be within 2″ of 180°, so close that the difference could be charged to errors of measurement. Hence the experiment was not decisive.

The provoking aspect of Gauss's triangle test is that even under the best of experimental conditions it could never prove that space is Euclidean, for even if the measured angle sum should be 180° allowance would always have to be made for errors of measurement and hence for the *possibility* that the true sum is less than 180°. Actually the test involved two unwarranted assumptions, either one of which could invalidate a conclusion drawn from it. The first of these was that the triangle formed by three mountain tops is large enough to be decisive. The second assumption was that the light rays which formed the sides of the triangle follow straight lines. The rays may actually curve slightly but imperceptibly.

While Gauss's test can be dismissed as an interesting but inconclusive experiment, the larger question of the applicability of non-Euclidean geometry still deserves attention. The surprising fact that emerges from all attempts to decide which of the two geometries fits physical space is that both fit *equally well*. We have already indicated that in a small triangle the new geometry calls for an angle sum close to 180° and the smaller the triangle the closer must the sum be to 180°. If we applied non-Euclidean geometry and accordingly used angle sums that are slightly less than 180° no harm would be done from the practical standpoint. Similarly, no harm would be done if we assumed that given a point P and a line L an infinite number of parallels to L exist through P and in the plane of P and L.

We might think that the new geometry cannot be applied to

the physical world because it asserts that similar triangles must be congruent. It certainly seems possible to construct two physical triangles that are similar but not congruent. In fact, one triangle could be made very large and the other very small. No matter how carefully the two triangles are constructed, however, we cannot be sure they really are similar – that is, that corresponding angles are exactly equal. The smaller triangle could have a larger angle sum, in accordance with the new, non-Euclidean geometry, but the difference might not be measurable. Hence for all practical purposes it would not matter whether we accepted the assertion of the new geometry. In other words, there is no way of deciding which geometry applies to physical space; either one could be used. Our prejudices and habits favour Euclidean geometry, which may also be somewhat simpler than the non-Euclidean geometry. But these reasons for preferring it do not discredit the applicability of the new geometry.

No doubt the reader is not satisfied. Perhaps he can be diverted with other, more pleasant arguments which show that non-Euclidean geometry can be applied to the physical world.

Let us return for a moment to Euclidean geometry. Imagine an enormous sheet of paper extending indefinitely in all directions. This sheet of paper is a physical illustration of the mathematical plane, the plane in which the theorems of Euclid's geometry hold. Now suppose we alter the shape of this vast sheet of paper by bending the left and right sides upward somewhat (fig. 80) so that it forms a curved surface, which, however, continues to extend indefinitely in all directions as did the plane. Such a surface is known as a cylindrical surface. As a result of the change in shape, most straight lines of the former plane become curves which, like the straight lines in the plane, are the shortest paths between the points they join on the surface. We shall call such curves *geodesics*. Two straight lines that were parallel in the plane become parallel geodesics, that is geodesics which do not meet on the surface. Triangles of the original plane become figures formed by arcs of geodesics on the surface. We shall call the new figures 'triangles' also. Circles on the plane give rise to figures we shall call 'circles'.

We come now to a very startling fact. Every axiom of Eu-

clidean geometry holds for the figures on the cylindrical surface
with the one provision – that we interpret the words line, triangle, and circle as just suggested above. Hence *the theorems of
Euclidean geometry, which follow from the axioms by deductive
reasoning, a process completely independent of the pictures we*

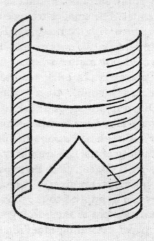

Figure 80. A new pictorial interpretation of Euclidean geometry

may draw, also hold for figures on the curved surface. To take
one instance, the sum of the angles of a 'triangle' on the surface is
180°.

The reader may object to the argument on the ground that
straight lines and figures defined in terms of straight lines no
longer have the proper meaning; they have lost their straightness.
Now, however, we take advantage of a fact first pointed out in
Chapter IV, namely, that the basic concepts of geometry such as
point and line are undefined. We use only the properties of these
concepts that are explicitly stated in the axioms. Hence if some
new physical picture of line, say, has the properties required by
the axioms, it is possible to adopt this new picture. Therefore it is
logically justifiable to associate a completely new physical picture
with all of Euclidean geometry.

The argument just made for the new physical interpretation of

Euclidean geometry applies as well to non-Euclidean geometry. And if we do take advantage of our liberty to choose the physical interpretation of the line and other figures we obtain an intuitively acceptable interpretation of the new geometry.

Fig. 81 illustrates a surface known as a *pseudosphere*. The curves on the pseudosphere which are the shortest paths between points on the surface – these special curves are also called geodesics – have the properties that straight lines possess in the

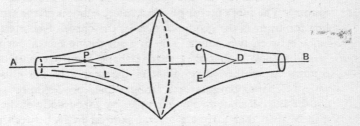

Figure 81. Pictorial interpretation of Lobatchevsky's and Bolyai's non-Euclidean geometry

axioms of Lobatchevsky and Bolyai. For example, the axiom that two points determine one and only one straight line applies to these geodesics. Two points on the pseudosphere (*C* and *D* in fig. 81) determine one and only one geodesic, or shortest path, between them. In like manner, the parallel axiom of Lobatchevsky and Bolyai, which says that through a point *P* not on a line *L* there is an infinite number of lines that do not meet *L*, applies to the geodesics on the pseudosphere.

Because the axioms of Lobatchevsky and Bolyai fit the geodesics on the pseudosphere, the theorems, as logical consequences of the axioms, must also apply. Thus the theorem that the sum of the angles of a triangle is less than 180° holds for triangle *CDE* formed by the arcs of geodesics. We have therefore obtained a visualization of the non-Euclidean geometry at the cost merely of a slight and justifiable change in the picture of the straight line.

Having made 'sense' of the new geometry let us return to our original question. Can the new geometry be descriptive of the physical world in which we live? The answer, as the reader may have anticipated, is that the geometry of physical space depends on the physical meaning we attach to the concept of straight lines. Experience tells us that if straight line is taken to be a stretched string, Euclidean geometry applies very well. It is neither necessary nor desirable, however, to allow straight line to mean a stretched string in all physical applications. Let us consider for a moment people who live in mountainous country and who are interested in the geometry of the surface of their country. The most useful physical interpretation of the straight line for them is the geodesic, that is, the curve of shortest distance between two points. The first surprising fact about these 'straight lines' is that they change shape from one part of the country to another, depending on the shapes of the hills and valleys. What axioms do these 'straight lines' obey? Almost surely not the Euclidean ones. For example, the topography of the area may be such that there are several shortest paths between some pairs of points. There may be many geodesics through a given point that do not meet a particular geodesic; and so forth.

In astronomical measurements, too, the stretched string is not the practical interpretation of the straight line. Here the light ray must serve instead. And what geometry fits best when light rays are used as straight lines? We shall leave this question with the reader until the next chapter. Meanwhile, we had better return to the mathematical account of non-Euclidean geometry. There are more mathematical worlds to be examined.

Lobatchevsky and Bolyai concentrated on Euclid's parallel axiom but accepted the other axiom in Euclid that is almost as questionable, namely, the axiom that says a line segment can be extended indefinitely in either direction. Here, too, is an axiom purporting to describe what happens in space trillions of miles from our Earth. How can we be sure of its truth, that is, its applicability to the physical world?

Not long after the work of Bolyai and Lobatchevsky on the concept of parallelism, the piercing glances of mathematicians lighted on the infiniteness of the straight line and sought to determine the wisdom of this axiom. The sickly, precocious Bernhard Riemann (1826–66), who had to beg his father, a

Lutheran pastor, for permission to abandon his training for the ministry so that he might study mathematics, undertook to pursue possible alternatives to this axiom.

One of his novel thoughts was that we must distinguish between endlessness and infiniteness. For example, the equator on the Earth is endless but finite. In view of this distinction Riemann proposed an alternative to Euclid's axiom on the infiniteness of the straight line, namely, the axiom that all lines are finite in length but endless.

This thought was then followed by reflections on the parallel axiom similar to those of Lobatchevsky and Bolyai but leading in this case to a different conclusion. As R moves to the left along L (fig. 82) and as Q moves to the right, both points must ultimately meet, for Riemann supposed the line L to be finite. The line PR

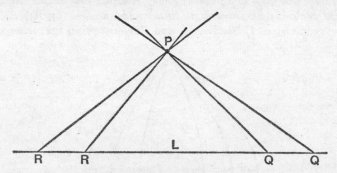

Figure 82. The geometrical basis for Riemann's parallel axiom

will, as a result, rotate around P into PQ without ever losing contact with L. That is, there should be *no* line through P parallel to L. Fig. 82 does not tell us how this complete rotation of PR about P is possible with our usual conception of the straight line, but the drawing is not intended to do more than suggest Riemann's thought. These reflections suggested to Riemann that he adopt along with the finiteness of the straight line an axiom to the effect that there are no parallel lines.

As if two radical departures from Euclid were not enough, Riemann proposed a third one: instead of requiring that two points determine one and only one line, Riemann adopted the axiom that two points may determine more than one line.

Before proceeding we remind the reader that these axioms are to be accepted for the moment purely as the basis for a logical development of a new geometry. The relation between this rather arbitrary system and the real world will be considered later.

The geometry of Riemann, like that of Lobatchevsky and Bolyai, has some theorems in common with those of Euclid. The theorem that vertical angles are equal and the theorem that angles opposite equal sides of a triangle are also equal hold in all three geometries, because these theorems depends only on axioms that are common to the three geometries.

Some of the theorems in Riemann's geometry that differ from Euclid's are striking. For example: *All the perpendiculars to a straight line meet in a point* (fig. 83). Another fact in this strange new world is that *two straight lines enclose an area* (fig. 84). As in the geometry of Lobatchevsky and Bolyai we find that *triangles*

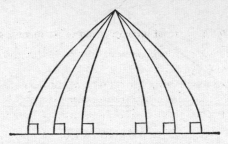

Figure 83. All the perpendiculars to a line meet in one point

Figure 84. Two straight lines enclose an area

which are similar are also congruent. Two other theorems might almost be anticipated. The first states that *the sum of the angles of any triangle is greater than 180°.* The second states that *of two triangles the one with larger area has the greater angle sum.*

We may now raise the same question that we raised with regard to the geometry of Lobatchevsky and Bolyai. Has Riemannian geometry any possible significance beyond that of an intellectual exercise for the mathematicians? Here again the answer is yes. It is possible to apply Riemann's geometry to the physical world with the usual understanding of straight line and never detect any differences between the assertions of the geometry and the physical situation. The argument here is precisely

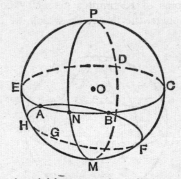

Figure 85. A pictorial interpretation of Riemann's geometry

the one that was made in connection with the geometry of Lobatchevsky and Bolyai.

Moreover, by changing the picture of the straight line we may find other, intuitively satisfying interpretations of Riemann's geometry. Just as we were able to picture Euclidean geometry on a cylindrical surface and Lobatchevsky's and Bolyai's geometry on a pseudosphere, so we may picture Riemann's geometry on the familiar sphere. The curve that connects two points on a sphere by the shortest route – that is, the curve that will be our picture of the straight line – is the arc of the great circle through the two points. By a great circle we mean one whose centre is also the centre of the sphere. Thus of the two circles through A and B (fig. 85), the circle *ABCDE* is the great circle whereas the circle *ABFGH* is not.

Let us see whether the axioms of Riemann's geometry apply to the sphere, provided we interpret straight line of the axioms to mean great circle on the sphere. In the first place, great circles are endless and finite in length. Second, there are no parallel lines on the sphere, for any two great circles meet. In fact, they meet not once but twice. For example, the great circles *ABCDE* and *MNPD* meet at *N* and *D*. The axiom that two points may determine more than one line is also fulfilled on the sphere. Two points such as *N* and *D* in fig. 85 have more than one great circle through them, while through two points such as *A* and *B* only one great circle passes.

Since the axioms of Riemann's geometry correctly describe facts about the sphere, the theorems, which are derived by valid

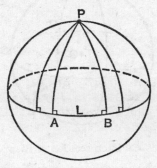

Figure 86. All the perpendiculars to a great circle on a sphere meet in one point

deductive reasoning, must also be true on the sphere. Let us check a few. One theorem states that all the perpendiculars to a straight line meet in a point. Taking the great circle *L* of fig. 86 as our straight line, we find that all the perpendiculars to *L* meet at *P*. If, for example, *L* were the equator on the Earth, *P* would be the North or South Pole.

Another theorem states that the sum of the angles of a triangle is greater than 180°. Since the straight lines of our axioms are great circles, a triangle is the figure formed by arcs of great circles. Such a triangle is illustrated by *ABP* in fig. 86. Since two of the angles of this triangle are right angles, the sum of the three angles is necessarily greater than 180°. This fact is true of every 'triangle' on a sphere.

It is not necessary to labour an already obvious point. *Every*

theorem of Riemann's geometry can be interpreted on the sphere
merely by thinking of the straight lines in the theorems as great
circles on the sphere. Hence we can give geometrical and intuitively satisfying meaning to Riemann's geometry. More than
that, this geometry supplies exact answers to practical and
scientific problems involving geometrical relationships *on the*
surface of the sphere. Hence it is, certainly to that extent, a
geometry of the physical world. In fact, every argument for the
theory that our physical world could be non-Euclidean in the
sense of Lobatchevsky's and Bolyai's geometry applies equally
well to Riemann's geometry. The applicability of non-Euclidean
geometries to the world in which we live will be discussed further
under the subject of relativity.

In retrospect, the history of the creation of non-Euclidean
geometries is the history of the blindness of human beings, great
and small. Man lives on the surface of the Earth. Suppose he
were to set about constructing a geometry to fit *this surface di-*
rectly instead of regarding it as a special surface in the three-
dimensional Euclidean world. What kind of geometry would he
develop? The 'line' in this geometry for the surface of a sphere
should obviously be the curve that joins any two points by the
shortest route, for this curve would be the most useful. This
curve, as we have seen, is the great circle joining the points. On
the other hand, the straight line in the familiar sense of Euclidean geometry would certainly not be chosen as the basic
curve, for it does not even exist *on* the surface of a sphere.

What axioms would a geometer choose for his great circles?
Why, none other than those Riemann selected, a system of
axioms in which no parallel lines exist and in which the line is
finite in length. In other words, the natural geometry, the practical geometry, the common-sense geometry for us earth-bound
mortals is Riemannian geometry.

For thousands of years this geometry has been right under the
feet of man. Yet during all those years the greatest mathematicians never once sought to test their attack on the parallel
axiom by checking with the geometry of the sphere. And as a
climax to these thousands of years the great Kant built his profound philosophy on the incontrovertible truth of Euclidean ge-

ometry, indeed on the impossibility of conceiving of any other geometry. Yet all this time he was living *on,* if not in, a non-Euclidean world.

How is it, then, that though geometry arose from measurements made on the Earth, Euclidean geometry was developed first? The answer is that to human beings living in a very limited region, the Earth does indeed appear flat, and the shortest distance on a flat surface is indeed the straight line in the commonly accepted sense of the term. With this stretched-string picture of the line the axioms and theorems of Euclidean geometry followed naturally. Once the geometry of flat surfaces was developed, the sphere had to be introduced within the framework of Euclidean geometry. No one, not even the Greeks, who were especially fond of the sphere, thought to approach the geometry of the sphere through a set of axioms designed to fit such a surface directly. This history shows that men are ruled as much by habits of thought as by physical habits, social customs, and conventions. Surely the unsuccessful precursors of Lobatchevsky and Bolyai did not lack technical skill or the ability to master difficult mathematics. They failed to solve the problem of the parallel axiom only because they were unable to break a habit of thought – Euclidean geometry. The history of this mental inertia is an excellent example of what Lecky, in his *History of Rationalism in Europe,* has described as that spirit, that *Zeitgeist* of an age, which predisposes people to points of view or beliefs independent of arguments for or against. So it was with Kant and all the mathematicians up to 1800. The belief in the truth, the unassailableness, the uniqueness of Euclidean geometry precluded anyone from even considering the possibility of another geometry, even though a non-Euclidean geometry lay right before them.

The importance of non-Euclidean geometry in the general history of thought cannot be exaggerated. Like Copernicus' heliocentric theory, Newton's law of gravitation, and Darwin's theory of evolution, non-Euclidean geometry has radically affected science, philosophy, and religion. It is fair to say that no more cataclysmic event has ever taken place in the history of all thought.

First, the creation of non-Euclidean geometry brought into clear light a distinction that had always been implicit but never recognized – the distinction between a mathematical space and physical space. The original identification of the two was due to a misunderstanding. Fleeting visitors to our minds, sensations of sight and touch, suggested that the axioms of Euclidean geometry were true of physical space. The theorems deduced from these axioms were checked with further sensations of sight and touch and, behold, they checked perfectly – at least as far as these sensations could reveal. Euclidean geometry was therefore held to be an exact description of physical space. This habit of thought became so well established over hundreds of years that the very notion of a new geometry failed to make sense. Geometry meant the geometry of physical space and that geometry was Euclid's. With the creation of non-Euclidean geometry, however, mathematicians, scientists, and laymen were ultimately compelled to appreciate the fact that *systems of thought based on statements about physical space are different from that physical space.*

This distinction is vital to an understanding of the developments in mathematics and science since 1880. We must say now that a mathematical space takes on the nature of a scientific theory. It is applied to the study of physical space as long as it fits the facts of experience and serves the needs of science. However, if one mathematical space can be replaced by another in closer agreement with the expanding results of scientific work, then it will be replaced just as the Ptolemaic theory of the motion of the heavenly bodies was replaced by the Copernican theory. Nor should the reader be surprised if he discovers that this possibility has materialized by the time he reaches the next chapter.

We should regard any theory about physical space, then, as a purely subjective construction and not impute to it objective reality. Man constructs a geometry, Euclidean or non-Euclidean, and decides to view space in those terms. The advantages in doing so, even though he cannot be sure space possesses any of the characteristics of the structure he has built up in his own mind, are that he can then think about space and use his theory in scientific work. This view of space and of nature generally does not deny that there is such a thing as an objective physical world.

It merely recognizes the fact that man's judgements and conclusions about space are purely of his own making.

The creation of non-Euclidean geometry cut a devastating swath through the realm of truth. Like religion in ancient societies mathematics occupied a revered and unchallenged position in Western thought. In the temple of mathematics reposed all truth, and Euclid was its high priest. But the cult, its high priest, and all its attendants were stripped of divine sanction by the work of the unholy three: Bolyai, Lobatchevsky, and Riemann. It is true that in undertaking their research these audacious intellects had in mind only the logical problem of investigating the consequences of a new parallel axiom. It certainly did not occur to them at the outset that they were challenging Truth itself. And as long as their work was regarded merely as an ingenious bit of mathematical hocus-pocus, no serious questions were raised. The moment men realized, however, that the non-Euclidean geometries could be valid descriptions of physical space, an inescapable problem presented itself. How could mathematics, which had always claimed to present the truth about quantity and space, now offer several contradictory geometries? No more than one of these could be the truth. Indeed, and even more disturbing, perhaps the truth was different from all these geometries. The creation of the new geometries, therefore, forced recognition of the fact that there could be an 'if' about all mathematical axioms. *If* the axioms of Euclidean geometry are truths about the physical world then the theorems are. But, unfortunately, we cannot decide on *a priori* grounds that the axioms of Euclid, or of any other geometry, are truths.

In depriving mathematics of its status as a collection of truths, the creation of non-Euclidean geometries robbed man of his most respected truths and perhaps even of the hope of ever attaining certainty about anything. Before 1800 every age had believed in the existence of absolute truth; men differed only in their choice of sources. Aristotle, the fathers of the Church, the Bible, philosophy, and science all had their day as arbiters of objective, eternal truths. In the eighteenth century human reason alone was upheld, and this because of what it had produced in mathematics and in the mathematical domains of science. The possession of mathematical truths had been especially comforting because they held out hope of more to come. Alas, the hope was blasted. The end of the dominance of Euclidean geometry was the end of the dominance of all such absolute standards. The philosopher may still claim the conviction of profound thought; the artist may passionately insist on the validity of the insight which his technical skill makes manifest; the religionist may fill the largest cathedral with the echoes of divine inspiration; and the romantic poet may lull our intellects into drowsy numbness and induce uncritical acceptance of his alluring composition. Perhaps these are all sources of truth. Perhaps there are others. But the rational person who has grasped the lesson of non-Euclidean geometry is at least wary of snares, and, if he accepts any truths, he does so tentatively, expecting at any moment to be disillusioned. Paradoxically, although the new geometries impugned man's ability to attain truths, they provide the best example of the power of the human mind, for the mind had to defy and overcome habit, intuition, and sense perceptions to produce these geometries.

The loss of the sanctity of truth appears to dispose of an age-old question concerning the nature of mathematics itself. Does mathematics exist independently of man, as do the mountains and seas, or is it entirely a human creation? In other words, is the mathematician in his labours unearthing diamonds that have been hidden in the darkness for centuries or is he manufacturing a synthetic stone? Even in the latter part of the nineteenth century, with the story of non-Euclidean geometry before him, the illustrious physicist, Heinrich Hertz, could say, 'One cannot escape the feeling that these mathematical formulas have an independent existence and an intelligence of their own, that they are wiser than we are, wiser even than their discoverers, that we get more out of them than was originally put into them.' Despite this opinion, mathematics does appear to be the product of human, fallible minds rather than the everlasting substance of a world independent of man. It is not a structure of steel resting on the bedrock of objective reality but gossamer floating with other speculations in the partially explored regions of the human mind.

If the creation of non-Euclidean geometry rudely thrust math-

Q

ematics off the pedestal of truth it also set it free to roam. The work of Lobatchevsky, Riemann, and Bolyai, in effect, gave mathematicians carte blanche to wander wherever they wanted. Because the non-Euclidean geometries, which were investigated originally for the sake of what seemed to be an interesting logical nicety, proved to have incomparable importance, it now seems clear that mathematicians should explore the possibilities in *any* question and in *any* set of axioms as long as the investigation is of some interest; application to the physical world, a leading motive for mathematical investigation, might still follow. At this stage in its history mathematics scrubbed the clay of earth from its feet and separated itself from science, just as science had broken from philosophy, philosophy from religion, and religion from animism and magic. It is possible now to say with Georg Cantor that 'The essence of mathematics is its freedom.'

The position of the mathematician before 1830 can be compared with that of an artist whose driving force is the sheer love of his art but who is compelled by the dictates of necessity to confine himself to drawing magazine covers. Freed from such a restriction the artist might give unlimited rein to his imagination and activities and produce memorable works. Non-Euclidean geometry had just this liberating effect. The tremendous expansion in mathematical activities as well as the increasing emphasis on aesthetic quality in the work of mathematicians since the middle of the last century bears witness to the influence of the new geometry.

Non-Euclidean geometry with its unparalleled importance in the history of thought was the culmination of two thousand years of dabbling in 'useless', logic questions. Mathematics thus provided one more example of the wisdom of abstract, logical thinking unmotivated by utilitarian considerations, and one more example of the wisdom of occasionally rejecting the evidence of the senses, as Copernicus asked us to do in his heliocentric theory, for the sake of what the mind might produce.

XXVII

The Theory of Relativity

Mad Mathesis alone was unconfined,
Too mad for mere material chains to bind,
Now to pure space lifts her ecstatic stare,
Now, running round the circle, finds it square.

Alexander Pope

THERE is an old bit of advice which says: Watch your friends; your enemies will take care of themselves. In the scientific métier, this saying goes: Suspect the obvious; the obscure truths will elude you anyway. Anyone who would challenge the obvious must nevertheless be daring, for the challenge is almost always regarded as an act of madness. Such daring is often displayed by genius, and perhaps for this reason genius does appear to be akin to madness as the popular phrase would have it.

The daring of genius is not pointless bravado, however. It has a goal which in the mathematical and scientific realms is a logically consistent account of the phenomenon under investigation. The passion for such an account is the earmark of the scientist; the ability to discern and the courage to follow the path of reason are the tests of his genius.

In modern times, one man, the creator of the theory of relativity, pre-eminently displayed such signs of greatness. With brilliance exceeded only by his modesty Albert Einstein attacked the obvious and revolutionized almost all branches of scientific and philosophic thought. The attack was directed against long accepted and, seemingly, the soundest concepts and assumptions of physical science.

Among the assumptions, one of the most firmly held was that space and figures in space obey the theorems of Euclidean geometry. It was, of course, true that at the time Einstein made his attack the non-Euclidean geometries had been in existence for about seventy-five years. It was also recognized that there was no guarantee as to the Euclidean character of physical space. Nevertheless, no one doubted but what the geometry for scientific work had to be Euclid's. The belief that physical space is Euclidean carries with it the belief that space is homogeneous, that is, that space on and near the Earth and in the region of the most distant stars possesses the same geometrical properties.

Nineteenth-century physics also rested upon certain metaphysical assumptions introduced by Newton and blithely adopted by later scientists. To appreciate the nature and role of these assumptions let us examine the most fundamental of physical processes, the measurement of length. Suppose a passenger walks from one position to another along the deck of a moving ship. What is the distance from his initial to his final position? The question is easily answered. The passenger can determine the distance by laying down a yardstick. Suppose now that this person's motion is in the direction in which the ship moves and that an observer on a ship anchored near by also measures the distance between the man's initial and final positions. He will find the distance to be greater than that obtained by the passenger himself because the moving ship has carried the passenger some distance.

Of course, no insurmountable difficulty is involved. The passenger measured a distance *relative to the ship*. The observer on the stationary ship measured the distance *relative to the sea*. If either one takes into account the motion of the moving ship, a correction can be made and the two observers will agree. Yet it must be recognized that the measurement of the distance varied from one person to the other. To speak of the distance from an initial to a final position is meaningless unless we specify *who* measured that distance.

Now the most important scientific laws involve distances either directly or indirectly as in the determination of velocities, accelerations, and forces. Hence, scientific laws should apparently depend on the observer whose measurements are used in the framing of that law. But this was not the usual understanding of a scientific law. Newton believed that our senses assure us of the existence of absolute space and absolute time and, therefore, he assumed that there are absolute laws even though we must be content with a formulation of these laws that depends on an observer on the moving ship called Earth. The absolute laws are known, he believed, to a superhuman observer, God, whose observations of space and time are absolute. And the ideal formulation of the mathematical and scientific laws of this universe are the laws God can obtain by his absolute measurements. It was only by knowing the motion of the Earth relative to the fixed observer, God, that man could translate his laws into the true form. We see, then, that Newtonian scientific thought rested ultimately on metaphysical assumptions involving God, absolute space, absolute time, and absolute laws.

One of the most firmly imbedded assumptions in the scientific thinking of the late nineteenth century was that a force of gravity exists. According to Newton's first law of motion a body at rest remains at rest and a body in motion continues in uniform motion along a straight line unless acted upon by a force. Hence, were there no force of gravity, a ball held in the hand and merely released would remain suspended in the air. Similarly, were there no force of gravity, the planets would shoot out into space along straight-line paths. Such strange phenomena do not occur. The universe acts *as if* there were a force of gravity.

Though Newton did show that the same quantitative law covered all the terrestrial and celestial effects of gravity's action, the physical nature of the force of gravity has never been understood. How does the sun, 93,000,000 miles away from the Earth, exert its pull on the Earth, and how does the Earth exert its pull on the variety of objects near its surface? Though there were no answers to these questions the physicists were not perturbed. Gravity was such a useful concept that they were content to ignore numerous objections to it. Indeed, if it were not for other more pressing questions and difficulties which arose around 1880, the complacency of physicists on the subject of gravity might not yet have been seriously ruffled.

Another problem raised by the introduction of the force of gravity had also been quietly thrust aside. In Chapter XIV we pointed out that every physical object possesses two apparently distinct properties, mass and weight. Mass is the resistance an object offers to a change in its speed or direction of motion. Weight is the force with which the Earth attracts an object. The mass of an object is constant, whereas its weight depends on how far the object is from the centre of the Earth. Though these two properties of matter are distinct, the ratio of weight to mass of *all* objects is always the same at a given place. This fact is as surprising as if the ratio of coal production to wheat production were exactly the same each year. Were coal and wheat production actually related in this way we should look for an explanation in the economic structure of the nation. In like manner, an explanation of the constant ratio of weight to mass was called for. Until Einstein's day none had been found.

One more physical assumption must be mentioned before we examine Einstein's work. Attempts to explain the nature of light date back to Greek times. Since the seventeenth century the most commonly accepted view of light regarded it as a wave motion much as sound is. Since it is not possible to conceive of a wave motion without a medium to carry the wave, scientists reasoned that there must be a medium which carries light waves. But the space through which light travels from the distant stars is a vacuum and hence contains no material substance to transmit the waves. Therefore, scientists had to assume the existence of a new 'substance', *ether,* which could neither be seen, tasted, smelled, weighed, nor touched. Moreover, for reasons unimportant to us, ether had to be a fixed medium coexistent with all space through which the Earth and other heavenly bodies moved. The introduction of ether to carry light waves drugged the scientists into a profound sleep which lasted over two hundred years. But by 1880 the properties that had to be assigned to ether were so contradictory that physicists began to doubt its existence altogether.

Despite the many dubious and poorly understood assumptions that lay at the foundations of late nineteenth-century physics, no group of scientists in any age was ever more cocksure that it had discovered the laws of the universe. The eighteenth century had been optimistic; the nineteenth was supremely confident. Two hundred years of partial success had so turned the heads of the scientists and philosophers that Newton's laws of motion and the law of gravitational attraction were declared to be immediate consequences of the laws of thought and pure reason. The word *assumption* did not appear in scientific literature despite the fact that, as Newton had expressly stated, the concepts of gravitation and ether were hypotheses, and, indeed, hypotheses not at all understood physically. But what was inconceivable to Newton was, to the nineteenth century, inconceivable otherwise.

The drastic overhauling of physics began inauspiciously enough when two American physicists decided in 1881 to check experimentally the conclusion that the Earth moves through a stationary ether. These two men, A. A. Michelson and E. W. Morley, devised an experiment based on a very simple principle.

A little arithmetic shows that it takes longer to row a given

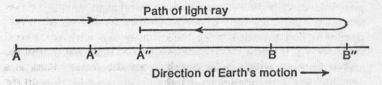

Path of light ray

A   A'   A"   B   B"

Direction of Earth's motion ⟶

Figure 87. The Michelson-Morley experiment

distance down a river and then back if there is a current than if there is no current. For example, if a man can row at the rate of 4 miles per hour in still water then, with no current present, he can go 12 miles down and then 12 miles back in 6 hours. If a current flows at the rate of 2 miles per hour, however, the man's progress downstream will be at the rate of 4 + 2 miles per hour while his rate upstream will be 4 − 2 miles per hour. At these rates his total time for the trip will be 2 + 6 or 8 hours. The principle involved here is that if a constant velocity, such as the velocity of the stream, hinders a motion for a longer time than it helps the motion (6 hours as against 2 hours in the example), the net result is a loss in time.

Michelson and Morley used this principle in the following way. From a point *A* (fig. 87) on the Earth a ray of light was sent to a mirror placed on the Earth at *B*; the direction from *A* to *B* was the direction of the Earth's motion around the sun. The ray was expected to travel through the ether to B at the usual velocity with which light travels and then be reflected back to A. Because of the Earth's motion, however, the mirror at B moves to a new position B' while the light ray is travelling towards it. Hence the Earth's motion delays the light ray in reaching the mirror. At B' the ray is reflected back towards A. While the ray is travelling *towards B*, however, the Earth carries A to A', and while the ray is travelling *back*, the Earth carries A' *to A''*. Therefore, the motion of the Earth helps the light ray in going from B' *to A''*. But the distance travelled from B' to A'' is shorter than the distance from A' *to B'*. In this way, the light ray is helped by the Earth's motion for a shorter time than it is delayed on the way out. In this situation the Earth's motion has the same effect as the velocity of the stream does in the example above. Hence, by the principle described in the preceding paragraph, the light ray should require *more* time to do the journey from A to B' to A'' than if it had travelled twice the distance AB with the Earth stationary in the ether. But despite the use of a very ingenious and delicate testing device known as an interferometer Michelson and Morley were unable to detect the increase in time. The motion of the Earth through the ether was apparently not taking place.

Physicists were faced with an inescapable dilemma. The ether that was needed to carry light had to be a fixed medium through which the Earth moved. Yet this condition was inconsistent with the result of experimentation. The failure of theory to agree with such a fundamental experiment could not be ignored. By this time physicists were convinced that their science needed some overhauling.

Though they were already sorely beset by fundamental problems, Einstein came along in 1905 to call attention to further difficulties in the basic concepts of simultaneity, length, and time. Einstein pointed out that under some circumstances it is theoretically impossible for two observers to agree on whether two phenomena or events are simultaneous and because of this the observers will not agree on the distance and time between events. Let us see why these disagreements must occur.

Suppose that a man in the middle of a long, very fast-moving train sees *simultaneously* two flashes of light, one of which emanates from a spot in the front car of the train and the other from the car at the rear. An observer standing alongside the track halfway between the front and rear of the train also sees the two flashes but *not simultaneously*. The one from the rear reaches him first. The question for consideration is: were the flashes *emitted* simultaneously?

Both observers would agree that they were not. As for the man on the ground, since he is exactly between the flashes, the two light rays must travel the same distance and, therefore, take the same amount of time to reach him. Since he saw the flash from the rear first, this flash must have occurred first. The man on the train would reason that relative to him the velocity of the light ray coming from the rear is the velocity of light minus the velocity of the train. On the other hand, relative to him the velocity of the ray from the front is the velocity of light plus the velocity of the train. Since both rays travelled half the train's length to reach him, and since the ray from the rear required more time, the flash from the rear must have been sent out first in order for the two to reach him simultaneously. There seems to be no difficulty whatever in this situation.

The two observers agreed on the order of the flashes because they both assumed that the man on the ground was at rest with respect to the ether while the man on the train was supposed to be in motion. Suppose, however, that the man on the train should take the unorthodox view that the train is at rest with respect to the ether and that the Earth is moving towards the rear of the train. According to this view the man on the train would correctly conclude that the flashes were emitted simultaneously. The man on the ground would undoubtedly prefer to stand by his previous position, namely, that he and the Earth are at rest with respect to the ether and that the flash from the rear car occurred first. We now have disagreement on the simultaneity of the two flashes arising out of disagreement about *who is at rest with respect to the ether*. Who is?

Unfortunately, the man on the train is as much entitled to the belief that the train is at rest with respect to the ether as the man on the ground is that the Earth is stationary in ether, for the Michelson-Morley experiment shows us that we cannot detect any motion through the ether. It follows that *two observers moving relative to each other must disagree on the simultaneity of two events*.

If two observers disagree about the simultaneity of two events, they must also disagree on the measure of distances. Suppose an observer on Mars and one on the Earth agree to measure the distance from the Earth to the sun. Since this distance is variable they must agree to measure it at a given instant. But for both observers to agree on the given instant, both must agree on the simultaneity of occurrences, such as the striking of clocks, which mark the instant. And since two observers moving relative to each other will not agree on the simultaneity of these occurrences, they will obtain different measures of the distance from the Earth to the sun 'at a given instant'.

Two observers moving with respect to each other will disagree not only on the measure of distances but also on the measure of time intervals. Otherwise, the observers would have to agree on the simultaneity of events that mark the beginning of the interval as well as on those marking the end. And this they cannot do.

The assumptions that space is Euclidean everywhere, that absolute lengths, absolute time, and absolute laws exist, that the force of gravity operates throughout the universe, that a fixed ether exists and carries light, as well as the problems these assumptions brought in their train, were in themselves becoming too numerous and weighty for science to carry with ease. When, in addition, it was recognized that simultaneity, time intervals, and lengths did not have a unique meaning, it also became apparent that no mere patchwork would resolve all the difficulties. A revolution in physical theory loomed, just as a political revolution does when the economic and social structure of a country fails to provide for the basic needs of its people.

In 1905, at the age of twenty-five, Einstein inaugurated the whole series of sweeping changes that were needed to re-establish physical theory. The Michelson-Morley experiment showed that the motion of the Earth does not affect the velocity of light relative to the Earth. Since science cannot run counter to experimental fact, Einstein accepted the basic assumption that the velocity of light is the same for *all* observers in the universe regardless of how they may be moving relative to each other. Hence, in one respect physical theory and experiment were made to agree. He accepted another axiom suggested by experience, namely, that no physical body has a velocity which exceeds that of light.

The concepts of absolute space and absolute time, which Newton needed to frame the true laws of the universe, Einstein discarded. Accepting the fact that two observers moving relative to each other will disagree on the measurements of space and time, he introduced the notions of *local length* and *local time*. Two observers *who are at rest relative to each other* will agree on the distance and time between two events. This distance and time are the local distance and local time for these observers. Two observers who are in motion relative to one another will obtain *different* measurements of the distance and time between the same two events. Each one's measurements are *his* local length and local time. In other words, men live in different space and time worlds.

If, for example, a Martian were to measure the distance and time intervals between two events on the Earth, he would find these quantities to be different from what our measurements would indicate. We in turn would find lengths on Mars and time intervals between events on Mars different from those the Martians would obtain.

It should be emphasized that we are not speaking of the effect of distance on sight or of optical illusions when we discuss the differences that different observers might obtain in the measurement of lengths. Even if Mars were to pass right by us as we measure lengths on it we should still find these lengths to be different from those measured by the Martians. Nor are we speaking of a psychological or emotional effect when we speak of disagreement on time intervals. The theory of local time says that two observers moving relative to each other and having identical clocks will record time intervals differently because these observers live in different time worlds.

To consider a numerical example, an observer on the Earth would find that a rocket ship moving at the speed of 161,000 miles per second relative to the Earth is half as large as a man on the ship would find it to be. Also, a clock on such a rocket would 'move half as fast' for the earth-bound observer as it would for the man in the rocket. An observer in the rocket would draw the same conclusions on size and time for objects and events on the Earth. And both sets of measurements are correct, each in its own space and time world.

We have in this doctrine of local length and local time one of the startlingly new assertions of the theory of relativity. The length of a room and the duration of our work day are not fixed quantities. They are one thing for us but different for an observer moving relative to us. The strangeness of these ideas should not blind us to the fact that they agree far more with experiment and the reasoning on simultaneity, which we examined above, than do the absolute notions of Newton. Indeed, if they did not, scientists would not hold them for a moment, relative or absolute.

In view of his abandonment of absolute space and absolute time, Einstein had to adopt a new concept of what constitutes a mathematical law of the universe. His conclusion was that there are no absolute laws in the sense of laws independent of observers. A law must be framed in terms of the measurements of a particular observer. If one observer formulates a law in terms of his measurements of space and time, then it is still possible to translate this law into the form given to it by another observer by means of formulas relating the length and time measurements of the two observers and involving their relative velocity. But in any event laws are tied to observers.

Though Einstein discarded absolute space, absolute time, and absolute laws, the question did arise whether any quantity connected with space and time measurements is the same for all observers. One very important quantity of this sort was discovered. Before we discuss it we must recall some notation from an earlier chapter. To represent points on the two-dimensional plane two coordinates, x and y, are used; to represent points in three-dimensional space three coordinates, x, y, and z, are used. To represent space and time measurements connected with events it is customary to use four letters, x, y, z, and t; the first three specify location in space and the fourth represents time. When discussing two different points or events, it is customary to use subscripts; thus x_1, y_1, z_1, t_1 represent the first event and x_2, y_2, z_2, t_2, the second one.

Now let us call upon a theorem of coordinate geometry. The distance between two points in a plane, one of whose coordinates is (x_1, y_1) and the other (x_2, y_2), is given by the expression

(1) $$\sqrt{(x_1 - x_2)^2 + (y_1 - y_2)^2}.$$

The distance between two points in space, (x_1, y_1, z_1) and (x_2, y_2, z_2), is given by

(2) $$\sqrt{(x_1 - x_2)^2 + (y_1 - y_2)^2 + (z_1 - z_2)^2}.$$

Concerning two events, (x_1, y_1, z_1, t_1) and (x_2, y_2, z_2, t_2), Einstein found that the quantity

(3) $$\sqrt{(x_1 - x_2)^2 + (y_1 - y_2)^2 + (z_1 - z_2)^2 - 186,000(t_1 - t_2)^2},$$

wherein it is assumed that distances are measured in miles and time in seconds, is the same for all observers. This absolute quantity is called the *space-time interval* between two events. It is obviously the analogue in the four-dimensional world of events of the quantities given by (1) and (2) above. The figure 186,000 is the velocity of light in miles per second.

Apparently, in order to find an absolute quantity, a quantity which is the same for all observers, an expression must be formed that involves both distance and time. And in this expression time measurements are treated no differently than spatial measurements. Now space and time had always been considered as different in kind and to treat time values like space measurements, as formula (3) does, seems to be an artifice designed specifically to produce an absolute quantity. Yet in 1908 H. Minkowski, a Russian mathematician, argued otherwise. It is true, he agreed, that we have harboured a notion of continuously flowing time which is independent of any notion of space. None the less, when we observe events in nature we experience time and space simultaneously. Moreover, time, itself, has always been measured by spatial means, for example, in terms of the distance moved by the hands of a clock, by the motion of a pendulum through space, or by the distance a shadow travels on a sundial. And our methods of measuring space necessarily involve time. Even during the simplest method of measuring distance, that of applying a rod, time elapses. No measurement is instantaneous. Hence the natural view of events should be in terms of a combination of space and time; that is, according to Minkowski, the world is a four-dimensional space-time continuum.

True, different observers may obtain different measures of the space and time components of the space-time interval between two events. But this is not surprising. Consider three-dimensional space itself. Two people in different parts of the globe see the same three-dimensional space but one analyses his experience of space into vertical and horizontal directions different from the vertical and horizontal directions of the other. Nevertheless, we continue to regard space as a three-dimensional whole rather than as an artificial combination of horizontal and vertical extents. Similarly, different observers may decompose space-time into different space and time components. This decomposition is as real and as necessary for the person who makes it as the distinction between horizontal and vertical is for a person walking down a flight of stairs. Yet it is man who does the differentiating; nature presents space and time together.

Einstein proceeded to utilize Minkowski's view that the universe should be regarded as a four-dimensional space-time world. The astounding innovations of Einstein's Special Theory of Relativity had not settled all of the difficulties enumerated at the beginning of this chapter. No explanation was as yet forthcoming in regard to just how gravity pulls objects to the Earth and 'holds' a planet in its course, or why mass and weight should always have a constant ratio at a given locale. Meanwhile, refinements in modern astronomical instruments began to put Newton to the test. These instruments were able to discover differences between the actual positions of the planet Mercury and the positions as predicted by the law of gravity. Consideration of these problems led Einstein to create and publish his

General Theory of Relativity. The new theory retained the major ideas of the earlier one but by extending them accomplished much more.

The idea that space and time should be regarded as a four-dimensional unity is employed in his general theory in the following way. Earlier we stated that formula (3) is to be regarded as an interval of space-time in a four-dimensional world. It is a generalization of formulas (1) and (2) which give the distances between two positions in a two-dimensional and a three-dimensional world, respectively. Now formulas (1) and (2) were derived on the basis of Euclid's geometry and are merely algebraic ways of expressing distance. Since formula (3) is essentially a generalization of (1) and (2), the space-time of Einstein's earlier theory is also Euclidean. (For our statement to be exact the minus sign in (3) should be a plus sign but this is a detail.)

Suppose, however, that we were to use instead of (3) an expression such as

$$(4) \quad \sqrt{2(x_1-x_2)^2+3(y_1-y_2)^2+7(z_1-z_2)^2-100{,}000(t_1-t_2)^2}.$$

If formula (4) were taken to be the numerical value of the space-time interval between the same two events whose coordinates (x_1, y_1, z_1, t_1) and (x_2, y_2, z_2, t_2) are involved in (3), then the value of the interval between these events would, of course, be different from that given by (3). In two and three dimensions the analogous procedure would amount to taking a number for the distance between two points different from that given by the formulas (1) and (2) of Euclidean geometry. What is the significance of altering the value of the distance or the space-time interval?

The choice of a formula for distance determines whether we have a Euclidean or non-Euclidean world. Let us see why this is so. Suppose we were to use the three-dimensional Cartesian coordinates discussed in Chapter XII to describe the positions of mathematical points corresponding to New York and Chicago. By using formula (2) to calculate the distance from New York to Chicago we should obtain what the formula is supposed to give, namely, the length of the straight-line segment joining the two cities. We could also use a different formula, for example, the one that gives the length of that arc of the great circle on the surface of the Earth between New York and Chicago.

Now suppose three cities, New York, Chicago, and Richmond, are brought into our discussion. These three cities are the vertices of a triangle. If we were to use formula (2) to calculate the sides of this triangle, we should obtain lengths that belong to a triangle formed by straight-line segments. On the other hand, if we were to use the formula for the great-circle arc length between each pair of vertices, we should obtain lengths that belong to a triangle formed by arcs of great circles on the surface of the sphere. That is, the choice of the formula for distance would determine whether we must think of our triangle as a plane triangle or as a triangle on a sphere. The properties of the two triangles differ, one being a triangle of Euclidean geometry, the other, a triangle in Riemann's non-Euclidean geometry. Thus the choice of the formula for distance determines the geometry that is used to describe the physical world.

Similarly, by adopting a formula such as (4) instead of (3) for the interval between two events in space-time we cause the geometrical figures in that four-dimensional mathematical world to possess properties different from those possessed by figures in Euclidean geometry; that is, we establish a non-Euclidean geometry in that space-time. We do not say that the new geometry will be the Lobatchevskian or Riemannian geometry examined in the preceding chapter but it will be non-Euclidean in the sense that it will differ from Euclid's.

The choice of a distance formula determines not only the geometry but also the shape of the geodesic, that is, the curve that gives the shortest distance between two points. In Euclidean geometry, the geodesic is the straight-line segment; in Riemann's geometry it is the arc of a great circle; in Lobatchevsky's and Bolyai's geometry it is the type of curve shown in figure 81 of Chapter XXVI. We have now to see how Einstein capitalized upon the choice of a 'distance' formula.

We should notice first that the location of a planet can be specified by using four coordinates, three for its position in space and the fourth for the time at which it occupies that position.

The successive locations lie on a curve in a four-dimensional mathematical world. Einstein's brilliant thought was to choose a formula for the space-time interval such that the 'path' of each planet is a geodesic in the resulting geometry.

What is accomplished by this ingenious mathematics? It will be recalled that the concept of a force of gravity was introduced to account for the fact that the planets move in ellipses instead of along the straight lines that Newton's first law of motion says they should follow. If now we revise Newton's first law of motion to read that bodies undisturbed by forces travel along the geodesics of Einstein's space-time, we have in just this revised first law the description of the motion of the planets around the sun *without having to introduce a fictitious force of gravity.*

But the force of gravity had also been used to account for the Earth's attraction of objects near it. Moreover, an apple dropping from a tree does not take the same path as do the planets. How does Einstein treat this phenomenon of gravitation? Here, too, he utilized the geodesics of space-time to eliminate the fictitious force. In his choice of a formula for the space-time interval he replaced the numbers 2, 3, 7, and – 100,000 in (4) by functions whose values vary from place to place in space-time *in accordance with the mass present.* Since the mass of the Earth differs from the mass of the sun, the structure of the geometric 'field' around the Earth differs from that near the sun. Consequently, the shapes of the geodesics vary from 'place' to 'place' in space-time. That is, by choosing the proper functions in his formula for the space-time interval, Einstein fashioned his space-time so that the presence of a mass in the physical world determined the character of that space-time and the geodesics around that mass, much as differences in the shapes of mountains in a range determine different geodesics on the surface of the Earth. In particular, objects near the surface of the Earth merely follow the geodesics of space-time in this region and again no force of gravity is needed to account for the paths.

The explanation of what were formerly considered gravitational effects in terms of the geometry of space-time disposes of another unsolved problem, namely, why the ratio of weight to mass is constant for all bodies on and near the Earth. Interpreted in the physical sense, this constant ratio is the acceleration with which all masses fall to the Earth* and which, according to Newtonian mechanics, is caused by the force of the Earth's gravitational pull on the masses. Hence the constant ratio of weight to mass means that all masses follow the same space and time behaviour in falling towards the Earth. Now in accordance with Einstein's reformulation of the phenomenon of gravitation, what was formerly regarded as the Earth's gravitational pull now becomes the effect of the shape of space-time near the Earth. All masses falling freely must, according to the revised first law of motion, follow the geodesics of space-time. In other words, all masses should show the same space and time behaviour near the Earth, and they do. Hence, the theory of relativity solves the problem of the constant ratio of weight to mass by eliminating weight as a scientific concept and by advancing an even more satisfactory explanation of the effects formerly attributed to weight.

As a climax to these accomplishments, the theory of relativity disposes of two other unsolved problems which had baffled scientists. The first of these concerns the motion of the planet Mercury. This planet does not pursue a purely elliptical path about the sun. Actually, the perihelion – that is, the point on the ellipse at which Mercury is closest to the sun – advances from one revolution to another. About a hundred years ago, the French astronomer Leverrier showed that part of this motion of the perihelion is due to the gravitational attraction of the other planets. A complete explanation eluded the scientists until the theory of relativity was created. The 'path' computed for Mercury in the space-time of the new theory agrees, within experimental error, with the observed motion. In other words, we attain more accurate computations of planetary motions by means of the new theory than with Newtonian theory.

The second of the problems that troubled scientists was the observed bending of light rays as they passed the sun on their way from the stars to the Earth. Such a bending might have been explained as a gravitational pull of the sun on the rays were it not for the fact that a light ray has no mass. If, in accordance with the revised first law of motion, we merely suppose that the light

* See Chapter XVI.

rays are following geodesics of the space-time region around the sun, the bending of the light rays is explained and the measured deviations from straight-line paths are in accord with computations based on the new theory.

Many a person who surveys the strange principles introduced by the theory of relativity and who finally realizes how complex its mathematical world is may be tempted to exclaim, 'Leave me alone with my ether, my gravitation, and my simple, intuitive, sense-satisfying Newtonian world. Your distorted construction may hew somewhat closer to experiment and to precise reasoning but it is too fantastic an account to be taken seriously.' Unfortunately, the person living today does not have this freedom of choice. Two predictions of the theory of relativity are now indispensable to science.

The first prediction is the relativity of mass. A ball held in a person's hand has, of course, a definite mass. If the person throws the ball up or out, then the new theory says that, as far as that person is concerned, the mass of the ball increases with its speed. This increase in mass of a moving body becomes considerable when the speed reaches any significant fraction of the speed of light, which is 186,000 miles per second. Such speeds are now common for electrons in hundreds of varieties of radio tubes and for electrons and other subatomic 'particles' in many types of atom-smashers. The theory of all these devices must take into account the relativistic increase in mass.

The other prediction of the theory that can no longer be ignored by any intelligent person of our century states that a given amount of energy is physically equal to a definite amount of mass; the energy in a light wave is essentially no different from that in a piece of wood. The precise quantitative expression, namely, that the energy in a given quantity of mass equals the mass times the square of the velocity of light (in suitable units), is now well known. In addition to establishing this formula, Einstein suggested that physicists examine the phenomenon of radioactivity to uncover a physical conversion of mass to energy. His suggestion proved to be sound. A few years ago man learned to control this conversion of matter into energy in the form of electromagnetic waves and produced the atomic bomb.

In spite of the astonishing and dramatic verifications of the theory, many people find its four-dimensional, non-Euclidean universe totally unpalatable. No one can visualize a four-dimensional, non-Euclidean world. But anyone who insists on visualizing the concepts with which science and mathematics now deal is still in the dark ages of his intellectual development. Almost since the beginning of work with numbers mathematicians have carried on algebraic reasoning that is independent of sense experience. Today they consciously construct and apply geometries that exist only in human brains and that were never meant to be visualized. Of course all contact with sense perception has not been abandoned. The conclusions about the physical world predicted by geometric and algebraic cogitations must be in accord with observation and experimentation if the logical structure is to be useful for science. But to insist that each step in a chain, even of geometrical reasoning, be meaningful to the senses is to rob mathematics and science of two thousand years of development.

Much more can be said in favour of the theory of relativity. In the preceding chapter we saw that the natural geometry of the *surface* of the Earth in a mountainous region, say the Rocky Mountain belt, would be non-Euclidean. On the surface of such a region there are no straight lines, no circles, and no other familiar paths. Moreover, whatever curve does give the shortest distance between two particular points may not serve similarly between two other points. Hence, the character of the geometry, which is determined by the nature of the geodesics, varies from place to place. This is precisely what happens in Einstein's general theory. Just as mountain masses cause the geometry of the Rocky Mountain belt to vary from point to point, so in relativistic space-time the character of the geometry and the shapes of the geodesics are affected by the presence of masses such as the Earth or sun.

We are asked in the new theory to accept the concepts of local space and local time, a relativity of time and space hitherto unknown. This much can certainly be said in defence of the conclusion that the time worlds of observers moving relative to each other are different. What might be called the subjective character of our experience of time has long been recognized. Were we to judge duration by our personal feelings on the subject there certainly would be marked disagreement on how much time elapses during any given interval. Hence only in regard to an artificial device, such as a clock, is the variation in time from one observer to another surprising. We have assumed that all observers using identical clocks would obtain the same result, but now we must recognize that even such a standardizing device will not serve to make time independent of the observer.

Once more thought should reconcile us to the radical suggestions made by Einstein. Consider the incredulity that must have prevailed when people were first told that the Earth is a round ball instead of a flat surface resting on some unknown foundation. What mathematical explanation could satisfy them that objects on the other side of the ball remain on the surface? Imagine their perplexity when they learned, in addition, that the Earth and the other planets whirl at tremendous speeds around the sun and rotate at the same time, contrary to the evidence of their senses. These assertions of Copernican theory, which are commonplace by now, must have been far more shocking to people of the sixteenth century than the sophisticated pronouncements of the theory of relativity are to us. Newton's explanation of why people remain on the Earth and the Earth in its path – the mysterious force of gravity – was not very satisfactory. Einstein, on the other hand, removes the need for this mysterious force and for other assumptions without actually contradicting the evidence of our senses.

Nor should we despair if Einstein's ideas do come hard to us despite all the arguments in their favour. It is not surprising that the average man, who cannot afford to spend much time speculating about the mysteries with which nature has surrounded us, has been much amazed and bewildered by the new mathematical and scientific ideas about space, time, matter, and gravitation. He may take consolation from the fact that his bewilderment is mild compared to the severe shocks that philosophers, who spend their lives building up sound thoughts on these subjects, have experienced.

We have often spoken of the close relationship between mathematics and philosophy, and we find in the theory of relativity an example *par excellence* of a mathematical creation which revolutionized modern philosophy.

The union of space and time and the influence of matter on space-time proposed by the theory of relativity, ideas that would have seemed outlandish to philosophers of the early 1900s, have now become embodied in a philosophy of nature more and more widely held. Nature presents herself to us as an organic whole with space, time, and matter commingled. Man has in the past analysed nature, selected certain properties that he regarded as most important, forgotten they were abstract aspects of a whole, and then regarded them thereafter as entirely distinct entities. He is now surprised to learn that he must reunite these supposedly separate concepts to obtain a consistent, satisfactory synthesis of knowledge.

Aristotle first formulated the philosophical doctrine that space, time, and matter are distinct components of experience. This view was subsequently adopted by scientists and used by Newton. We, following him, have become so accustomed to thinking of space and time as fundamental and distinct components of our physical world and separate from matter that we no longer recognize this view of nature as man-made and as only one of a number of possible views. Of course, the philosophers of the contemporary scene, among them the late Alfred North Whitehead, do not argue that this analysis of nature is useless. On the contrary, it has proved quite valuable and even essential. But we should be aware that it is artificial, and we should not mistake our analysis for nature itself any more than we should mistake the organs observed by dissection of the human body for the living body itself.

The theory of relativity upsets one of the fundamental philosophical assumptions of science, the relationship of cause and effect. Under the usual conception of this relationship the cause of an effect must certainly precede it. In accordance with the new theory, however, the order of two events is no longer an absolute affair. When we discussed the question of simultaneity we found that the order of the two flashes of light depended on the observer. If these two flashes were replaced by events that appeared to be cause and effect to some observers, there might, nevertheless, be other observers who could not view the events in that relation, for to them the event called the effect might occur *before* the cause. Revision of the concept is obviously in order.

The existence of free will seems to stand or fall with cause and effect. Free will implies that a voluntary act of the mind can cause a subsequent act of the body. To the person exercising his 'free will' this may indeed be the order of events; but for some observers the order of events may be reversed in time so that the act of the body would appear to be making up the person's mind. The latter view may remind us of the modern theory of emotions which says, for example, that we are afraid because we run away from danger rather than that we run away because we are afraid. The question of whether human beings possess free will must evidently be reconsidered in the light of the theory of relativity.

The revolutionary doctrines of the new theory have focused attention on our tendency to accept patterns of thought just because we grow up with them. Newton taught us to think in terms of a force of gravity which reached out from the sun millions and hundreds of millions of miles away and kept the planets in their orbits. This concept was acclaimed in the eighteenth century because it permitted accurate predictions. We have been following that lead unquestioningly. Youngsters two or three generations in the future will no doubt laugh at our naïveté and credulity.

Another mental process to which the work on relativity calls attention and which is obstructive to progress is that of making assumptions unconsciously. We are guilty of glibly and uncritically assuming, for example, that time, distance, and simultaneity are the same for all people in this universe. Mathematicians and scientists now realize that more attention must be paid to assumptions made implicitly than to those that have been recognized and explicitly stated.

Perhaps the author is also guilty of an unwarranted assumption at this point, the assumption that the reader has been able to swallow and digest so rapidly not only the major ideas of the theory of relativity but its philosophical implications as well. Let us therefore review the main events briefly: The physics of the nineteenth century was built upon the foundation of Euclidean geometry, the ideas of absolute lengths, absolute time, and absolute simultaneity of events, Newton's laws of motion, the force of gravity, and the concept of ether. Each of these foundation stones involved assumptions about the physical world believed to be well warranted. The Michelson-Morley experiment showed that an inconsistency in physical theory was involved in the use of ether as the carrier of light waves. Einstein then showed that the assumptions of absolute length, time, and simultaneity were also unjustified. A revolution in physical thought followed. The notions of local length, local time, and local order of events replaced the absolutes. The search for new absolute quantities ended with the realization that we must combine space and time to produce them. Minkowski then made us appreciate that the universe was naturally a four-dimensional, space-time unity, and that it was unreal, even though sometimes necessary for practical reasons, to separate space and time. Einstein followed up this idea by developing a non-Euclidean geometry that explains the effects of Newton's force of gravity in terms of the natural paths of bodies in the new space-time.

With these developments a broad trend in the history of mathematics and science comes to a grand climax. We spoke in an earlier chapter of the mathematization of science. A most significant step in this direction was taken when scientists decided in the seventeenth century to fashion their thoughts and procedures in terms of quantitive relationships. Then phenomena of motion, forces, sound, light, and electricity were all successfully studied and applied only after this transmutation into mathematics was accomplished. Thereupon, many domains of science became merely extensions of the mathematics of number.

It is now possible to appreciate how much of science has become mathematized in the form of geometry. Since the days of Euclid the laws of physical space have been no more than theorems of geometry. Then Hipparchus, Ptolemy, Copernicus, and Kepler summarized the motions of the heavenly bodies in geometrical terms. With his telescope Galileo extended the application of geometry to infinite space and to many millions of heavenly bodies. When Lobatchevsky, Bolyai, and Riemann showed us how to construct different geometrical worlds, Einstein seized the idea in order to fit our physical world into a four-dimensional, mathematical one. Thereby gravity, time, and matter became, along with space, merely part of the structure of geometry. The belief of the classical Greeks that reality can be best understood in terms of geometrical properties and the Renaissance doctrine of Descartes that the phenomena of matter and motion can be explained in terms of the geometry of space have received sweeping affirmation.

The theory of relativity is but one of the twentieth century developments in mathematics that are decisively shaping our civilization and culture. To be fair to our century we should investigate a related and perhaps even more influential development – quantum theory. Whereas the theory of relativity has been most useful in the treatment of phenomena involving great distances, times, and velocities, quantum theory has enabled scientists to treat the minute world inside the atom. Hence, the science of the vast universe and the infinitesimal realms have both been revolutionized. Unfortunately, the course of twentieth century science is departing farther and farther from 'common sense', from intuitively accessible concepts, and from simple, physical pictures. More and more science is resorting to complicated mathematics for which the physical account is either incomplete or even inconsistent despite the fact that this account is real enough to design and produce atomic bombs. It is impossible, therefore, in a survey as brief as this one, to attempt any account of quantum phenomena. We regret that we must merely mention in passing this second major development of a century that is only half over.

XXVIII

Mathematics: Method and Art

The Science of Pure Mathematics, in its modern developments, may claim to be the most original creation of the human spirit.

Alfred North Whitehead

In the preceding chapters we have examined some of the ideas of mathematics proper, the origins of these ideas in their contemporary setting, and their influences on branches of our culture. In modern times these ideas have multiplied at an almost fantastic rate. Correspondingly, the influences of mathematics have grown in number, depth, and complexity. It would be possible to take any one of the fields that enjoyed close contact with mathematics in some period we have examined and trace the continuation and extension of that association right down to the present day. There is neither space nor time, however, to permit a comprehensive account of the relation of mathematics to art, science, philosophy, logic, the social sciences, religion, literature, and a dozen other major human activities and interests. It is hoped that enough has been said to support the thesis of this book, namely, that mathematics has played a predominant role in the formation of modern culture.

One subject has, however, been thus far slighted. Mathematics is, itself, a living, flourishing branch of our culture. Several thousand years of development have produced an imposing body of thought whose essential characteristics should be familiar to every educated person. Though the nature of modern mathematics was somewhat foreshadowed by the contributions of the Greeks, the events of the intervening centuries and the creation of non-Euclidean geometry in particular have radically altered the role and character of the subject. An examination of the nature of twentieth century mathematics will not only redress a wrong but will perhaps make it evident why the subject has gained in power and stature.

More than anything else mathematics is a method. The method is embodied in each of the branches of mathematics, such as the algebra of real numbers, Euclidean geometry, or any non-Euclidean geometry. By examining the common structure of these branches, the salient features of this method will become clear.

Any one branch or system of mathematics deals with a class of concepts pertinent to it; for example, Euclidean geometry deals with points, lines, triangles, circles, and so on. Precise definitions of the concepts belonging to a system are all-important foundation stones on which the delicate superstructure is built. Unfortunately not every concept or term can be defined without entering upon an unending succession of definitions. It is true that the meanings of the undefined terms are suggested by physical examples. Addition, one of the undefined terms of algebra, can be explained by talking in terms of the number of cows that would be obtained by forming one herd from two separate herds. But such explanations in physical terms are not part of mathematics, for the subject is logically independent and self-sufficient. Of course, some concepts can be defined by appealing to the undefined ones, just as *circle* can be defined in terms of point, plane, and distance by describing it as the set of all points in a plane at a fixed distance from a given point.

If some terms are undefined and the physical pictures and processes we customarily associate with these terms are not part of mathematics proper, what facts about them can we use in the reasoning? The answer is to be found in the axioms. These assertions about the undefined and defined terms, which we accept without proof, are the sole basis for any conclusions that may be drawn about the concepts under discussion.

But how do we know what axioms to accept, especially in view of the fact that they involve undefined terms? Are we not in the position of dogs chasing their own tails? As in the case of the undefined terms, experience usually supplies the answer. Men accepted the axioms about number and the axioms of Euclidean geometry because experience with collections of objects and with physical figures vouched for these axioms. Here, too, we must caution against including the physical experience as part of mathematics. Mathematics begins with the statement of the axioms regardless of where they are obtained. Experience was the sole source of axioms until the nineteenth century. The investigations in non-Euclidean geometry, however, were motivated by a desire to use a parallel axiom different from Euclid's. In these cases mathematicians were deliberately going contrary to experience.

Though the axioms of non-Euclidean geometry appeared to be contrary to ordinary human experience they yielded theorems applicable to the physical world. In view of this fact it would seem that there should be considerable latitude in the choice of axioms. This is a partial truth, for the axioms of *any one* branch of mathematics must be consistent with each other, or else only confusion results. Consistency means not only that the axioms must not contradict each other but that they must not give rise to theorems which contradict each other.

The requirement of consistency has begun to take on great significance in recent years. As long as mathematicians regarded their axioms and theorems as absolute truths, it did not occur to them that contradictions could ever arise, except through an error in logic. Nature was consistent. Since mathematics phrased facts of nature in its axioms and deduced other truths, whether or not immediately apprehended in nature, mathematics also had to be consistent. The creation of non-Euclidean geometry, however, caused the mathematicians to see that they must stand on their own feet. They were not recording nature; they were interpreting. And any interpretation might not only be wrong but might also be inconsistent. The problem of consistency was further emphasized by the discovery that followed in the train of Cantor's contributions.

It may be possible to determine by direct examination of a set of axioms that no one of them contradicts another. But how can we be sure that no one of the hundreds of theorems which may be deducible from the axioms will ever contradict another? The answer to this question is lengthy and, it must be confessed, not entirely satisfactory at the present time. Much recent work in mathematics has been directed towards establishing the consistency of the many mathematical branches. Mathematicians, however, have been baulked, at least thus far, in their efforts to prove that the mathematical system which comprises the axioms and theorems about our ordinary real number system is consistent. The situation is extremely embarrassing. In recent years consistency replaced truth as the god of mathematicians and now there is a likelihood that this god too may not exist.

In addition to being consistent with each other, the axioms of a branch of mathematics should be simple. The reason for this requirement is clear. Inasmuch as axioms are accepted without proof, we should be aware of precisely what we are agreeing to. Simplicity insures this understanding. It is preferable, though not essential, that the axioms of a mathematical system be independent of each other. That is, it should not be possible to *deduce* an axiom from one or more of the others. The axiom that can be so deduced is better affirmed as a theorem, for we thereby reduce to as small a number as possible the statements accepted without proof. Finally, the axioms of a mathematical system must be fruitful; like carefully selected seeds they must yield a valuable crop, for one objective of mathematical activity is to obtain the new knowledge and insight implicit in the axioms. Euclid's contribution to mathematics was valuable because he chose a simple set of axioms that yielded hundreds of theorems.

Granted that a set of axioms fulfilling all the necessary and desirable conditions has been selected, how does the mathematician know what theorems to prove and how does he go about proving them? Let us consider these questions in turn.

There are many sources of possible theorems. Of such sources experience is by far the most fruitful. Experience with physical or real triangles suggests many likely conclusions about mathematical triangles. Deduction from the axioms then either establishes these conclusions as theorems of mathematics or discredits them. Of course experience must be understood in a broad sense. Random observations sometimes suggest possible theorems.

Scientific problems arising in laboratories or observatories and the artistic problem of depicting depth on a flat surface have led to precise theorems.

To a large extent mathematics generates its own problems. Many a possible theorem arises as a generalization of observations about numbers and geometrical figures. Anyone who has played with integers, for example, has doubtless observed that the sum of the first *two* odd numbers, that is, $1 + 3$, is the square of *two*; the sum of the first *three* odd numbers, that is, $1 + 3 + 5$, is the square of *three*; and similarly for the first four, five, and six odd numbers. Thus simple calculation suggests a general statement, namely, that the sum of the first *n* odd numbers, where *n* is any positive integer, is the square of *n*. Of course this possible theorem is not proved by the calculations above. Nor could it ever be proved by such calculations, for no mortal man could make the infinite set of calculations that would be required to establish the conclusion for *every n*. The calculations do, however, give the mathematician something to work on.

Consider another instance of generalization as a source of suggestions for theorems. A triangle is a polygon of three sides. Now in Euclidean geometry the sum of the angles of a triangle is 180°. Is it not natural to ask whether any general theorem could be found about the sum of the angles of any polygon? This question is answered by a very old theorem. The sum of the angles of any polygon is found by subtracting two from the number of sides and then multiplying the result by 180°.

We have already seen how the purely logical problem of deducing the assertion contained in Euclid's parallel axiom from more acceptable axioms led to non-Euclidean geometry. Once the idea of such geometries was grasped numerous suggestions for theorems were obtained by seeking the analogues of theorems that held in Euclidean geometry. For example, what is the analogue of the theorem that the sum of the angles of a quadrilateral is 360°?

These few indications of how the mathematician secures suggestions for theorems do not tell the whole story. Even if we add the more fortuitous sources such as pure chance, guesswork, and blundering about until a theorem is found, we still have left out the most valuable source of possible theorems – the imagination, intuition, and insight of creative genius. Most people could look at a quadrilateral indefinitely without becoming aware that if the midpoints of the four sides are joined (fig. 88), the figure formed is a parellelogram. Such knowledge is not the product of logic but of a sudden flash of insight.

In the domains of algebra, calculus, and advanced analysis especially, the first-rate mathematician depends on the kind of

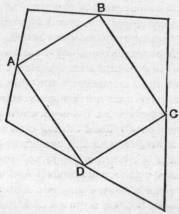

Figure 88. The lines joining the midpoints of the sides of any quadrilateral form a parallelogram

inspiration that we usually associate with the composer of music. The composer feels that he has a theme, a phrase which, when properly developed and embroidered, will produce beautiful music. Experience and a knowledge of music aid him in developing it. Similarly the mathematician divines that he has a conclusion which will follow from the axioms. Experience and knowledge may guide his thoughts into the proper channels. Modifications of one sort or another may be required before a correct and satisfactory statement of the new theorem is achieved. But essentially both mathematician and composer are moved by a divine afflatus that enables them to 'see' and 'know' the final edifice before one stone is laid.

Knowing *what* to prove is inextricably involved with knowing *how* to prove it. The mathematician may be convinced from an examination of the known facts in a situation that it should be possible to prove a certain theorem. But until he can give a deductive proof of this theorem he cannot assert or apply it. The distinction between conviction that a theorem should hold and proof of the theorem is made clear by many classic examples. The Greeks proposed the three famous problems of doubling a cube, trisecting an angle, and squaring a circle by means of a straightedge and compass. Over a period of two thousand years many a mathematician was convinced that it was impossible to perform these constructions under the conditions stated, but it was not until definite proofs of the impossibility were given in the nineteenth century that the problems were considered settled.

An excellent example of a conjecture, the truth of which seems indubitable, is that every even number is the sum of two prime numbers. A prime number, let it be remembered, is a whole number divisible only by itself and 1; thus 13 is a prime number but 9 is not. In accordance with this conjecture 2 is $1 + 1$; 4 is $2 + 2$; 6 is $3 + 3$; 8 is $3 + 5$; 10 is $3 + 7$; and so on. We could continue to test even numbers indefinitely and we should find that the conjecture holds. This conjecture is not a theorem of mathematics, however, because no proof of it has thus far been given.

A theorem must be established beyond question by deductive reasoning from axioms, and mathematicians literally work thousands of years to obtain such proofs. In our daily use of the phrases 'mathematical exactness' and 'mathematical precision' we pay homage to this unrelenting search for certainty.

Evidently much mathematical work must be done to find methods of proof even after the question of what to prove is disposed of. This point needs no emphasis for those readers who have struggled with exercises in geometry wherein the statement of what to prove is given and the student is expected to take over from that point. In the search for a method of proof, as in finding what to prove, the mathematician must use imagination, insight, and creative ability. He must see possible lines of attack where others would not, and he must have the mental stamina to wrestle with a problem until he has succeeded in finding a solution. Just what goes on in his mind while he works on the problem we do not know, any more than we know exactly what thought processes inspired Keats to write fine poetry or why Rembrandt's hands and brain were able to turn out paintings that suggest great psychological depth. We cannot define genius. We can only say that creative ability in mathematics calls for mental qualities of unusual excellence.

Perhaps we have been riding our Pegasus too hard and too high. Having anticipated a theorem and having then established it, has the mathematician really learned something new? After all, he derives from the axioms only what he puts into them, since all the conclusions that follow are logically implicit in the axioms. Mathematicians adopt axioms and spend centuries deducing theorems that are actually no more than elaborations of what the axioms say. In the words of the philosopher, Wittgenstein, mathematics is just a grand tautology.

But how grand! It is literally correct to describe the logical structure of mathematics as tautology, but this statement is about as adequate as saying that Venus de Milo is just a big girl. The description of mathematics as a tautology says that the choice of a set of axioms is like the purchase of a piece of mining land – the riches are all there. This description omits, however, the patient, hard digging which must be performed, the careful sifting of the precious metal from the base rock, the value and beauty of the treasure obtained, and the pleasure and exhilaration of accomplishment.

The divination and establishment of theorems complete the structure of a branch of mathematics. Such a branch, then, comprises terms, undefined and defined, axioms, and theorems proved on the basis of these axioms. This analysis of a mathematical system describes the structure of the mathematics of number and the structure of each of the geometries. It thus seems to epitomize the nature of mathematics. But a fuller appreciation of our subject calls for a little deeper investigation.

Every mathematical system contains undefined terms: for example, the words *point* and *line* in a geometrical system. In our discussion of the non-Euclidean geometries we found that we can attach physical meanings to the word *line* that differ considerably from the stretched string mathematicians had in mind in constructing these geometries. The fact that we can take such liberties with undefined terms, that we can give them seemingly unwarranted interpretations, suggests some deeper significance in the existence of undefined terms than has heretofore been made evident.

Let us forget mathematics for a moment and concern ourselves with the less logical field of diplomacy. A statesman at an international congress was faced with the delicate task of forming committees to perform various functions and decided that it would be tactful to form these committees in accordance with the following conditions:

(a) Any two nations should appear on at least one committee.
(b) Any two nations should not appear on more than one committee.
(c) Any two committees should have at least one nation in common.
(d) Every committee should have at least three nations on it.

Though these conditions seemed wise to the statesman, he was somewhat afraid that they might lead to undesirable complications which he could not foresee. He consulted a mathematician who at once pointed out some of the consequences.

(1) Any particular combination of two nations will appear on one and only one committee.
(2) Any two committees will have one and only one nation in common.
(3) There will be many combinations of three nations that will not appear on any committee.

The mathematician was able to state these conclusions at once because he recognized the conditions on nations and committees were precisely like the following statements about points and lines:

(a′) Any two points appear on at least one line.
(b′) Any two points do not appear on more than one line.

(c') Any two lines have at least one point in common.

(d') Every line contains at least three points.

The only difference between the two sets is that the words *point* and *line* take the place of nation and committee. Then the very theorems the mathematician had once deduced about points and lines from the conditions (a') to (d') carried over to nations and committees because *only the facts (a') to (d') had been used to establish the theorems*. The mathematician had but to replace point and line by nation and committee in the mathematical theorems to obtain the consequences he presented to the states-man. Thus the absence of well-defined meanings for the un-defined terms *point* and *line* proved to be a great advantage.

A fact of great consequence should now be clear: *In deductive proof from explicitly stated axioms the meaning of the undefined terms is irrelevant*. The mathematician of today realizes that any physical meaning can be attached to point, line, and other undefined terms as long as the axioms involving these terms hold for the physical meanings. If the axioms do hold, then the theorems also apply to these physical interpretations.

It would seem that our new conception of the nature of math-ematics robs it of all its meaning. Instead of being inseparably related to definite physical concepts and giving us insights into the physical world, it now appears to be concerned with empty words 'signifying nothing'. But the reverse is true. Mathematics is far richer in meaning, vaster in scope, and more fruitful in application than had ever been suspected before. In addition to the physical meanings which were formerly associated with mathematical concepts and can still be retained, an unlimited variety of new meanings may be found that satisfy the axioms of mathematical systems. In such new situations the theorems of these systems have new meanings and hence new applications.

Yet pure mathematics itself is not immediately or primarily concerned with the special meanings that may be given to the undefined terms. Rather it is concerned with the deductions that can be made from the axioms and the defined concepts. Applied mathematics, on the other hand, is concerned with those physical meanings of the concepts of pure mathematics that render the theorems useful in scientific work. The transition from pure to applied mathematics usually goes unnoticed. The statement that the area of a circle is πr^2 is a theorem of pure mathematics. The statement that the area of a circular field is π times the square of a certain physical length is a theorem of applied mathematics.

The distinction we have drawn between pure and applied mathematics is precisely what Bertrand Russell had in mind when he made the seemingly flippant but entirely justified remark that pure 'mathematics is the subject in which we never know what we are talking about, nor whether what we are saying is true'. Of course many a person has entertained such thoughts about mathematics without encouragement from Russell. He may not have known, however, how true they were or how to justify them. Mathematicians do not know what they are talking about because pure mathematics is not concerned with physical meaning. Mathematicians never know whether what they are saying is true because, as pure mathematicians, they make no effort to ascertain whether their theorems are true assertions about the physical world. Of such theorems we may ask only whether they were obtained by correct reasoning.

The abstract character of mathematical systems and their relation to physical meanings may be illustrated by a comparable situation in music. Beethoven composed the Fifth Symphony. Lesser mortals contributed the interpretations. Hope, despair, victory, defeat, man's struggles against fate, all these themes are read into Beethoven's creation. The music, like mathematics, exists without any such 'applications', however.

We might be inclined to believe that the process of deducing conclusions from axioms about undefined terms is peculiar to pure mathematics. But a moment's digression will convince us that this type of reasoning is not at all unusual. Let us consider the typical thought process of lawyers. The lawyer accepts as an axiom, although he prefers to call it a principle or rule, the fact that every sovereignty has police power. New York State is, according to the definition of a state in our union, sovereign in local matters. Industries conducted wholly within New York State are purely local matters. Hence New York has police power over industries located wholly within the state borders. The employment of elevator operators in buildings in New York is, by legal definition, an industry conducted entirely in New York. Hence New York State has police power over the employment of elevator operators in buildings within the state and, in particular, over the employment of women elevator operators.

By using some axioms about concepts or terms the lawyer has arrived at a conclusion. Let us notice, however, that no definition of police power was given or used in the reasoning. Our lawyer has utilized only the axiom that every sovereignty has it. Hence the term police power was used as an undefined term just as the mathematician uses point and line. More than that, while assenting to the reasoning above the reader inexperienced in the law may have associated police power with policemen. The usual legal interpretation of police power, however, is the power to provide for health and general welfare. As a matter of legal history, police power did not at one time include the fixing of minimum wages for women, so that our reasoning would then have led to the conclusion that New York State could not fix minimum wages for women elevator operators. Later, however, a court decision declared that police power did include the fixing of minimum wages for women. Hence, by this interpretation of police power New York State can fix minimum wages for women who operate elevators in New York buildings. Thus the undefined term police power can be given completely contradictory interpretations and yet the conclusion obtained by the reasoning above applies under either interpretation.

We see from this example that the lawyer, like the mathematician, engages in chains of deductive reasoning about undefined terms and often gives concrete meaning to these terms only when he is ready to apply his conclusions. Also, just as the mathematician gives varying, even contradictory, physical meanings in different situations to an undefined word such as line, so the court gives contradictory meanings at different times to an undefined term such as police power.

The analogy between mathematical and legal procedure extends beyond the use of undefined terms in chains of deductive reasoning. Principles of law are not merely axioms; they belong to systems as do the axioms of mathematics, and different systems may contain contradictory principles. For example, the legal right of the individual to engage in private enterprise is a principle in a capitalistic system of government just as the Euclidean parallel axiom is an axiom in the Euclidean system of geometry. The differences among fascist, democratic, and communistic forms of government stem from differences in fundamental principles, just as the differing theorems in the several geometries stem from different axioms. And just as each geometry attempts to treat physical space, each political system attempts to treat the social order.

Not only lawyers within legal systems but also politicians within political parties use the mathematical scheme of things that we have been describing. Before each election campaign the politicians dare to be logical. The leading members of each party draw up a platform, each plank of which is, in a real sense, an axiom of that party's political creed. From the statements of this platform it should be possible to deduce a party's position on future legislation. So far so good. What the politicians fail to point out, much less stress, is the free use in their platforms of undefined terms such as liberty, justice, Americanism, democracy, and the like. Needless to say the use of undefined terms in this connection is deliberate.

Our discussion of the significance of undefined terms in mathematical systems should help us appreciate the abstractness of mathematical thinking. This abstractness results from the fact that mathematics proper drops the physical meanings originally associated with the undefined terms. Mathematical method is abstract in another sense as well. Out of the medley of experiences proffered by nature, mathematics isolates and concentrates on particular aspects. This is abstraction in the sense of delimiting the phenomenon under investigation. For example, the mathematical straight line has only a few properties compared to those of the straight lines made by the edge of a table or drawn with pencil. The few properties the mathematical line possesses are stated in the axioms; for example, it is determined by two points. The physical lines, in addition to this property, have colour and even breadth and depth; moreover, they are built up of molecules each of which has a complicated structure.

It would seem offhand that an attempt to study nature by concentrating on just a few properties of physical objects would fall far short of effectiveness. Yet part of the secret of mathematical power lies in the use of this type of abstraction. By this means we free our minds from burdensome and irrelevant detail and are thereby able to accomplish more than if we had to keep the whole physical picture before us. The success of the process of abstracting particular aspects of nature rests on the divide-and-conquer rule.

In addition to delimiting the problem being studied there are further advantages in concentrating on a few aspects of experience. The experimental scientist, because he deals so directly with physical objects, is usually limited to thinking in terms of objects perceived through the senses. He is chained to the ground. Mathematics, by abstracting concepts and properties from the physical objects, is able to fly on wings of thought beyond the sensible world of sight, sound, and touch. Thus mathematics can 'handle' such 'things' as bundles of energy, which perhaps can never be qualitatively described because they are apparently beyond the realm of sensation. Mathematics can 'explain' gravitation, for example, as a property of a space too vast to visualize. In like manner mathematics can treat and 'know' such mysterious phenomena as electricity, radio waves, and light for which any physical picture is mainly speculative and always inadequate. The abstractions, that is the mathematical formulas, are the most significant and the most useful facts we have about these phenomena.

Abstracting quantitative aspects of physical phenomena often reveals unsuspected relationships because the quantitative laws turn out to be the same for apparently unrelated phenomena. This statement is nowhere better illustrated than by Maxwell's discovery that electromagnetic waves and light waves satisfy the same differential equations. It suggested at once that light and electromagnetic waves possess the same physical properties, a relationship confirmed a thousand times since. As Whitehead says:

Nothing is more impressive than the fact that as mathematics withdrew increasingly into the upper regions of ever greater extremes of

abstract thought, it returned back to earth with a corresponding
growth of importance for the analysis of concrete fact. . . . The paradox is now fully established that the utmost abstractions are the true
weapons with which to control our thought of concrete fact.

Those who, admitting the paradox, still deplore the fact that to
achieve success the physical sciences have to pay the price of
mathematical abstractness must reconsider what it is they would
look for in a scientific exposition of the nature of the physical
world. Eddington's answer is that a knowledge of mathematical
relations and structure is all that the science of physics can give
us. And Jeans says that the mathematical description of the universe *is* the ultimate reality. The pictures and models we use to
assist us in understanding are, to him, a step away from reality.
They are like 'graven images of a spirit'. We go beyond the mathematical formula at our own risk.

We have been discussing mathematics as a method, a method
applied to the study of quantitative and spatial relationships and
to concepts arising from these original fields of investigation.
The province of mathematics is no longer clearly delimited, however. The creation of non-Euclidean geometry, as we have seen,
released the mathematician from the bondage of producing
truths and set him free to adopt axioms and to investigate ideas
that may have no apparent usefulness in mastering or understanding the physical world. And so the mathematician is compelled to ask himself what guides his choice of subject matter and
what motivates his activity. What distinguishes his work from
cheap riddles, crossword puzzles, or even mere nonsense? (The
reader who would like to answer this question at once may be a
bit too hasty.) For about a hundred years now mathematicians
have come to recognize what was felt and asserted by the Greeks
but had been lost sight of in the intervening centuries: mathematics is an art and mathematical work must satisfy aesthetic
requirements.

No doubt many people feel that the inclusion of mathematics
among the arts is unwarranted. The strongest objection is that
mathematics has no emotional import. Of course this argument
discounts the feelings of dislike and revulsion which mathematics induces in some people. This argument also undervalues

the delight experienced by creators of mathematics when they succeed in formulating their ideas and in erecting ingenious and masterful proofs. Even the student of elementary mathematics is pleased by his success in proving stereotyped exercises and by his ability to see light, meaning, and order where formerly there was obscurity and confusion.

Nevertheless, it is true that mathematics generally appeals to the emotions less than music, painting, and poetry do. And a person is logically able to insist that the primary function of art is to arouse emotions and stir feelings. According to this concept of art, however, a dramatic photograph that catches at our hearts would be considered more artistic than numerous great paintings; abstract painting and much contemporary sculpture would probably be disregarded and there would be doubt about the status of architecture and ceramics. The still-life paintings of Picasso, impressionistic studies, such as Monet's, of atmospheric and light effects, the work of Seurat and Cézanne, and the 'arrangements' of the Cubists would also fail to satisfy the requirement. In fact, the pure art of modern times puts emphasis on the theoretical and formal side of painting, on the use of line and form, and on technical problems. Such work appeals much more to the intellect than to the emotions (see Plate xxvii). Whereas most Renaissance paintings, despite the intellectual studies involved in their composition, act directly on the emotions, the works of modern artists must first be 'figured out'. The requirement that an art must stir the emotions would seem to be especially inapt today.

An art must provide an outlet for the creative instinct of man. A glance backwards at the growth of our number system, the improvements in methods of calculation, the initiation and expansion of new branches inspired by the problems of the arts, sciences, and philosophy, and the refinements in standards of rigorous reasoning shows that mathematicians create. The determination of the precise assertions contained in the theorems, and the proofs which establish those theorems, are acts of creation. As in the arts each detail of the final work is not discovered but composed.

Of course the creative process must produce a work that has

design, harmony, and beauty. These qualities, too, are present in mathematical creations. Design implies the presence of structural patterns, of order, symmetry, and balance. Many mathematical theorems reveal just such a design. Consider, for example, the following theorem of plane geometry: Of all n-sided polygons with the same area, the regular n-sided polygon, that is, the one with equal sides and equal angles, has least perimeter. So far, then, mathematics tells us that a regular polygon requires less perimeter than a non-regular one with the same area and same number of sides. And now, of regular polygons with *different* numbers of sides but with the same area, which has least perimeter? The answer is that among regular polygons with the same area the one with the greatest number of sides has least perimeter. We can, of course, form regular polygons with any number of sides. Which figure then requires least perimeter for a given area? Here even an intuitive feeling for design suggests the answer. As the number of sides of a regular polygon increases· the figure approaches a circle in shape. The circle then should require least perimeter. And this is a theorem of mathematics. Such theorems are the essence of order and design.

Design is not merely accidental in mathematics. It is necessarily present in any logical structure. Only through conscious design was it possible for Euclid to produce the entire development of Euclidean geometry from the few axioms he adopted at the outset.

An excellent example of design utilized as a principle in mathematical creation may be found in the construction of higher dimensional geometries. Since $x^2 + y^2 = r^2$ is the equation of a circle in a plane and $x^2 + y^2 + z^2 = r^2$ is the equation of a sphere in three-dimensional space, $x^2 + y^2 + z^2 + w^2 = r^2$ is taken to be the equation of a hypersphere in four-dimensional space. Thus the design of two- and three-dimensional coordinate geometry is deliberately carried over to higher dimensions.

In any artistic creation the relation of the parts to each other and of the parts to the whole must be harmonious. The harmony in mathematical creations is partly intellectual in the form of logical consistency. The theorems of any one mathematical system must be in complete accord with each other. There are, however, other harmonies. The entire structure of Euclidean geometry is in harmony with the mathematics of number. By means of coordinates it is possible to interpret geometrical concepts and theorems algebraically. Conversely, algebraic equations have a geometric interpretation. Thus the two creations are harmonious with each other.

Major mathematical themes have been harmonized with each other. In our brief survey we have touched on four distinct branches of geometry – Euclidean, projective, and two non-Euclidean geometries. As we have viewed these subjects they appear distinct and in some cases inconsistent with each other. Nevertheless, one of the most satisfying mathematical contributions of recent times has shown that it is possible to erect projective geometry on an axiomatic basis in such a way that the theorems of the other three geometries result as specialized theorems of projective geometry. In other words, the content of all four geometries are now incorporated in one harmonious whole.

Mathematics offers still another kind of harmony. The plan that mathematics either imposes on nature or reveals in nature replaces disorder with harmonious order. This is the essential contribution of Ptolemy, Copernicus, Newton, and Einstein.

It is, of course, quite possible for a creation to possess all the formal characteristics of a work of art and yet fail to belong to that category. Many of the people who have listened to modern music or looked at modern painting would hold this to be true of the art being produced today. The ultimate test of a work of art is its contribution to aesthetic pleasure or beauty. Fortunately or unfortunately, this is a subjective test and depends on the cultivation of a special taste. Hence the question of whether mathematics possesses beauty can be answered only by those who have studied the subject.

As a matter of fact, the search for aesthetic pleasure has always influenced and prompted the development of mathematics. Out of a host of themes and patterns which suggest themselves the mathematician chooses those that satisfy a conscious or unconscious sense of beauty. The Greeks of the classical period investigated geometry because its forms and logical structure were beautiful to them. They valued the discovery of geometrical relations in nature not because such discoveries helped them to master nature but because they revealed her beautiful structure. Copernicus, we saw, advocated the new view of planetary motions because the mathematics of his theory gave him aesthetic pleasure. And Kepler too valued the heliocentric theory for this reason. 'I have attested it as true in my deepest soul,' he said, 'and I contemplate its beauty with incredible and ravishing delight.' Inspired by the work of Copernicus, Kepler himself spent most of his life searching for aesthetically satisfying mathematical laws. Newton also was genuinely concerned with beauty as the ultimate sanction of his mathematical and scientific work. He speaks of God as interested in the preservation of cosmic harmony and beauty. We can find similar remarks and views in the writings of most mathematicians.

Indeed the aesthetic sense of the true mathematician is more demanding than the most shrewish wife. Much research for new proofs of theorems already correctly established is undertaken simply because the existing proofs have no aesthetic appeal. There are mathematical demonstrations that are merely convincing; to use a phrase of the famous mathematical physicist, Lord Rayleigh, they 'command assent'. There are other proofs 'which woo and charm the intellect. They evoke delight and an overpowering desire to say, Amen, Amen.' An elegantly executed proof is a poem in all but the form in which it is written.

The tantalizing and compelling pursuit of mathematical problems offers mental absorption, peace of mind amid endless challenges, repose in activity, battle without conflict, 'refuge from the goading urgency of contingent happenings', and the sort of beauty changeless mountains present to senses tried by the present-day kaleidoscope of events.

The appeal offered by the detachment and objectivity of mathematical reasoning is superbly described by Bertrand Russell:

Remote from human passions, remote even from the pitiful facts of nature, the generations have gradually created an ordered cosmos, where pure thought can dwell as in its natural home and where one, at least, of our nobler impulses can escape from the dreary exile of the actual world.

Even laymen have been convinced of the artistic character of mathematical works. Thoreau says: 'The most distinct and beautiful statements of any truth must take at last the mathematical form.' The reader who remains unimpressed may at least find the attitudes and efforts of mathematicians more intelligible by knowing that these men have sought beauty.

It would appear from the analysis above that the usual criteria of an art are satisfied by mathematics. Nevertheless many people refuse to grant the subject the status. Unconsciously, however, they do acknowledge it. No one speaks of a talent or a gift for history or for economics or even for biology. But most everyone speaks of a talent or genius for mathematics, if only by way of regretting its absence. Mathematical ability is thereby classed with artistic ability.

We regret that we cannot pursue further our investigation of the subject matter, nature, and influences of mathematics. If time would permit the investigation of the more advanced branches of mathematics, we could then explore many more of the contributions mathematics has made to our culture. Unfortunately it requires years of study to master mathematical ideas and there is no royal road that materially shortens the process. It is hoped that the material presented here has at least dispelled the impression that mathematics is a closed book, a story told in Greek times, and a minor chapter in the history of mankind, and that it has conveyed some understanding of the position mathematics holds in our civilization and culture.

Unfortunately, mathematics does not solve all the problems man faces. Reason, the axiomatic method, and quantitative analysis do not serve as the approach to all phases of life. The artist may use mathematical perspective but correct perspective is not in itself art. Though the eighteenth-century thinkers were sure they could discover the laws of society and solve all social problems by means of mathematics, the social order is unfortunately even more confused today than it was in the eighteenth century. Nor would we recommend mathematics as the means for solving the problems of romance and marriage, though anthropologists at a recent symposium did seriously urge the application of mathematics to just these problems. The scope of mathematics is limited and the reason that it is limited is succinctly stated in the phrase that man is a rational animal. His rationality is merely a qualification of his animality. And inasmuch as man's desires, emotions, and instincts are part of his animal nature and often unsatisfied by and even opposed to his reason, reason alone will not suffice to guide and control all of man's activities. Of course these remarks are not intended to imply that the application of reason to the affairs of man has by any means reached the point of surfeit.

Mathematics is variously described as a body of knowledge, as a practical tool, as a cornerstone of philosophy, as the perfection of logical method, as the key to nature, as the reality in nature, as an intellectual game, as an adventure in reason, and as an aesthetic experience. Our survey of mathematics should have indicated the grounds for these descriptions. When we consider the number of fields on which mathematics impinges and the number of these over which it already gives us mastery or partial mastery, we are tempted to call it a method of approach to the universe of physical, mental, and emotional experiences. It is the distillation of highest purity that exact thought has extracted from man's efforts to understand nature, to impart order to the confusion of events occurring in the physical world, to create beauty, and to satisfy the natural proclivity of the healthy brain to exercise itself. We, who live in a civilization distinguished primarily by achievements owing their existence to mathematics, are in a position to bear witness to these statements

Selected References

Armitage, Angus: *Copernicus*, W. W. Norton, New York, 1938.
 Sun, Stand Thou Still, Henry Schuman, New York, 1947; paperback under the title, *The World of Copernicus*, New American Library, New York, 1951.

Ball, W. W. Rouse: *A Short Account of the History of Mathematics*, Macmillan, New York, 4th ed., 1908; reprint, Dover Publications, New York, 1960.

Becker, Carl L.: *The Declaration of Independence*, Harcourt, Brace, New York, 1922.
 The Heavenly City of the Eighteenth Century Philosophers, Yale University Press, New Haven, 1932.

Bell, Arthur E.: *Newtonian Science*, Edward Arnold, London, 1961.

Bell, Eric T.: *Men of Mathematics*, Simon & Schuster, New York, 1937; paperback, Penguin Books, 1953.

Berkeley, George: *Three Dialogues Between Hylas and Philonous*, Open Court Publishing Co., Chicago, 1929.

Berry, Arthur: *A Short History of Astronomy*, John Murray, London, 1898; reprint, Dover Publications, New York, 1961.

Bohm, David: *Causality and Chance in Modern Physics*, Routledge & Kegan Paul, London, 1957.

Born, Max: *Einstein's Theory of Relativity*, Dover Publications, New York, 1962.

Bridgman, P. W.: *The Nature of Physical Theory*, Princeton University Press, Princeton, 1936; reprint, Dover Publications, New York, 1949.

Bronowski, J., and Mazlish, B.: *The Western Intellectual Tradition*, Harper & Bros., New York, 1960; paperback, Penguin Books, London, 1963.

Bunim, Miriam: *Space in Medieval Painting and the Forerunners at Perspective*, Columbia University Press, New York, 1940.

Burtt, Edwin A.: *The Metaphysical Foundations of Modern Physical Science*, 2nd ed., Routledge & Kegan Paul, London, 1932; paperback, Doubleday, New York, 1964.

Bury, J. B.: *A History of Freedom of Thought*, Oxford University Press, London, 1952.
 The Idea of Progress, Macmillan, New York, 1932; reprint, Dover Publications, New York, 1955.

528   *References*

Bush, Douglas: *Science and English Poetry*, Oxford University Press, New York, 1950.

Butterfield, Herbert: *The Origins of Modern Science*, Macmillan, New York, 1951; paperback, Collier Books, New York, 1962.

Cardan, Jerome: *The Book of My Life*, E. P. Dutton, New York, 1930; reprint, Dover Publications, New York, 1962.

Carr, Herbert Wildon: *Leibniz*, Constable, London, 1929; reprint, Dover Publications, New York, 1960.

Caspar, Max: *Kepler* (translated), Abelard-Schuman, New York, 1960; paperback, Collier Books, New York, 1962.

Cassirer, E.: *The Philosophy of the Enlightenment*, Princeton University Press, Princeton, 1951.

Chamberlin, Wellman: *The Round Earth on Flat Paper*, National Geographic Society, Washington, D.C., 1947.

Childe, V. Gordon: *Man Makes Himself*, Watts & Co., London, 1936; paperback, New American Library, New York, 1951.

Clark, Kenneth: *Piero della Francesca*, Phaidon Press, London, 1951.

Cohen, Morris R., and Nagel, E.: *An Introduction to Logic and Scientific Method*, Harcourt, Brace, New York, 1934; paperback, Harcourt, Brace & World, New York, 1962.

Coolidge, Julian L.: *The Mathematics of Great Amateurs*, Oxford University Press, London, 1949; reprint, Dover Publications, New York, 1963.

Crum, Ralph B.: *Scientific Thought and Poetry*, Columbia University Press, New York, 1931.

Dampier-Whetham, William C. D.: *A History of Science and Its Relations with Philosophy and Religion*, Cambridge University Press, London, 1929.

Dantzig, Tobias: *Number, The Language of Science*, 3rd ed., Macmillan, New York, 1939; paperback, Doubleday, New York, 1964.

Descartes, René: *Discourse on Method*, E. P. Dutton, New York, 1912; paperback, Penguin Books, London, 1960.

Dijksterhuis, E. J.: *The Mechanization of the World Picture* (translated), Oxford University Press, New York, 1961.

Dreyer, J. L. E.: *History of the Planetary Systems from Thales to Kepler*, Cambridge University Press, London, 1906; reprint, Dover Publications, New York, 1953, under the title *A History of Astronomy from Thales to Kepler*.

Duhem, Pierre: *The Aim and Structure of Physical Theory* (translated), Princeton University Press, Princeton, 1954; paperback, Atheneum, New York, 1962.

Eddington, Sir Arthur S.: *The Nature of the Physical World*, Macmillan, New York, 1928; paperback, University of Michigan Press, Ann Arbor, 1958.

Space, Time, and Gravitation, Cambridge University Press, Cambridge, 1920; paperback, Harper & Row, New York, 1964.

Farrington, Benjamin: *Francis Bacon: Philosopher of Industrial Science*, Henry Schuman, New York, 1949; paperback, Collier Books, New York, 1961.

Greek Science, 2 vols., Penguin Books, London, 1944 and 1949.

Freund, John E.: *Modern Elementary Statistics*, 2nd ed., Prentice-Hall, Englewood Cliffs, 1960.

Fry, Roger: *Vision and Design*, Penguin Books, London, 1937; Meridian Books, New York, 1956.

Galilei, Galileo: *Dialogue on the Great World Systems*, University of Chicago Press, Chicago, 1953.

Gibson, James J.: *The Perception of the Visual World*, Houghton Mifflin, Boston, 1950.

Gilson, Etienne: *Painting and Reality*, Pantheon Books, New York, 1957; paperback, Meridian Books, New York, 1959.

Gombrich, E. H.: *Art and Illusion*, 2nd ed., Pantheon Books, New York, 1961.

The Story of Art, Oxford University Press, New York, 1950.

Hall, A. Rupert: *From Galileo to Newton*, Collins, London, 1963.

The Scientific Revolution, Longmans, Green, New York, 1954; paperback, Beacon Press, Boston, 1956.

Hall, Everett W.: *Modern Science and Human Values*, D. Van Nostrand, New York, 1956.

Hamilton, Edith: *The Greek Way to Western Civilization*, Norton, New York, 1930; paperback, New American Library, New York, 1948.

Hardy, G. H.: *A Mathematician's Apology*, Cambridge University Press, London, 1940.

Hazard, Paul: *The European Mind*, Yale University Press, New Haven, 1953; paperback, Penguin Books, London, 1964.

European Thought in the Eighteenth Century, Yale University Press, New Haven, 1954; paperback, Penguin Books, London, 1965.

Ivins, William M., Jr: *Art and Geometry*, Harvard University Press, Cambridge, 1946; reprint, Dover Publications, New York, 1964.

Jeans, Sir James H.: *The Growth of Physical Science*, Cambridge University Press, London, 1951; paperback, Fawcett, New York, 1958.

Science and Music, Cambridge University Press, London, 1937; paperback, 1961.

Jones, Richard Foster: *The Seventeenth Century*, Stanford University Press, Stanford, 1951.

Kline, Morris: *Mathematics and the Physical World*, T. Y. Crowell, New York, 1953; paperback, Doubleday, New York, 1963.

Mathematics: A Cultural Approach, Addison-Wesley, Reading, Mass., 1962.

Kuhn, Thomas S.: *The Copernican Revolution*, Harvard University Press, Cambridge, 1957.

Lawson, Philip J.: *Practical Perspective Drawing*, McGraw-Hill, New York, 1943.

Lecky, William E. H.: *History of the Rise and Influence of the Spirit of Rationalism*, 2 vols., Longmans, Green, London, 1882; reprint, George Braziller, New York, 1955, in one volume.

Levinson, Horace C.: *The Science of Chance*, Rinehart, New York, 1950; reprint, Dover Publications, New York, 1963, under the title *Chance, Luck and Statistics*.

Miller, Dayton C.: *The Science of Musical Sounds*, 2nd ed., Macmillan, New York, 1926.

More, Louis T.: *Isaac Newton*, Charles Scribner's Sons, New York, 1934; reprint, Dover Publications, New York, 1962.

Mortimer, Ernest: *Blaise Pascal: The Life and Work of a Realist*, Harper & Bros., New York, 1959.

Nicolson, Marjorie Hope: *Newton Demands the Muse*, Princeton University Press, Princeton, 1946.

Ore, Oystein: *Cardano: The Gambling Scholar*, Princeton University Press, Princeton, 1953.

Panofsky, Erwin: *Meaning in the Visual Arts*, Doubleday, New York, 1955; paperback, Penguin Books, London, 1971.

Peierls, R. E.: *The Laws of Nature*, Charles Scribner's Sons, New York, 1956; paperback, 1962.

Pope-Hennessy, John: *The Complete Works of Paolo Uccello*, Phaidon Press, London, 1950.

Randall, John H., Jr: *Making of the Modern Mind*, rev. ed., Houghton Mifflin, Boston, 1940.

Reichmann, W. J.: *The Use and Abuse of Statistics*, Oxford University Press, New York, 1962; paperback, Penguin Books, London, 1964.

Russell, Bertrand: *The ABC of Relativity*, Harper & Bros., New York, 1925; paperback, New American Library, New York, 1959.

Our Knowledge of the External World, George Allen & Unwin, London, 1926; paperback, New American Library, New York, 1960.

A History of Western Philosophy, Simon & Schuster, New York, 1945; paperback, 1957.

Sampson, R. V.: *Progress in the Age of Reason*, Harvard University Press, Cambridge, 1956.

Schrödinger, Erwin: *Science and the Human Temperament*, G. Allen & Unwin, London, 1935; reprint, Dover Publications, New York, 1957, under the title *Science, Theory and Man*.

Singer, Charles: *A Short History of Scientific Ideas to Nineteen Hundred*, Oxford University Press, New York, 1959.

Smith, Preserved: *A History of Modern Culture*, 2 vols., Henry Holt, New York, 1930.

Snow, C. P.: *The Two Cultures and a Second Look*, Cambridge University Press, London, 1963; paperback under the title *The Two Cultures and the Scientific Revolution*, New American Library, New York, 1964.

Stephen, Leslie: *The English Utilitarians*, 3 vols., G. P. Putnam's Sons, New York, 1900.

Strong, Edward W.: *Procedures and Metaphysics*, University of California Press, Berkeley, 1936.

Taylor, Henry Osborn: *Ancient Ideals*, 2 vols., 2nd ed., Macmillan, New York, 1913.

Thought and Expression in the Sixteenth Century, 2 vols., 2nd ed., Macmillan New York, 1930; paperback in 5 volumes, Collier Books, New York, 1962: Bk 1, *The Humanism of Italy*; Bk 2, *Erasmus and Luther*; Bk 3, *The French Mind*; Bk 4, *The English Mind*; Bk 5, *Philosophy and Science in the Sixteenth Century*.

Taylor, Lloyd William: *Physics, the Pioneer Science*, Houghton Mifflin, Boston, 1941; reprint, Dover Publications, New York, 1959.

Toulmin, Stephen: *The Philosophy of Science*, Hutchinson's University Library, London, 1953; paperback, Harper & Row, New York, 1964.

and June Goodfield: *The Fabric of the Heavens*, Harper & Bros., New York, 1961; paperback, Penguin Books, London, 1963.

Whitehead, Alfred North: *Science and the Modern World*, Macmillan, New York, 1925; paperback, Penguin Books, London, 1938.

Willey, Basil: *The Seventeenth Century Background*, Chatto & Windus, London, 1934; paperback, Penguin Books, 1962.

The Eighteenth Century Background, Chatto & Windus, London, 1940, paperback, Penguin Books, 1962.

Nineteenth Century Studies, Chatto & Windus, London, 1949, paperback, Penguin Books, 1964.

Wood, Alexander: *The Physics of Music*. Methuen, London, 1944; reprint, Dover Publications, New York, 1962.

Index

Mathematics in Penguin

GEOMETRY AND THE LIBERAL ARTS

Dan Pedoe

For an architect like Vitruvius and artists like Leonardo and Dürer, the fascination of geometry lay in its contribution to solving problems of order, proportion and perspective. Taking these three great practitioners as his principal examples, Professor Pedoe traces the lasting appeal of geometry to artists, scientists and philosophers and its effect on artistic achievement.

a Peregrine book

THE PSYCHOLOGY OF LEARNING MATHEMATICS

Richard R. Skemp

Many of us have had teachers to whom mathematics was obvious – but we only learned rules without reasons. Richard Skemp looks at the ways in which we understand mathematics, and discusses the emotional factors in teaching. It will be of enormous help to those for whom mathematics is a closed book, and will broaden the insight of those who comprehend (and even teach) intuitively.

CONCEPTS OF MODERN MATHEMATICS

Ian Stewart

'Modern maths' looks askance at rules of thumb and Euclid (with his eternal triangle) learnt by rote. With plenty of humour and a number of anecdotes. Dr Stewart shows here how the new approach, once grasped, encourages a genuine understanding of mathematics.

and forthcoming

THE PENGUIN BOOK OF TABLES
Penguin Education

UIN MATHEMATICAL AND STATISTICAL TABLES
David Nelson
a Reference book